SOCIETAL RISK ASSESSMENT
How Safe is Safe Enough?

PUBLISHED SYMPOSIA

Held at

General Motors Research Laboratories

Warren, Michigan

1979   R. C. Schwing, W. A. Albers, Jr., eds., *Societal risk assessment: How safe is safe enough?* Plenum Press, New York, 1980.

1978   J. N. Mattavi, C. A. Amann, eds., *Combustion modeling in reciprocating engines,* Plenum Press, New York, 1980.

1978   G. G. Dodd, L. Rossol, eds., *Computer vision and sensor-based robots,* Plenum Press, New York, 1979.

1977   D. P. Koistinen, N. -M. Wang, eds., *Mechanics of sheet metal forming: Material behavior and deformation analysis,* Plenum Press, New York, 1978.

1976   G. Sovran, T. A. Morel, W. T. Mason, eds., *Aerodynamic drag mechanisms of bluff bodies and road vehicles,* Plenum Press, New York, 1978.

1975   J. M. Colucci, N. E. Gallopoulos, eds., *Future automotive fuels: Prospects, performance, perspective,* Plenum Press, New York, 1977.

1974   R. L. Klimisch, J. G. Larson, eds., *The catalytic chemistry of nitrogen oxides,* Plenum Press, New York, 1975.

1973   D. F. Hays, A. L. Browne, eds., *The physics of tire traction,* Plenum Press, New York, 1974.

1972   W. F. King, H. J. Mertz, eds., *Human impact response,* Plenum Press, New York, 1973.

1971   W. Cornelius, W. G. Agnew, eds., *Emissions from continuous combustion systems,* Plenum Press, New York, 1972.

1970   W. A. Albers, ed., *The physics of opto-electronic materials,* Plenum Press, New York, 1971.

1969   C. S. Tuesday, ed., *Chemical reactions in urban atmospheres,* American Elsevier, New York, 1971.

1968   E. L. Jacks, ed., *Associative information techniques,* American Elsevier, New York, 1971.

1967   P. Weiss, G. D. Cheever, eds., *Interface conversion for polymer coatings,* American Elsevier, New York, 1968.

1966   E. F. Weller, ed., *Ferroelectricity,* Elsevier, New York, 1967.

1965   G. Sovran, ed., *Fluid mechanics of internal flow,* Elsevier, New York, 1967.

1964   H. L. Garabedian, ed., *Approximation of functions,* Elsevier, New York, 1965.

1963   T. J. Hughel, ed., *Liquids: Structure, properties, solid interactions,* Elsevier, New York, 1965.

1962   R. Davies, ed., *Cavitation in real liquids,* Elsevier, New York, 1964.

1961   P. Weiss, ed., *Adhesion and cohesion,* Elsevier, New York, 1962.

1960   J. B. Bidwell, ed., *Rolling contact phenomena,* Elsevier, New York, 1962.

1959   R. C. Herman, ed., *Theory of traffic flow,* Elsevier, New York, 1961.

1958   G. M. Rassweiler, W. L. Grube, eds., *Internal stresses and fatigue in metal,* Elsevier, New York, 1959.

1957   R. Davies, ed., *Friction and wear,* Elsevier, New York, 1959.

SOCIETAL RISK ASSESSMENT
How Safe is Safe Enough?

Edited by
RICHARD C. SCHWING and WALTER A. ALBERS, Jr.
General Motors Research Laboratories

PLENUM PRESS • **NEW YORK-LONDON** • **1980**

Library of Congress Cataloging in Publication Data

Main entry under title:

Societal risk assessment.

Proceedings of an international symposium held Oct. 8–9, 1979 at the General Motors Research Laboratories, Warren, Mich. and sponsored by the Laboratories. Includes bibliographical references and indexes.

1. Technology assessment – Congresses. 2. Risk – Congresses. I. Schwing, Richard C. II. Albers, Walter A. III. General Motors Corporation. Research Laboratories.

T174.5.S6   363.1   80-23833

ISBN 0-306-40554-7

Proceedings of the General Motors Symposium on Societal Risk Assessment, held in Warren, Michigan, October 7–9, 1979.

© 1980 Plenum Press, New York
A Division of Plenum Publishing Corporation
227 West 17th Street, New York, N.Y. 10011

PREFACE

This volume constitutes the papers and discussions from a symposium on "Societal Risk Assessment: How Safe is Safe Enough?" held at the General Motors Research Laboratories on October 8-9, 1979. This symposium was the twenty-fourth in an annual series sponsored by the Research Laboratories. Initiated in 1957, these symposia have as their objective the promotion of the interchange of knowledge among specialists from many allied disciplines in rapidly developing or changing areas of science or technology. Attendees characteristically represent the academic, government, and industrial institutions that are noted for their ongoing activities in the particular area of interest.

The objective of this symposium was to develop a balanced view of the current status of societal risk assessment's role in the public policy process and then to establish, if possible, future directions of research. Accordingly, the symposium was structured in two dimensions; certainty versus uncertainty and the subjective versus the objective. Furthermore, people representing extremely diverse disciplines concerned with the perception, quantification, and abatement of risks were brought together to provide an environment that stimulated the exchange of ideas and experiences. The keys to this exchange were the invited papers, arranged into four symposium sessions. These papers appear in this volume in the order of their presentation. The discussions that in turn followed from the papers are also included. It is our hope that the issues and ideas presented in this volume will serve to identify the needs for further development of useful societal risk assessment and to clarify the role that risk assessment can play in the public policy process.

Many people played an important role in the planning, executing, and hosting of the symposium as well as in the preparation of this volume, and we wish to acknowledge our gratitude. An advisory group consisting of Dr. Lester B. Lave (Brookings Institution), Professor Howard Raiffa (Harvard), and Dr. Paul Slovic (Decision Research) were instrumental in organizing the program in addition to

W. A. Albers, Jr.

their active participation as either speakers or session chairmen. Dr. Chauncey Starr (Electric Power Research Institute), Dr. Allen Kneese (Resources for the Future), and Professor William D. Rowe (The American University) served admirably as session chairmen. The encouragement, guidance, and advice of Mr. John D. Caplan and Dr. Nils L. Muench, both of the General Motors Research Laboratories, was invaluable. Among many other colleagues at the General Motors Research Laboratories who have been of assistance, we should like to especially thank R. Thomas Beaman, Linda E. Meyer, and David N. Havelock for the symposium arrangements, secretarial needs, and manuscript layout and art work, respectively. And finally, each of the authors was most cooperative and responsive in putting together papers that were requested and supplying manuscripts for publication.

Richard C. Schwing and Walter A. Albers, Jr.

CONTENTS

Preface... v

SESSION I: The Risks We Run and The Risks We "Accept"
Chairman: Chauncey Starr,
    Electric Power Research Institute.................................... 1

The Nature of Risk
    William W. Lowrance, Stanford University 5
        References.. 14
        Discussion 14

The Uncertain Risks We Run: Hazardous Materials
    Marvin A. Schneiderman, National Cancer Institute.................... 19
        References.. 37
        Discussion 38

The Known Risks We Run: The Highway
    Barbara E. Sabey and Harold Taylor,
    Transport and Road Research Laboratory............................. 43
        References.. 62
        Discussion 66

Perceptions of Risk and Their Effects on Decision Making
    Raphael G. Kasper, National Academy of Sciences 71
        References.. 80
        Discussion 80

SESSION II: "Acceptability" with Fixed Resources
   Chairman: Allen Kneese, Resources for the Future.....................   85

On Making Life and Death Decisions
   Ronald A. Howard, Stanford University.............................   89
      References...   106
      Discussion   106

Economic Tools for Risk Reduction
   Lester B. Lave, Brookings Institution   115
      Discussion   124

Trade-offs
   Richard C. Schwing, General Motors Research Laboratories..............   129
      References...   141
      Discussion   142

Risk Spreading through Underwriting and the
Insurance Institution
   J. D. Hammond, Pennsylvania State University   147
      References...   175
      Discussion   176

SESSION III: "Acceptability" in a Democracy - Who Shall Decide?
   Chairman: William D. Rowe, The American University.................   179

Facts and Fears: Understanding Perceived Risk
   Paul Slovic, Baruch Fischhoff and Sarah Lichtenstein,
   Decision Research, A Branch of Perceptronics   181
      References...   212
      Discussion   214

Ethics, Economics and the Value of Safety
   William D. Schulze, University of Wyoming   217
      References...   229
      Discussion   230

Problems and Procedures in the Regulation of
Technological Risk
   Dorothy Nelkin and Michael Pollack, Cornell University...............   233
      References...   247
      Discussion   248

The Role of Law in Determining Acceptability of Risk
   Harold P. Green, George Washington University 255
      References 266
      Discussion 267

SESSION IV: Directions and Perspectives of Societal Risk Assessment
   Chairman: Howard Raiffa, Harvard University 271

Aesthetics of Risk: Culture or Context
   Michael Thompson, Institute for Policy and Management Research 273

Witches, Floods, and Wonder Drugs:
Historical Perspectives on Risk Management
   William C. Clark, International Institute for
   Applied Systems Analysis 287
      References 311
      Discussion 313

Prospects for Change
   Melvin Kranzberg, Georgia Institute of Technology 319
      References 332

Miscellaneous Discussion 333

Concluding Remarks
   H. Raiffa, Harvard University 339

Symposium Summary
The Safety Profession's Image of Humanity
   C. W. Churchman, University of California 343
      References 346

Participants 347

Proper Name Index 353

Subject Index 355

SESSION I
THE RISKS WE RUN
AND THE RISKS WE "ACCEPT"

Session Chairman
Chauncey Starr

Electric Power Research Institute
Palo Alto, California

INTRODUCTORY REMARKS

SESSION I

Chauncey Starr

Electric Power Research Institute
Palo Alto, California

First I want to congratulate Paul Chenea and Dick Schwing and Walter Albers for putting together such an interesting agenda and inviting so many people who are actively interested in this relatively new area of risk assessment. We are trying to create a discipline out of a very uncertain subject and I hope that his conference will be a productive step in bringing together some of the insights to it. I am not going to take very long to introduce this subject but I would like to comment on some of the general background.

Decision making in general, of course, is a difficult art. In technical areas it's difficult enough to decide whether new technologies will work or not work. Of course, the people who promote the technologies always tell you how well it will perform *when it works.* They rarely tell you the second element of decision making, the vulnerability of the technology to failure. The more astute research managers are now familiar with that second aspect — the off design characteristics which prevent new technologies from really hitting the targets. A third element, of course, is the risk of the new technologies to people. That's very rarely discussed, and it's only recently, perhaps a decade or so, that we have become deeply concerned with that part of the decision making process.

Risk arises from an abnormal event. It requires analysis under off design conditions that are not part of our normal structure of thinking. Now there are certain basic truisms about risk, and you and I know that everybody getting into this subject discovers these truisms. One is that everyone knows that personal risk is an accepted part of life. Living is fun but it is also dangerous. People recognize this and accept it. Its the quantification of *how* dangerous which gets to be difficult. Secondly, everyone reacts differently to risks they take voluntarily and to risks that are imposed by some outside group, the involuntary exposure. Our individual behavior patterns and responses are quite different as to whether we decide to do something or someone decides for us. A third characteristic is that group decisions by other people imposing involuntary risks on us are being made all the time. They've been made throughout history. We may or may not approve of the basis on which its done but its being done. The result is as we reach the fourth truism, that because we react differently to involuntary and our own choices, we face a conflict when a group decision imposes a risk on us. Historically that has always occurred and will continue to occur, and creates a conflict which will never be completely resolved to everybody's satisfaction. It is a characteristic of an organized society that conflict between the balance of voluntary and involuntary exposure is inherent to societal function. It's unavoidable, and the whole idea that society can avoid risk conflict by some method is basically unsound, it's just not possible. The conflict will always be there and so we face a problem in this symposium as to how to handle this inherent conflict.

I would like to explore why we study risks. What questions are we trying to answer? The first question has to do with the process. How should group processes proceed to minimize social costs? The processes in our societies range from those that go from complete anarchy where there is no definitive process to complete dictatorship where the process is pretty clear. In between we have a mix, and in most of the industrial world, we face that mix. The process for decision making is itself a major issue.

Secondly, when you talk about social costs, the question immediately evident is what costs are included and how are they weighted, a very difficult task. That's one of the questions we have to address. It's obvious that if you have a decision process, if you knew how to weight and value social costs, you'd still have a problem with the full disclosure of all the social costs. What is full disclosure? How far do you go in seeking all the ripples that exist in society from any one event before you have finished full disclosure, and do you include the options of risk management as part of full disclosure, i.e. the alternatives to what you can do about risk? What is full disclosure in practice? Do you talk about near time events, far time events, the people who get the benefits, or the people who bear the costs? You have geographic distributions, time distributions, demographic distributions — all of these things are involved in the term full disclosure. Where do you draw the boundries? That's one of the questions we want to answer.

A third question is who makes the decisions, even in an organized society — who is responsible? Decisions are not formed by institutions; the decision process involves people. The government is constantly making agencies and committees (as you and I know because we have served on many of these) so that, in fact, a few people in the agencies, a few people on the committees, really decide what happens. So how do you pick out those who are responsible for making decisions that are going to affect all society? Where do you allocate the responsibility? That's one of the questions.

And then, of course, the fourth and ultimate question is the value systems. When we establish the social costs and we weigh them how do we value the result? How do we set our priorities? How do we determine the relative merits of things? What value system should one use? That itself is a subject for a separate symposium because, of course, value systems depend on cultures, background, economic status, all kinds of things.

Now, these are the type of questions we try to answer in a subject of this nature. I hope that as the discussion proceeds in the next two days some of the questions, as well as the answers, will be elucidated.

Now part of the problem in the whole subject of risk assessment is a set of confusions that exist even in the discussion of the subject, the confusions that involve the reality of situations, the analysis of situations, and the perceptions. The reality is what actually happens or what will actually happen down the road. The analysis is usually based upon very limited data, the collection of cases and statistics which may or may not be correct and, based upon this, we set up models of one sort of another and predict an outcome down the road. So we end up with facing the fact that there may be real situations that we may or may not predict correctly. We use historical data to give us statistics which may or may not be correct, we used models to predict which may or may not be correct. Finally we have the question of just what is the intuitive perception of the people that we deal with. The risk takers of the country are not ourselves but the great population of the country and they have perceptions of risk. Their perceptions may be so far from reality that you and I know that they're absurd, but that's how they feel about it and that's the way they perceive things. So, in discussing the subject, we really have to distinguish between the reality of what may or may not occur, the analysis of it, and the perception of it. Similar confusions exist, incidentally, on social costs and social benefits all of which are involved. I can ask the rhetorical question, "Who in the year 1900 could have predicted reasonably well the social costs and benefits of the automobile, which General Motors, our host, is concerned with?" I've looked at the perceptions of the situation in the year 1900, and they are quite different from what the automobile has done and is doing to our society today.

I have to end my comments with just one little personal incident. You know I live in California. When I first moved there I lived in an old house that had steam heat with wall radiators. They were on a clock. We had some guests from Chicago, and when they arrived in the evening we said to them as we led them to the bedroom, that the heat goes off automatically about 11:00 at night and goes on again at 7:00 in the morning and when it does the radiators tend to clank, so don't be disturbed by this. Well, the next morning about 7:00 our guests from Chicago not only heard the clanking but the bed shook, the walls rattled, and the pictures on the walls really began to shake! It was quite a set of tremors. The disturbance settled after a while. When they came down to breakfast they said to me, "You really ought to do something about getting that heating system fixed," and I said to them, "Well, you know that's probably true, but you happened to have been here just at the time this morning when we had one of the greatest earthquakes we've ever had." That day they took the next plane back to Chicago with the notion that California has earthquakes all the time.

Now the point of telling this story is that peoples' perception of probabilities reaches absurdities. You will get questions today from people who will ask you what the Three Mile Island accident proves on probability. Well, it tells you something, but not very much about probabilities. The inadequacy of single events giving you probability numbers is something which any of you experts can explain, but the public response and the public perception is based on these single events and their personal relation to them. So that even if we as a group come up with what we consider intelligent answers, we're going to have a hard time explaining them to the public.

Well, those are my introductory comments. These are the questions I think we ought to try to cover. These are the kind of problems we face. With this preamble I'm going to ask our first speaker, Dr. Bill Lowrance, to get started.

THE NATURE OF RISK

William W. Lowrance

Stanford University, Stanford, California

ABSTRACT

"Risk" may be defined as a compound measure of the probability and magnitude of adverse effect. Important distinctions can be made among six major classes of hazard (infectious and degenerative diseases; natural catastrophes; failure of large technological systems; discrete, small-scale accidents; low-level, delayed-effect hazards; and sociopolitical disruptions). Decisions about risks meet with four kinds of limitations: of empirical analysis of "the facts", of social value appraisal, of "risk management", and of the assignment of rights and responsibilities.

INTRODUCTION

We begin to be taught about risks from a very early age: "Don't run; you'll skin your knees!" "Don't go near the water!" "Don't chase your ball into the street!" "Stay away from the stove!" We are taught moderation: "Don't swing so high; you'll fall out." We learn that Nature is capricious: "Bring your umbrella; it may rain." "Don't aggravate the dog; he'll bite you." At the beach, the tide comes in and, to our tears, washes away our sandcastles. We learn that technology has its limits: higher and higher we stack our building blocks, until they topple. Our board-bridge over the creek works fine the first summer, but the second summer it becomes rotten, cracks, and dumps somebody in the water. Generality is imparted by classical teachings: the Careless Little Pig builds his house of straw, but the Wise Little Pig builds his of brick.... We get safety instruction in swimming class. We are taught to be careful in using household appliances and tools. As we become of independent age we learn about the risks in driving automobiles and fooling around with sex, alcohol, and drugs. In all of this we learn about particular risks and become sensitized to the nature of risk-taking. Then, of course, as we grow into adults, it all gets terribly complicated.

References p. 14.

WHAT IS "RISK?"

I prefer to define "risk" as a compound measure of the probability and magnitude of adverse effect.

Thus a statement about "risk" is a description of the likelihood and consequences of harmful effect. We are familiar with parallels from financial investment and from gambling: "There is a very small chance that you will win the jackpot, but a large chance that you will simply contribute to that jackpot", and from weather prediction: "The chance of a snowstorm is 80%".

Risk may be expressed in many ways: number of lives lost per year, or average shortening of lifespan, or degree of loss of hearing, or frequency of chromosomal mutations.

Once the size of a risk is estimated, whether by intuitive guestimate or by formal empirical analysis, larger personal and social decisions have to be made about whether to bear that risk, taking into account many normative factors. I often find it useful to enquire into the "acceptability" of a risk, with "acceptability" having several meanings.

From our dreary experience with personal misjudgments and technological surprises, it is only too clear that we are not able to size up all threats with equal precision and accuracy. At one extreme are untoward events that are so familiar that they can be expressed actuarially: the classic example is highway fatalities, the statistics on which are tallied regularly and which do not vary in number or pattern very much from year to year. (Of course, although we know these numbers for groups of people, and can use them in calculating the odds of misfortune for given individuals, we are helpless to know for sure what will happen to particular people.) At this extreme, then, for risks that have a long recorded history of repeated events, we know the odds. That is, this knowledge is recorded within our social ledgers; individuals may or may not avail themselves of this knowledge, and even those who bother to enquire about the numbers may not be able to grasp what the numbers really mean or how they compare with other such risk numbers. At the other extreme are risks about which we have almost no knowledge. Surely we in this room are exposed to unknown or barely-known hazards at this very moment: trace chemicals unknown to the best chemists, antigens not yet identified, ultrasonic vibrations, microscopic flora breeding luxuriously in the carpets and in the dark crannies of the ventilating system, weak electromagnetic radiation.... Even hazards we partially understand, we have trouble predicting with specifically: how bad the 'flu epidemic will turn out to be, when the earthquake will occur.

Notice that across this entire range, from well-known hazards to ones only imagined, estimates of risk can be made using the tools of science: empirical knowledge, developed systematically within the prevailing orthodoxy of the scientific community, subject to the tests of repeatability, control, and the other guides of Western science, and evaluatable in retrospect on grounds of predictive power. Refinement of the notion of "scientific objectivity" has been one of the major accomplishments of this century, but this concept still is not widely understood. I will risk slighting this complex development by simply arguing that objective, scientific "truth" is knowledge that is subjectively endorsed by the scientific community. I recognize full well that scientists define their own "community", that this community is self-perpetuating, that it sets its own rules, and that the very essence of science is to pursue its orthodoxies so hard, to extend them further and further, that those orthodoxies take a pratfall and have to be replaced. Good science is science that "works": science that can predict with consistency and generality and accuracy what will happen in the physical and social world.

The point I want to make by this is that estimates of risk, whether made by scientists or by laypeople, cannot escape containing elements of "subjectivity", of human opinion. Subjectivity enters into the very defining of the questions, and into the designing of the experiments used in assembling evidence, and then into the weighing of the social importance of the risk. Therefore we should not be surprised when scientists disagree among themselves or when the lay public views a risk differently from the experts. Non-scientist officials are fond of ridiculing scientists for failing to reach internal agreement, and sometimes make the assumption that lack of scientific consensus allows the conclusion that the hazard is negligible: "Danger I don't know, must be trivial". Science is a matter of voting, directly as experts debate, or indirectly as they adopt, refine, or ignore putative "facts". Organizations of scientists are not much different, as they exert collective judgment and develop communal biases. We will be hearing more about this later.

The U.S. Weather Bureau performs a useful educational function as it announces its forecasts. These predictions are naturally the butt of many jokes — surely this is an occupational hazard of the meteorological profession — but I would argue that the basic approach is sound and is a model for other risk predictions. What does it mean to say that "there is a 30% chance of rain"? This statement is defined to mean that for a given geographic area, on 30% of the days for which such a forecast is made, it actually does rain a predefined minimum amount; on seven out of ten such days it should not be expected to rain. Contrary to cynics' beliefs it is possible for the weatherfolks to find themselves "wrong". Periodically the National Weather Records Center's computers evaluate individual weather stations' forecasting records and recommend adjustments: "On such-and-such kinds of days you tend to underestimate snow by 10%". Thus what is striven for is predictive validity.

Unfortunately, especially with infrequent or "freak" events, many people have trouble interpreting risk estimates. There are those who criticize what they call hysteria and overanxiousness in the 1976 rush campaign to immunize all Americans against swine 'flu. The single early swine influenza fatality at Fort Dix did not turn out to presage an epidemic, or, worse, a repeat of the 1918 pandemic that took twenty million lives around the world. I don't approve of the particular actions the Ford Administration took. But our hindsight — the knowledge that the pandemic did not in fact materialize — does not at all give us license to say that it was foolish for public health experts, the President, and others to become quite apprehensive about the disease and seek precautions; had the 1976 epidemic slaughtered the number of people it did in 1918, the 1976 precautions would have seemed irresponsibly slight.

More complicated issues surround the nuclear fiasco at Three Mile Island. Toward the end of the crisis period, as the reactor cooled down and the hydrogen bubble shrank, two extreme views emerged. One view was that the accident was a terrible event that showed that the nuclear technocrats had gone entirely too far, had failed to take account of human operator error and incompetence, had come within a hair's breadth of totally losing control of the machine, and had finally given the nation reason to close down the entire nuclear industry. Opposed to that was a conviction that the system had "worked", had vindicated the prior risk assessments, had demonstrated that the reactor could be kept under control even through a series of unprecedented and unanticipated extreme disturbances. Which opinion people favor seems to depend a lot on whether they like nuclear power in the first place. I admit that I took advantage of the Three Mile affair's extraordinary coincidence with Jane Fonda's movie, and adopting columnist Herb Caen's witticism, entitled several lectures on the accident, "Best Supporting Reactor". But I was quite offended by those who pooh-poohed it as simply a "media event", as Edward Teller did and as the *National Enquirer* did with its screaming headline, "EXCLUSIVE: NUCLEAR PLANT CRISIS A HOAX." How serious the threat was depends on how close to disaster the accident came, how many people might have been hurt how badly, how quickly help could have been rendered, and so on. What should not be overlooked is the probabilistic nature of the whole affair. What also should not be overlooked, in the large view, is how the risks of the entire nuclear industry compare with the risks of alternative energy sources.

CLASSIFICATION OF HAZARDS

Let us for a moment take note of six classes of hazard and point out distinguishing features of each. The six categories I would point to are:

1. infectious and degenerative diseases;
2. "natural" catastrophes;
3. failure of large technological systems;
4. discrete, small-scale accidents;
5. low-level, delayed-effect hazards; and
6. sociopolitical disruptions.

Obviously these overlap.

Infectious and Degenerative Diseases — Some infectious diseases, such as tetanus, syphilis, and pneumonia, are perennials. Others, such as influenza, recur, but in slightly varying form. Still others, such as bubonic plague, surface quite sporadically. The outcome of the duel between our immune systems and the microbial invaders is not easy to predict. Immunizing against, and deploying antibiotic chemicals against, pathogens may itself be hazardous. Most of the degenerative diseases, such as cirrhosis and arthritis and heart disease, are poorly understood; the main generalization that must be made is that these diseases will always, by definition, administer the *coup de grace* — if people live long enough to succumb to them. And as medical science or "clean living" overcome each such finality, other terminal conditions will surface. Just as our conquering of pneumonia and tuberculosis left us vulnerable to cancer (albeit at a later age on average), the conquering of cancer will leave us vulnerable to other ways of shuffling off this mortal coil.

Natural Disasters — This year, which has on average been like most years, we have had floods in Texas and other parts of the West and Midwest, earthquakes in New England and in the Southeast strong enough to bring into question the earthquake resistance of nuclear power plants there, the usual earthquakes in California, tornadoes in many places, and in the Gulf states, male-named hurricanes. Other countries have had even worse problems. These Acts of God are sporadic and hard to predict, although their mechanisms and effects are pretty well understood. We have some ability to influence the courses of some of them, such as floods and hurricanes, but even then our risk-aversion attempts are often weak. All we can do is assume that such episodes will recur and take measures to limit and buffer damage.

Failure of Large Technological Systems — Failure of dams, power plants, airplanes, and ships, from structural malfunctioning, abuse, or amplification of the effects of natural disruptions, are typified as being of low probability but potentially high consequence. Assessing such risks depends upon elaborate analytic modelling, based on tenuous basic assumptions. In essence, analysts study the known and inferred failure possibilities of the parts, then the larger subsystems, then the interactions of the subsystems, then the backup safeguards, then eventually cumulate these sub-odds into overall failure potentials; then they consider the site context, then the effects local geology and weather can exert on the system, then the profile of the human population at risk, and so on. In this long chain of surmises there is plenty of room for misjudgment. Such nightmares as independent simultaneous failure of several parts of the structure are hard to predict. Small machines not wedded to a site can be tested all the way to failure, perhaps repeatedly; but huge structures such as dams can never be pushed through such overload tests. Nobody knows how to predict the success of evacuations. An Achilles Heel of engineering analysis will always be human operator error: numerous shipwrecks attest to the power of operator error to mis-steer the machine despite everything that designers can do, and such accidents as the 1975 Browns Ferry fire, ignited by a technician using a candle to test for an air leak, and the fire that took the lives of the three astronauts of Apollo I while it was still on the launching pad, remind us that we simply commit the sin of pride when we think we have been so smart as to have forestalled absolutely every possibility of failure. Further, to strive to ensure against human error by making an engineering system fully automatic simply submits that system to the host of troubles that infernal devices get into when left to themselves.

Discrete Small-Scale Accidents — Discrete highway, workplace, and sports accidents take an enormous toll. They have the property of being, in general, fairly well understood: things break, children act like children, drunks drive like drunks. Usually, too, the risk can be reduced if there is sufficient will to do so. Such recur-

References p. 14.

ring accidents have a "numbing" effect, and although people may genuinely deplore a hazard, they may tend to understate its magnitude. One feature of, say, consumer product hazards is that because the threat is discrete and identifiable, society is easily tempted to take quite specific action on the hazard, such as regulating its design or use; although in and of itself this may be desirable, it tends to leave us with a huge and unwieldy set of regulatory strictures and to lead, in the aggregate, to hazard-reduction investments that are not "rationally" apportioned.

Low-Level Delayed-Effect Hazards — One of the most frustrating categories of risk is that of low-level, delayed-effect hazards. Such examples as asbestos, vinyl chloride, and other chemicals come to mind, as do radiation, noise, and psychological stress. Analysis of such effects requires assessing exposure, proving that exposure leads to harmful effect, and then gauging the extent of the effect across society. Tests, which often involve epidemiological surveys or experiments on animals, are made difficult in many cases by long latency before onset of effect, by the need to administer massive doses in order to be able to observe effect, by the very large number of test animals required for statistical significance, and so on.

Sociopolitical Disruption — The last category of risk I would mention is sociopolitical disruption. Under this cumbersome rubric I include the terrorism and nuclear-weapons-proliferation risks associated with nuclear power, the oil-embargo risks of petroleum dependency, and the like. Such risks are difficult to predict, of course, and hardly lend themselves to quantitative analysis.

LIMITATIONS OF RISK DECISIONS

Risk decisions meet with four kinds of limitations: of empirical analysis, of social value appraisal, of "risk management", and of the assignment of risks and responsibilities.

Empirical Analysis — One set of limitations has to do with empirical analysis, which many of this symposium's speakers will be discussing. The goal of techniques labelled "risk—benefit analysis" or "decision theory" is usually to map out the probabilities, consequences, and social worth of alternative possibilities, with the intent of constructing some kind of social "balance sheet" that can provide guidance to decisionmakers. The analysis is thus a problem of handicapping what will happen and intercomparing quantities that are rarely expressible in common-denominator terms.

With some (few) well defined projects for which goals and constraints are agreed upon by the major affected parties, and for which health and environmental risks, costs, and benefits are well known and understood, not only in magnitude but in social distribution, over both the near and long term, risk—benefit accounting has proven itself useful. Under such (rare) circumstances of certainty, commonsensical back-of-the-blueprint estimates as well as more formal analyses derived from operations research are applicable. These tend to be favored by specialists, technical or otherwise, who have been given a job to accomplish. The occasional "suc-

cessful" application of such techniques — and, one suspects, also the all-embracing ring of their title — is increasingly leading legislators, administrators, and
judges to call for their use. Undoubtedly versions of "risk—benefit analysis" will
continue to be employed, since all decisions, even our private ones, must involve
the weighing of good against bad, gain against loss.

Uncertainties arise because technical experts simply are not gifted with omniscience: nobody knows how to predict the timing and force of tornadoes or earthquakes. Scientific "proof" is still circumstantial; smokers are statistically more
likely than non-smokers to develop some kinds of cancer, but this sort of proof still
leaves many people unconvinced. From personal experience and from many dramatic accidents we realize that it is very difficult to take account of human error:
"You can make it foolproof, but not damn-fool-proof"; humans make human
mistakes. It is hard to anticipate the risks from abuse of technology, such as terrorism, theft, or sabotage abuse of nuclear power plants. Some of the most chastening
problems are those with which we simply have not had much experience: with
recombinant-DNA research, what is at question in part is the hazard of the very
experiments that are necessary in order to appraise the hazard.

Surely humility is called for on the part of analysts. I would point out that for
both of the prime contending energy systems, coal and nuclear, after all the mining
risks, transportation risks, powerplant accident risks, and known and speculated
health risks are estimated, state-of-the-art appraisal is seriously worried that potentially the most hazardous effects of these technologies are at the same time those
that are the least quantifiable. For coal, the Department of Energy and a panel of
advisors headed by Dr. Gordon MacDonald warned in June, "It is the sense of the
scientific community that carbon dioxide from unrestrained combustion of fossil
fuel is potentially the most important environmental issue facing mankind" [1].
For nuclear, the 1977 Ford Foundation/MITRE Corporation review concluded,
"In our view, the most serious risk associated with nuclear power is the attendant
increase in the number of countries that have access to technology, materials, and
facilities leading to a nuclear weapons capability" [2].

Social Value Appraisal — The second set of limitations on risk decisions are
those of social value appraisal. Again, this audience knows all too well the problems of "weighing" risks against benefits, predicting costs, assigning monetary
value to amenities and intangibles, dealing with distributional effects, and even
pricing human life itself.

Whereas some earlier laws called for protection of human health "without
regard to cost", more recent legislation has tended to require that costs be considered explicitly. This is the case, for example, with the Occupational Safety and
Health Act, the Toxic Substances Control Act, and the Safe Drinking Water Act.
Unfortunately, the legislation has not provided much guidance on how the
appraisals are to be conducted, mostly specifying only that "unreasonable" risks
be avoided.

For some purposes it is useful to try to take some measure of the levels of risk
society is willing to accept — with "accept" having a range of denotations, from
willing endurance, to passive acquiescence, to fatalistic stoicism. We will hear of

various approaches to this.

In an important test case, this spring the Food and Drug Administration proposed a new scheme for setting the standards for exposure to the hormone diethylstilbestrol (DES), used for fattening beef cattle but known to induce human cancer when ingested. The Commissioner argued that "the acceptable risk level should (1) not significantly increase the human cancer risk and, (2) subject to that constraint, be as high as possible in order to permit the use of carcinogenic animal drugs and food additives as decreed by Congress . . . a risk level of 1 in 1 million over a lifetime meets these criteria better than does any other that would differ significantly from it." The agency is still pursuing that proposal, which would if adopted, be one of the first regulations explicitly to acknowledge a specific contribution to the overall human carcinogenic burden [3].

This month a landmark case will come before the Supreme Court regarding the necessity for the Occupational Safety and Health Administration to "demonstrate" that the net social benefit from tightening the standards for occupational exposure to benzene by an order of magnitude (as proposed by OSHA) exceeds the net cost of that regulatory action. No matter how it comes down, this high-court ruling is bound to set precedent for regulatory analysis, even beyond the occupational domain. [4].

There definitely are situations in which formal analyses can help: in clarifying the questions, in making underlying assumptions explicit, in anticipating consequences, and in describing the options and trade-offs. Even if these analyses do not themselves decide the public debates, they can inform and assist. A not inconsequential aspect of them is that they usually are embodied in documents that can be criticized. Their biggest liability is overreaching.

Risk Management — The third cluster of limitations has to do with what I will call, for shorthand, social "risk management". Society has available to it numerous ways of "dealing" with risks: it can reduce risks, buffer and share them, and remedy and redress their damage. For these purposes it employs instruments of market economics, government regulation, legal injunction and redress, and insurance and indemnification.

"Regulation" has become one of the cursewords of our times. We hear constant complaint that life in our country is over-regulated; what people usually mean is that personal and economic freedoms have been restricted, that bureaucratic structures are clumsy and wasteful, and that progress is unstable and innovation stifled. The trend of effort, from all sides, seems to be to learn how to simplify; how to select the most important cases, work them through, and establish benchmarks; how to streamline procedures; and how to deregulate where possible. For example, the Administration has been developing ways of assigning toxic chemicals to categories by level of risk and seeking generic approaches to their regulation, rather than risking regulatory catatonia by continuing the piecemeal approach. The latest National Academy of Sciences report on food safety, published in March, urged generic approaches:

> The report recommends: (a) that there be a single policy applicable to all foodstuffs, food additives and food contaminants and that the public official

responsible for implementation of that policy be given sufficient flexibility to factor risks, benefits and other considerations into account when making a decision concerning a material that has been called into question, (b) that there be available to that official options other than decisions simply to ban or not ban, and (c) that to facilitate such implementation, materials under consideration first be categorized as exhibiting low, moderate or high risk [5].

The same thing is happening with pesticides; a committee of the National Research Council will soon deal with this in a report entitled, "How to regulate 35,000 pesticides". The Nuclear Regulatory Commission has established "Generic Environmental Standards for Mixed Oxides", covering the oxides of uranium and plutonium that are coming into increased use as fuels. The National Earthquake Hazards Reduction Act of 1977 mandated extensive work on contingency planning, seismic building codes, earthquake prediction, and the like; these efforts, along with others, are being consolidated into a new Federal Emergency Management Agency. Spurred in part by the spectacular rupture of Teton Dam in 1976, the federal government has reviewed the country's dams and concluded that "at least 20,000 of the 2,100 Federal and 47,000 private dams in the nation are so located that failure could result in the loss of human life and appreciable property damage"; an ongoing review — which has already reported over 300 unsafe dams to state governors — continues. Again, the movement is toward systemization.

Assignment of Rights and Responsibilities — The fourth cluster of limitations are limitations on the assignment of rights and responsibilities. Many risk decisions hinge on issues of rights: the right of a terminally ill patient to attempt his own cure, even if that treatment is not sanctioned by officialdom (as with this June's Supreme Court decision that laetrile may not be sold to cancer victims); the right of government to force certain risks upon the public (as with national security actions); the right of government to require risk-aversive actions (such as mandatory wearing of motorcycle helmets); the right of an individual to refuse to avoid risks (as with refusing to wear automobile seatbelts, or refusing to quit smoking); the right of a minority to embrace practices that can threaten the majority (as with the question of whether Amish children, many of whom are currently afflicted with polio, have a right to remain unvaccinated); the right of those now alive to leave hazardous bequests to future generations (as with current decisions about long-lived nuclear waste, or synthetic genes); or the right of one country to transfer its hazards onto another (as may happen when experimental pharmaceuticals are tested on developing-country subjects).

Other issues are those of responsibility: the responsibility of technical people to take cautionary or protective action on behalf of the larger society; the obligation of manufacturing firms to look out for the health of their workers and clients; the duty of governments toward their citizens; the international stewardship of nations and firms. In this regard one of the most troublesome growing issues concerns the increasing diffusion of responsibilities through the ranks of corporations and the layers of bureaus. It now appears that the DC-10 wreck will be traced, not to one single careless action, but to a combination of systems weaknesses and regulatory

References p. 14.

and inspection oversights for which many different organizations and individuals were responsible in a general but diffuse sense.

ENVOI

Thus the agenda, for society and for this Symposium, is full. I wish us good fortune with it.

REFERENCES

1. P. Shabecoff, "Increase of carbon dioxide in air alarms scientists," New York Times, June 9, 1979.
2. Nuclear Energy Policy Study Group (sponsored by the Ford Foundation and administered by the MITRE Corporation), Nuclear Power Issues and Choices, Ballinger Publishing Company, Cambridge, Massachusetts, 1977, page 22.
3. U.S. Food and Drug Administration, "Chemical compounds in food-producing animals. Criteria and procedures for evaluating assays for carcinogenic residues," 44 Federal Register, 17070-17114, March 20, 1979.
4. "American Petroleum Institute versus Occupational Safety and Health Administration," U.S. Supreme Court docket number 78-1036.
5. Committee for a Study on Saccharin and Food Safety Policy, National Academy of Sciences, Food Safety Policy: Scientific and Societal Considerations; this summary paragraph appears in the prefacing letter of transmittal to the U.S. Secretary of Health, Education, and Welfare, March 1, 1979.

DISCUSSION

E. V. Anderson (Johnson and Higgins)

You've mentioned risk management by government. Isn't the larger problem risk management by the individual and the degree of knowledge he is given so he may take action?

W. Lowrance

I'm sure of that, but all I can comment on is that individuals are not likely to do that very systematically, as you know, and I wonder whether we're getting any better at it as a society. I'm not sure, I think it's the nature of human beings to jump on one issue after the other and I have not sat down myself to think systematically about all cases, I simply haven't thought about it very much. But I do think there will be sporadic outcroppings of new decisions for society and the individuals participating within it.

C. Starr *(Electric Power Research Institute)*

You mentioned two examples of scientific analysis. One was the CO_2 problem and the other was the nuclear proliferation problem. I just happen to be involved deeply in following these and I think they illustrate a very important point. The statement that came out of the National Academy committee on CO_2 I consider a gross technical exaggeration based upon very flakey — proper term — technical information. This is the same group, incidently, that predicted that the supersonic transport would have a deleterious ozone affect. The latest data show that the effect is just the opposite of what this group predicted. The fact is that one of the hazards we run is premature conclusions based upon very elementary scientific indications and the drawing of cataclysmic consequences. In this particular case, the whole concept of depriving the world of the use of fossil fuels is not only cataclysmic but one which the world does not obviously accept, though there is a very serious issue in what I consider bad information.

In the case of the nuclear proliferation issue, the Ford/MITRE study has been very severely critizèd by those who know the subject well, simply because that study was written with a presumption that the plutonium economy was dangerous, and in fact all the experimental evidence is that nuclear weapon technology and military technology is proceeding apace without having anything to do with the nuclear power technology. So that in both cases the so-called scientific community that's involved in making a decision is leading the world either if not totally astray certainly confusing the issue. I think it does create one of the problems in risk analysis; that there is a tendency, based upon very limited information, to jump to policy conclusions which may not be warranted.

W. W. Lowrance

W. Lowrance

Well, I agree with you on the last point. We often jump to conclusions that are not warranted.

I would like to speak to both of your points. On the CO_2 question, I am not able — not being a geophysicist — to follow all of the research. I have read the papers. It seems to me that when Gordon MacDonald and others raised this issue, it was responsible to foster debate, because in the worst scenario one can imagine, the CO_2 build-up could be terrible for the world. To raise the issue was proper, given the fact that we only have something like five years or so to *focus on the question.* I agree that a lot of people jump to hasty conclusions and make hasty warnings, but I'm glad the issue has been raised and I happen to think we ought to continue to put millions of dollars into that research and ask the questions even better. So I don't find it irresponsible science.

On your second question, I, too think that nuclear weapons proliferation is not easily amenable to scientific, or even systematic, analysis. I don't agree that nuclear weapons research will proceed quite independent of the spread of civilian nuclear technology. I also must take exception to your saying that those who conducted the Ford/MITRE reports simply didn't know what they were talking about. The problem is that those on the "other side" are people who have known nuclear power all too well over the years, and a lot of people from the nuclear community did meet with that panel. I don't agree with all of the report's conclusions either, but I don't think we can write off the proliferation threat as being un-important just because it doesn't lend itself to numerical analysis.

C. Starr

Just one quick point. I didn't say that they didn't know what they were talking about. I said they started with the presumption that the plutonium economy was inherently dangerous.

W. Lowrance

I agree that they probably had that prejudice.

C. Starr

And in the case of CO_2 I think everyone who studied it agrees that intensive geophysical research, world wide, has to go on in terms of the effect. I don't want to prolong the discussion but it illustrates a specific point. You made a statement that if the CO_2 levels go up that this would be cataclysmic. It is a fact that the kinetics of the geosphere itself lead to climatic changes. There are people who've looked at the problem who said that the climatic changes might be beneficial.

W. Lowrance

I Agree.

C. Starr

So you can't just draw the conclusion that this would be cataclysmic. What I am cautioning against is the tendency of the scientific and technical community to draw specific conclusions rather than a range of possibilities from very inadequate data. And we've done it in the past and we're likely to do it in the future and it is something I think we ought to be aware of.

THE UNCERTAIN RISKS WE RUN:
HAZARDOUS MATERIALS

Marvin A. Schneiderman

National Cancer Institute, Bethesda, Maryland

ABSTRACT

The problems of risk assessment are discussed from the dual point of view of:
1. Who is exposed — and to how much and
2. Given the estimated levels of exposures, what estimates of risk can be developed.

The major issue of concern is that in general there is inadequate data on who is exposed, and almost no data on how much exposure there is. Individual exposure monitoring is rarely done (except for such things as radiation, through radiation badges) making it impossible to associate individual exposure with subsequent disease — if any. It is often difficult to know how many employees are exposed to a given agent. Before monitoring is undertaken an agent usually has to come under suspicion, thereby making inevitable unmonitored exposures to materials later found to be hazardous.

The mathematical models for extrapolation from species to species vary enormously in the predictions they yield, even when one accepts all the assumptions about species-to-species jumping. Recent work derived from Whittemore and Keller would seem to imply that in a multi-stage cancer process, the initiator stage dose-effect may pile up early in the life history of exposure, implying that several short term exposures to different carcinogens may create more risk than one long-term exposure. Promoting agents, according to this model, on the other hand, seem to have a different mode of action — with far more rapidly diminishing residual effects — so that the promoter may "have to be there" for its effect to be manifest. These different modes of action have implications for research and applications for prevention.

INTRODUCTION

In the past several weeks the newspapers have carried accounts of environmental hazards that seem both frightening and essentially unquantifiable. What health damage could we expect (i.e. quantify) from that PCB transformer spill in Montana that led to destroying millions of eggs and chickens in the thousands? What were the health risks created by that sloppy factory in Arizona that led to school lunches, and school children, becoming contaminated with radioactivity? What's going to happen here in southern Michigan, now that "chemical contamination may be so widespread . . . that . . . we may find it cheaper to simply write off the ground-water supplies of large portions of southern Michigan." What should we do? Let'em drink scotch whiskey? That's been found to contain nitrosamines — known carcinogens for several animal species. Use more surface water? Surface waters usually contain more contaminants than ground water. Have the authorities over-reacted: Are these things really not harmful? Not very harmful?

There are people who argue that the harm has been exaggerated. If we look at the increases in lifespan in this century and the reduction in death rates from most diseases other than cancer (Fig. 1, Table 1) give credence to the argument that some aspects of our environment must be getting better. I think we can safely say that humans have handled, and improved (for humans, at least), both our bacteriological environment, and much of our physical environment. Good heating in the winter and cooling in the summer have surely helped reduce the swings in winter-to-summer mortality from respiratory diseases [1], and, with better food handling, have also reduced things like "summer diarrhea" in children. What we are still struggling to handle as well is our chemical environment and perhaps those parts of our physical environment associated with ionizing radiation.

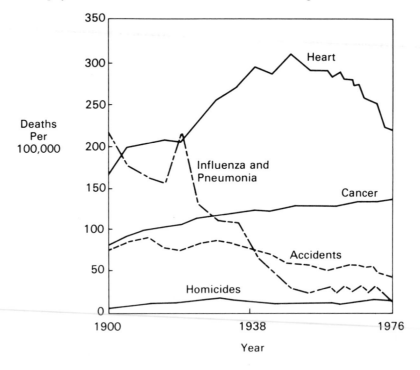

Fig. 1. Age adjusted death rates for United States.

TABLE 1

Age-Adjusted Death Rates for Selected Causes of Death,
United States, 1900-1978

Year	Cause of Death				
	Malignant Neoplasms	Disease of the Heart	Influenza and Pneumonia	Accidents	Homicides
1900	79.6	167.3	209.5	75.3	1.2
1905	90.9	198.7	175.5	85.4	2.1
1910	97.0	201.7	163.0	88.4	4.5
1915	100.8	206.3	154.7	77.4	6.0
1920	104.9	203.6	213.1	74.0	6.9
1925	112.5	229.6	128.1	81.9	8.5
1930	113.4	252.7	108.2	84.6	9.2
1935	117.5	269.0	109.2	80.7	8.6
1940	120.3	292.7	70.2	73.1	6.3
1945	119.9	282.4	45.6	68.7	5.8
1950	125.4	307.6	26.2	57.5	5.4
1955	125.8	287.5	21.0	54.4	4.8
1960	125.8	286.2	28.0	49.9	5.2
1961	125.4	278.6	22.1	48.1	5.2
1962	125.6	282.7	23.7	49.7	5.4
1963	126.7	285.4	27.7	50.9	5.5
1964	126.7	276.9	22.8	52.1	5.7
1965	127.9	275.6	23.4	53.4	6.2
1966	128.4	275.8	23.8	55.6	6.7
1967	129.1	267.7	20.8	54.8	7.7
1968	130.2	270.0	26.8	55.1	8.2
1969	129.7	262.3	24.6	55.3	8.6
1970	129.9	253.6	22.1	53.7	9.1
1971	130.7	252.0	19.3	52.0	10.0
1972	130.7	249.3	20.8	52.0	10.3
1973	130.7	244.4	20.1	51.7	10.5
1974	131.8	232.7	16.9	46.0	10.8
1975	130.9	220.5	16.6	44.8	10.5
1976	132.3	216.7	17.4	43.2	9.5
1977	133.0	210.4	14.2	43.8	9.6
1978	133.2	207.3	15.4	45.3	9.9

SOURCE: U.S. Dept. of HEW, National Office of Vital Statistics, *Vital Statistics Special Reports*, Vol. 43, Nos. 1-31, October 1956.

U.S. Dept. of HEW, NCHS, *Vital and Health Statistics*, Series 20, No. 16, March 1974.

U.S. Dept. of HEW, *Vital Statistics of the United States*, Vol. II, Mortality, Part A, 1970-75.

Monthly Vital Statistics Report, Final Mortality Statistics, 1976, Vol. 26, No. 12, Suppl. 2, March 1978.

References p. 37.

People are now questioning whether we really need to do much about our chemical environment. They argue that no great harm has yet shown itself from our living in an environment in which synthetic chemicals are much more common. Cancer mortality and incidence rates do continue to increase, slowly, however, while almost all other diseases show decreasing mortality rates. Much (but not all) of that increase is related to cigarette smoking, the major personal pollutant that we have added to our environment in the last 60 or so years. Some people would like us to believe that most of the increase is due to our own personal inadequacies and indulgencies — smoking too much, drinking too much, eating too much of the wrong things, etc. However, I (and others) have estimated that if we remove all of the cigarettes' contribution to cancer rates, for example, we will still be left with increases in cancer mortality [2]. Many people ascribe this increase to the increasing "chemicalization" of our environment. We must consider that as a real possibility. But if we are to protect ourselves from risks possibly arising out of the chemical environment we must recognize that they are still uncertain risks. We don't really know their magnitude, and the question is, will we find ways to protect ourselves in a pre-intraocular traumatic era? That is, before the data hits us between the eyes.

Eric Ashby, in the book *Reconciling Man with the Environment* [3], asks what we will do with the fact that all future generations of mankind "will have to live with problems of population, of resources, of pollution" that we are creating today (or created yesterday). Unlike the man who asked "Why should I do anything for posterity? What has posterity ever done for me?", Ashby wants to know if man can "adapt himself to anticipate environmental constraints? Or will he (like other animal societies) adapt himself only in response to the constraints after they have begun to hurt?" In our society it is obvious that there is a lot of reacting, and a little anticipating. The anticipating I see as cause for optimism. The Clean Air Act, the Safe Drinking Water Act, and TOSCA show humans as anticipating animals that probably can respond to constraints before being hit between the eyes. How we use these laws, how we deal with possible hazards, will show how much basis there really is for optimism.

Anticipation in a positive sense requires knowledge. Negative anticipation, worry, involves fear of the unknown, which feeds on the lack of knowledge, which of course derives from a lack of information. Unfortunately, this paper deals with much of what we don't know, which then may hamper our successful anticipation, and leave us frightened. I see us faced with problems of operating in the face of limited information. I indicate here, how limited some of that information is.

LIMITATION OF INFORMATION

By way of example of our difficulties in effective anticipation I will talk about nine materials, not all new synthetic organic chemicals, that have some possible link with cancer. I will talk about what we know about these materials, what we don't know, and how that could hinder us in a completely rational positive anticipation of the kind that Ashby talks about. Six of the materials have been extensively reviewed by IARC in their series of volumes on potential (or actual) carcinogens, and I have drawn heavily on their summaries. Table 2 summarizes some of the pertinent material. For three of the materials — trichlorethylene, ethylene dibromide and perchlorethylene — I have not used the IARC as a source, but rather the NCI carcinogenesis testing program documents, the reference numbers of which are listed in the table.

The Figs. 2 through 10 show changing production or use data for these materials from about the time of World War II to the mid-1970's.* In at least one case there are no recent published production, or use, data.

Table 2 may be used to indicate what has been suggested as a basis for concern. From it we may be able to pinpoint some of our difficulties in anticipating the effects of exposures to these materials. These materials are ones which have been found to be carcinogens in animal systems, and, in some cases, humans. The concerns expressed are related to either the continued high use of these materials, the substantially increased use of some of them since World War II, and, in the case of arsenic, the lack of information about production, or use.

The argument about risk that Figs. 2 through 10 imply is an indirect one. If a material is a carcinogen, and more of it is being produced or used now than some time in the past, then there is reason to be concerned that it may be producing more cancer. That more material is in use today of course does not mean that more people are necessarily exposed. Where human data have shown carcinogenic effects, it is necessary to examine current (and recent past) exposures to see if they are similar to the exposures that led to the discovery of the carcinogenic effect in the first place. Thus it can be argued that WW II exposures to asbestos were probably very much higher than present-day exposures. Although asbestos use has apparently doubled in the United States since WW II, it is possible that work place exposure may be decreased, because of better work practices, exposure regulation, and changes in product mix. Thus risks should now (or soon) be lower than in the past. This may be true. But how can we see the effects? Asbestos-related cancers have a long latent period, and (for lung cancer, at least) are also related to cigarette smoking. The establishment of work exposure standards in 1968 (since lowered, with recommendations for further lowering) and the anti-smoking campaigns of recent years should not be expected to show much effect for some years to come. For another material, nickel, data I have received from Norway [4] seem to show that refinery workers hired after 1960, by which time the refining process had been substantially changed for many years, are not at the increased risk of lung cancer that earlier workers were. So increased use or consumption is not necessarily in one-to-one correspondence with increased exposure. This raises questions of how we can use both the relative risk data developed in the past and the production data to give us reasonable estimates of future risk.

---

*I am indebted to Mr. Brian Magee of the Environmental Law Institute for preparing these data.

References p. 37.

TABLE 2

Some Potentially Dangerous Materials: Use and Possible Worker Exposure

Material	Use Production Levels Estimated Number of U.S. Workers Exposed (NOHS-1972/74)	Carcinogenicity Mutagenicity	Notes: Some Restraint and Regulatory* Action in the United States
Acrylonitrile and co-polymers	Piping and fittings. Automobile components. Electric & electronic equipment and appliances. 1940-45: near zero 1976-77: 1700x10⁶ lbs. Workers: 184,000	Carcinogenic in rats (for stomach, brain, zymbal gland). No data on several co-polymers. Humans: (preliminary) increased lung and large intestine cancer (synthetic fibres plant). Mutagenic.	May not be used in packaging for alcohol-containing materials (foods) or for beverage containers.
Arsenic and inorganic arsenic compounds (many forms)	Alloys, pesticides, insecticides. Electronic devices, glass, wood. Medicinal, defoliant, herbicide. No U.S. production data published since 1960. Workers: 951,000 NIOSH 545,000 OSHA	Animal data generally negative. Humans show inc. skin cancer and lung cancer (smelter workers, also exposed to other chemicals).	Most medicinal, herbicide and pesticide uses sharply decreased in recent years. Permissible exposure limits for certain forms. Protective clothing.
Asbestos	Insulation material. Water pipes (asbestos-cement). Floor and ceiling tiles. Brake & clutch linings. Shipbuilding. Use in U.S.: 1940-45 - 400,000 short tons 1970-75 - 800,000 short tons Workers: 1,500,000 to 2,500,000	All commercial forms tested are carcinogenic in mice, rats, hamsters, rabbits. Carcinogenic in humans: lung cancer, gastro-intestinal cancer, mesothelioma (interaction with cigarette smoking: lung cancer).	Limited use in insulation, public buildings. Permissible exposure limits established — 1968; reduced in 1975; further reductions recommended.

TABLE 2 (continued)

Material	Use Production Levels Estimated Number of U.S. Workers Exposed (NOHS-1972/74)	Carcinogenicity Mutagenicity	Notes: Some Restraint and Regulatory* Action in the United States
Benzene	Blends in gasoline. Chemical intermediates. Plastics & polymers. Solvent in paints and rubber. 1940-45: 200x10^6 gallons 1970-75: 1300x10^6 gallons Workers: 1,900,000 to 3,000,000	Mouse data appears inadequate. Case reports and case control studies in humans suggest leukemogenesis.	Limitations and standards now being litigated. Permissible limits established. Protective clothing, monitoring.
Chromates (ores, plus many compunds, lead, potassium, sodium, etc.) Chromium compounds	Alloys of steel. Corrosion inhibitor. Pigment. Magnetic tapes. Plating. Ores & concentrates: 1940-45: 500x10^3 short tons 1970-75: 1250x10^3 short tons Workers: 540,000 to 4,400,000	Various compounds are carcinogenic in rats and lung cancer excess among workers. No evidence of non-occupational exposure as a cancer hazard.	Ban on lead chromate paints containing >0.06% lead. Permissible exposure limits for several compounds.
Nickle & Nickle Compounds	Transportation (alloys of steel). Chemicals. Electrical goods. Textile industry. Edible oil hardening. U.S. consumption: 1945: 100x10^3 short tons 1970-75: 175x10^3 short tons Workers: 700,000 to 1,500,000	Some compounds of nickle carcinogenic in rodents upon inhalation, or subcutaneous injection. Humans: excess lung cancers in refinery workers, possibly modified by new refinery processes.	

TABLE 2 (continued)

Material	Use Production Levels Estimated Number of U.S. Workers Exposed (NOHS-1972/74)	Carcinogenicity Mutagenicity	Notes: Some Restraint and Regulatory% Action in the United States
Trichlorethylene	Industrial solvent, PVC, drycleaning. Food processing, fumigant. Anesthetic-analagesic. 1940-45: 100×10^6 lbs. 1970 (PEAK): 600×10^6 lbs. 1976: 300×10^6 lbs. Workers: over 3,600,000 (IARC: 283,000)	Hepatocellular cancer in mice of both sexes. No significant effects among rats (NCI-CG-TR-2). No human carcinogenesis data reported by IARC (Vol. 11, 1976). Operating room effects(?)	Restrictions in decaffinated coffee. NIOSH has recommended standards. U.S.: TLV = 100 ppm (535 mg/M³). USSR: 10 mg/M³. General concerns about volatile anesthetics.
Ethylene Dibromide (1, 2-Dibromoethane)	Gasoline production. Soil and grain fumigant solvent. Chemical intermediate. 1940-45: 50×10^6 lbs. 1970-76: 275×10^6 lbs. Workers:	Carcinogenic to Osborne, Mindel rats and B6C3F1 rats: for stomach hepatocellular cancer in female rats. (NCI-CG-TR-86).	Reduced use because of reduced consumption of leaded fuels.
Perchlorethylene (Tetrachlorethylene)	Drycleaning. Other cleaning and degreasing. Paint remover. Medicinal. 1940-45: $<50 \times 10^6$ lbs. 1974: 734×10^6 lbs. Workers:	Mice, rats-increased hepatocellular carcinoma (early deaths in rats). Liver carcinogen in both sexes of B6C3F1 mice (NCI-CG-TR-13).	(Found in air, water, food.) See Trichlorethylene.

Data from IARC publications – Regulatory information may not be current.

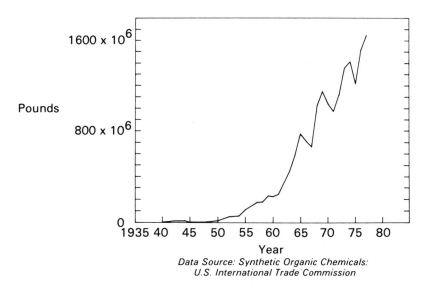

Fig. 2. Acrylonitrile.

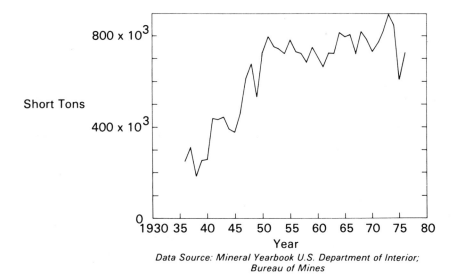

Fig. 3. Asbestos.

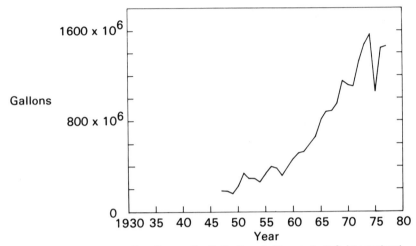

Data Source: Synthetic Organic Chemicals: U.S. International
Trade Commission; U.S. Imports for General Consumption and
General Imports; U.S. Department of Commerce; U.S. Exports;
U.S. Department of Commerce.

Fig. 4.   Benzene (U.S. apparent consumption).

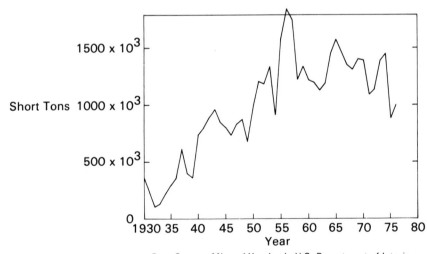

Data Source: Mineral Yearbook, U.S. Department of Interior;
Bureau of Mines.

Fig. 5.   Chromites: ores and concentrates.

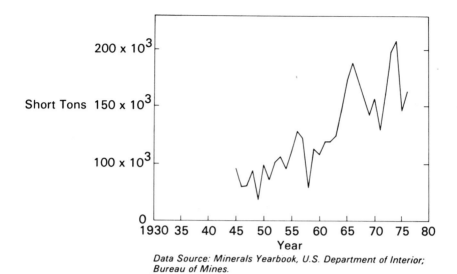

Data Source: Minerals Yearbook, U.S. Department of Interior;
Bureau of Mines.

Fig. 6.　Nickel, U.S. consumption.

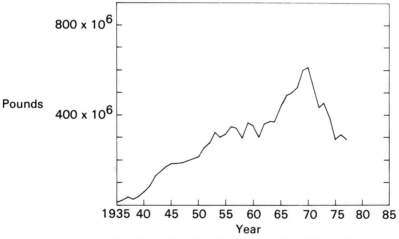

Data Source: Synthetic Organic Chemicals: U.S. Production and
Sales; U.S. International Trade Commission.

Fig. 7. Trichlorethylene.

References p. 37.

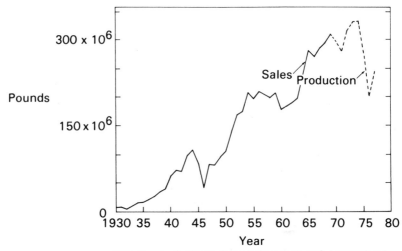

Data Source: Synthetic Organic Chemicals: U.S. International
Trade Commission (1969 - 1977); Restricted Source (1930 - 1968).

Fig. 8.   Ethylene dibromide.

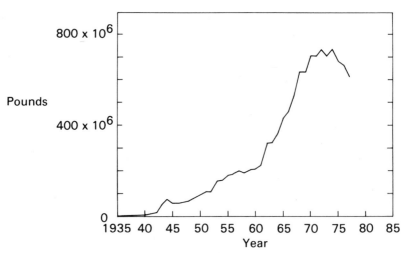

Data Source: Synthetic Organic Chemicals: U.S. Production and
Sales; U.S. International Trade Commission.

Fig. 9. Perchlorethylene.

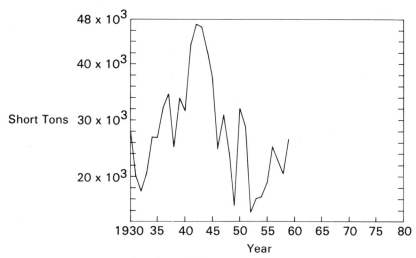

Data Source: Mineral Yearbook; U.S. Department of Interior;
Bureau of Mines. Note: All production data after 1960 is withheld
by the Bureau of Mines to avoid disclosing individual company
confidential data.

Fig. 10. White arsenic.

FROM DATA TO CANCER RISK ESTIMATION

Many of our problems in rational anticipatory behaviour lie in not having what we need to put together to do our risk computations to quantify our anticipation. A logical way to proceed from production or use data to eventually estimating risks might be to:

1. Find out who is exposed, and how.
2. Find out how much the exposees are exposed to.
3. Find out what excess risk that much exposure leads to.
4. Find out what else they are exposed to, and then find out how exposure to that "what else" changes the risk to the first exposure or to the things we are concerned about.

Some of the difficulties in doing this become obvious as we try to go through this process.

Who Is Exposed and How? — In general we have some ideas about who is exposed, but no unchallengeable data. On the face of it data on worker exposure should be the best we have, or at least, the easiest to get. The National Occupational Health Survey (NOHS) sample survey conducted for NIOSH from 1972 to 1974 [5] led to estimates about who might possibly be exposed. The samplers had to be able to identify the materials to which the worker was exposed, of course. Where trade secrets were involved, this information was sometimes not available. The NOHS Survey sample contained 5000 industrial units, not all of which were actually covered. Engineers, with some knowledge of manufacturing processes, estimated who was exposed but not how much or for how long. It should not be surprising that estimates of workers exposed made by different agencies vary enormously. For examples, NIOSH estimates 10x as many workers potentially exposed to trichlorethylene as does IARC. The nature of exposure (constant, sporadic, full time, part time) is generally not known or given. For some materials there are several routes and sources of exposure on the job, in the ambient environment, in air, water, food. In general we know only approximately who is exposed. We know even less *how* he(she) is exposed.

How Much Is He Exposed To? — We know less how much anyone is exposed to of a suspect material. In the epidemiologic studies which have been used to establish the reality of the risks that we are concerned with, exposure information beyond "yes or no," or "a lot or a little" is generally not available. Even with asbestos, a material that has been very extensively studied, only very few studies have been able to give any exposure, or dose-response information. The NOHS reports do not give exposure information, and, in fact, include part time workers as well as full time in the totals. Yet these are the best country-wide estimates of exposure now available, although they are only on a "yes-no" basis.

How Much Risk Does that Exposure Entail? — Risk is undoubtedly dose-related, with higher dose usually (but not invariably) producing higher risks. Since the cancer process requires that an altered or transformed cell must survive to reproduce itself to eventually produce cancer, doses that kill cells may be less carcinogenic than doses that do not kill cells. If there are protective (detoxification) systems operating, doses that overwhelm these systems will be carcinogenic, while lower doses will not. If the protective systems have common pathways, and there are many exposures, as in man, then this generalization may not hold.

Uncertainty From Epidemiological Studies — In estimating risk from human population studies two potential sources of error in the data operate in opposite directions. For something to be identified as a risk in an epidemiologic study, usually a large difference, or a large relative risk must be found. Because of this, published findings may overstate true relative risks. If exposure levels have declined (through regulation, etc.) since these studies were done, then, using the relative risks from these studies will even further overstate current and future risks. On the other hand in computing relative risks, comparisons are often made with populations unexposed to this material, but who may be exposed to other materials which increase their risk. These comparisons will understate the risk to the material we are interested in. Finally we usually have to deal with the "healthy worker effect" which usually leads to understating risks in epidemiologic studies.

As noted earlier, in the studies establishing risk for work exposure to certain materials, doses to the exposed workers are rarely known so that dose response relationships can not be established, other than in the grossest terms. Thus, if standards for exposure are established it will usually not be possible to develop firm estimates of how much illness will develop at the (lower) doses permitted under the standards. Information on adherence to standards is sparse (but is slowly being developed by OSHA), so that even where there are standards, one can not be sure they are adhered to. Responses in the non-worker population, and/or to exposures different in time, delivery, and duration than exposures in the population from which the risk was established, are usually not known, or easily estimated.

Uncertainty From Animal Studies — If we have such difficulty with using epidemiologic data to estimate risk, could animal or other laboratory data give a better indication of risk? Extrapolation from animal studies to man may have several serious limitations.

   a. Assumptions about the relative susceptibility of humans and the animal species. Usually it is assumed, for safety sake perhaps, that humans are as sensitive as the most sensitive animal species tested. This sensible suggestion can lead, however, to both overstatements and understatements of risk: overstatements where some specially sensitive animal species has been tested, particularly where man has evolved in some way to render him less sensitive (e.g. perhaps with respect to exposure to alcohol ingestion which is much more common in human societies than in animal societies, particularly special laboratory strains. Man may have a tougher liver than the mouse.) Understatements come when the appropriate (most sensitive) animal species has not yet been tested; e.g. a material like β-napthalamine which was not shown to be a carcinogen in the usually tested rodent species.

   b. Metabolic differences among species. This is particularly important for materials that have to be converted within the body to a "proximate carcinogen" from the material to which individuals are exposed. The fact of "metabolic activation" is important when we attempt to estimate the dose that should be entered into the mathematical model used for extrapolation. Dose-response models appropriate biologically for the appropriate metabolite but not for the administered dose of the material, will seem to be in error, because the wrong dose is entered into the model.

   c. Choice of the mathematical model (i.e. the shape of the dose-response curve at doses outside those observed in experimentation). Several mathematical models are available with more or less biological validity. Frequently, in practical problems many of the models fit the data equally well, although they yield substantially different extrapolations at the (usually) lower doses in which we are interested. In an attempt to find the effects of different models on the extrapolations made to low doses, or to estimate doses needed to produce low response levels, four different mathematical models were fitted (Fig. 11) to some animal data from Maltoni on vinyl chloride [6]. Aiming for a response of 10^{-8}, given the dose level which produced a 1 in 100 response rate, the fractions of that dose required by each of four models were:

One hit model	10^{-6}
Logit (slope: -3.454)	10^{-4}
Probit (Mantel-Bryan)	
slope:1	10^3 to 10^{-4}
Two hit model	10^{-3}

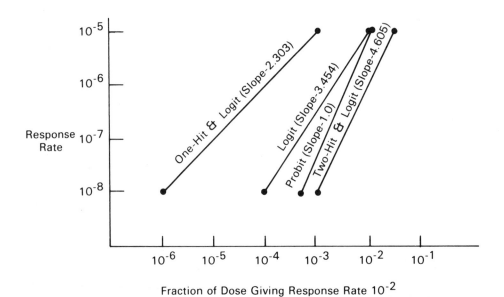

Fig. 11. "Response" to vinyl chloride.

Similar findings were recorded recently in the NAS report on "Saccharin: Technical Assessment of Risks and Benefits" [7]. (Table 3-8 "Estimated Human Risks 0.12 g/day ingestion of saccharin.") (Rat dose adjusted to human dose by surface area rule.)

Model	Lifetime Cases Per 10^6 Exposed
One hit model (Hoel)	1200
Two stage model (Hoel)	5
Multi-hit model (Food Safety Council)	.001
Probit (Mantel/Bryan)	450

Still different results are gotten if the rat dose is adjusted to humans on a mg/kg/day basis or on a mg/kg/lifetime basis. The smallest ratio between the highest and lowest estimates of risk at the 0.12 gm/day dose was in the mg/kg/day computation: 2×10^5:1.

Other Exposure — What else are people exposed to? What do these other materials do to modify incidence?

Cancer appears to be a multi-stage disease, perhaps with different materials contributing to the different stages. Exposure to two or more materials sometimes multiplies the cancer-causing effects of each. Thus the cigarette smoking asbestos worker is at substantially higher risk (30 to 90 fold) than the non-smoking non-asbestos worker [8]. Similar excesses have been reported for cigarette-smokers who also consume large quantities of alcohol [9]. The increases with age of the over-all age specific cancer rates are consistent with a multi-stage model of cancer [10]. With the further assumption that the effects of several carcinogens are additive in dose, we are led to a mathematical dose-response model at small incremental doses that is linear in response against dose; i.e. the same as the one-hit model [11]. For increasing dose the multi-stage model sometimes leads to a dose-response curve that curves upward with increasing dose that would give the relatively low response rates that one usually sees in humans. Mathematically, this usually leads to a dose response curve with three terms, one representing "background," one showing increase directly with dose, and the third showing increase with the square of the dose. In the saccharin estimation problem noted above, the multi-stage model led to risk estimates intermediate between those given by the one-hit and the multi-hit models.

Concerns about multiple exposures have led to a substantial amount of research on "initiators" and "promoters," or, as Richard Peto [12] prefers to call them, early-stage carcinogens and late-stage carcinogens. The control of late stage carcinogens (e.g. cigarette smoke, post-menopausal estrogens) has promise of leading to relatively rapid reduction in the appearance of cancer. The argument is that if an ultimate, or penultimate stage of cancer development can be interfered with, then cancer will not develop (unless there are other late stage carcinogens in the environment that operate in the same manner as the removed materials). Currently these mechanisms are not well understood, and there is research going on to attempt to explain the action of the late stage carcinogens, that would account for why lung cancer incidence fell off so rapidly (in British physicians) when cigarette smoking was reduced, or why the incidence of endometrial cancer so closely followed the increase and decrease in prescriptions written for post-menopausal estrogens.

In the several pages above I have talked about some of the problems that make risk estimation difficult. I have confined myself largely to materials with some known carcinogenic effects, and have tried to show that even with these known materials, not very much is known of what we need for computing the risks to a working population or, by extension, to the population as a whole. For exposure information, I have confined myself to worker exposure, noting the wide ranges in estimates for even worker exposures. For materials that "escape" into the general environment, exposure estimates both in terms of numbers and intensity are still harder to come by.

RISKS OF (NON-CANCER) TOXIC EFFECTS

There are several issues not talked about here that deserve at least passing mention. The only risk I have referred to is cancer. There is general awareness that this is not the only toxic effect that we need be concerned with, although sometimes regulation and arguments about regulation go on as if this were the only thing worth paying attention to. I have tried to show how risk estimation for cancer for a population is necessarily crude; i.e. uncertain. For most other chronic conditions, ranging from MCD (minimum cerebral dysfunction) to heart disease, things are even worse. For them there seems to be no way to make any reasonable estimates. For some possible toxic environmental exposures looking at only the cancer consequences is surely inappropriate, and at best, inefficient. Closer monitoring of fetal damage, or birth defects, might tell us a great deal more, and sooner.

In cancer, current testing systems (life-time testing in intact animals) are expensive, slow, and not particularly sensitive. Better, quicker systems are needed. Systems are needed that can tell us how differences in mode and duration of exposure modify responses. Systems are needed that can tell us something about the effects of multiple exposures.

Time and place affect our perceptions, and our behavior, very strongly. What Eric Ashby saw only a few years ago as a growing concern for the integrity of the environment can easily give way to a growing concern for energy production with a concomitant sacrifice of air quality, and other aspects of the environment, and health. Although man is a planning animal, what he sees or perceives in the near term seems to have much greater effect on his planning than what he perceives of what might possibly happen to him in the long term.

Although cancer, and the "obstructive" lung diseases, if one lumps them together, are the only major diseases that have increased all through this century, (even after we have taken population age changes into account) the rates of increase have not been great, and some very important cancers have been declining, rapidly. Partly because we know how to test for carcinogenesis, we have been discovering more and more materials that cause cancer, at least in our test systems. Because our test systems have also confirmed what we first noticed in man, we have begun to develop more faith in the test systems, and are now looking to them to help turn around the processes that may be adding carcinogens to our environment. We want to discover things in test systems, so that we might reduce or eliminate man's exposure to them, in order that we might never see their deleterious effects in man.

SUMMARY

To the degree that we develop tools that allow us to anticipate the risks that humans may run, in the sense that Ashby has indicated, the more likely we are to live longer, better lives. Clearly we need to develop better systems for gathering the data necessary to make risk estimates. Better, and perhaps more frequent, inventories of worker exposure are certainly needed. Estimates of exposure to the general population seem to me still harder to accomplish. I have troubles with the science fiction concept of all of us wearing personal monitors which transmit (automatically using a proved non-harmful section of the electro-magnetic wave band) data to an automated analytic laboratory which then feeds into my personal life-time exposure data bank which then correlates my illnesses with my exposures, etc. And I am not concerned only because such a data gathering is likely to be inefficient. What is "efficient" is not necessarily "good." What is good lies in the class of truth and beauty. It lies in the eye of the beholder, with the beholder's vision modified by the place and time, the morality and ethic in which he lives, and to which he subscribes. We want to do things rationally. We want to quantify so that we can compare. And we see that as "good." At today's level of information (and knowledge) I see substantial difficulties in quantifying risks. I have not dealt at all with the non-commensurable aspects of risk, or the social-equity issues involved in risk estimation, or risk taking. I remember reading that Pythagoras said "All is number." I think he was wrong. And I think as humans we are going to have to anticipate risks, and act on them often without definitive numbers from our estimates. If we know all our numbers are likely to be crude, and behave that way, we may even behave sensibly. At least we will have an escape route to follow when we discover we have made a wrong decision. If our numbers are too firm, or if we believe they are firm when they are not, I am afraid we may bully ourselves into doing bad things, efficiently.

REFERENCES

1. M. Momiyama and K. Katayama, "Deseasonalization of Mortality in the World," Int. J. Biometeor., 16(4): 329-342, 1972.
2. J. E. Enstrom, "Rising Lung Cancer Mortality Among Non-Smokers," J. Natl. Ca. Inst., 62: 755-760, 1979.
3. E. Ashby, Reconciling Man with the Environment, Stanford University Press, Stanford, California, 1978.
4. Personal Correspondence with Dr. Knut Magnus, Norwegian Cancer Registry, 1979.
5. National Occupational Hazard Survey, Vols. I-III, National Occupational Safety and Health Administration, 1977.
6. M. A. Schneiderman, N. Mantel and C. C. Brown, "From Mouse to Man — Or How to get from the Laboratory to Park Avenue and 59th Street," Annals of the New York Academy of Sci., 246: 237-248, 1975.
7. Saccharin: Technical Assessment of Risks and Benefits. National Academy of Sciences/National Research Council, Washington, D.C., 1978
8. E. C. Hammond, I. J. Selikoff and H. Seidman, Asbestos Exposure: Cigarette Smoking and Death Rates. IN PRESS in Annals of the New York Academy of Sciences, 1979.
9. K. Rothman and A. Keller, "The Effect of Joint Exposure to Alcohol and Tobacco on Risk of Cancer of the Mouth and Pharynx," J. Chron. Dis., 25: 711-716, 1972.
10. P. Armitage and R. Doll, "A Two-stage Theory of Carcinogenesis in Relation to the Age Distribution of Human Cancer," Brit. J. of Ca., 11: 161, 1957.
11. K. S. Crump, "Fundamental Carcinogenic Processes and their Implications for Low Dose Risk Assessment," Ca. Res., 36: 2973, 1976.
12. R. Peto, Epidemiology, Multi-stage Models, and Short-term Mutagenicity Tests. IN: Origins of Human Cancer. H. H. Hiatt, J. D. Watson, J. A. Winsten (Eds.), Cold Springs Harbor Symposium, 1403-1428, 1977.

DISCUSSION

W. D. Rowe *(The American University)*

As I looked at your data I got perhaps a little different conclusion. Maybe six or seven years ago NCI came out and said perhaps only 10% of cancer was caused by viruses. Suddenly everybody came to the conclusion that the other 90% were environmental. Your age adjusted cancer rate, which is only going up about 1% a year or maybe even a bit more, does not correlate with the introduction of these chemicals. Even assuming latency periods, the age adjusted rate hasn't changed. Perhaps a good proportion of cancer is just diseases of old age, not caused by introduction of large amounts of chemicals.

M. A. Schneiderman

M. A. Schneiderman

NCI never said 10% of cancers were due to viruses. Eric John Boyland of Great Britain did, in one paper, come to a number like 90%, by subtraction — eliminating possible virus cancer and possible "genetic" cancers. There are people taking comfort from the fact that we've not seen very large increases in the overall cancer mortality rate and therefore perhaps the introductions of these materials will not constitute a great hazard and hence should not be a great concern. They say perhaps we ought not worry ourselves unduly about them. I would like then to take these data I've shown and play with them. Some people say that when you do that you are manipulating the data. That's the perjorative way of saying it. I'm doing exploratory data analysis. "Exploratory data analysis" is a nice way and "manipulate" is a perjorative way of talking about examining data carefully. I'd like to take cancer of the uterine cervix out of these mortality data, for example, when I talk about trends in cancers possibly related to industrial exposure. I don't think that there's any indication that uterine cervix cancer is related to any industrial or related exposure except for one rather limited industry in this country. I might like to take stomach cancer out of these data too, although there's some indication that some stomach cancer may be related to chemical exposures or industrial exposures. If I took those out, then I am left with substantial increases in the remaining cancer rates. So other people say to me, "Hey, Schneiderman, take out lung cancer too. You know most lung cancer's due to cigarette smoking." So, I'll take out some of lung cancer from the total. I'm not going to take out as much as most people suggest because of the possible interaction with industrial exposure. After I got done with all these appropriate adjustments, I still found some increases in cancer incidence — perhaps a little more than 1% a year. I don't know what to make of this except that cancer is the only major disease in addition to one obstructive lung disease that is increasing in this country now. For me that is a cause for some concern. I don't like to have people going around saying that there's a cancer epidemic. I think they've overstated it. But there are increases. I hope that we will not have any extensive health repercussions of these large increases in the manufacture or use of synthetic organic chemicals. I wanted to show you those increases to show you what it is that is concerning some people, and the fact that we have rather limited information on who is exposed and to how much they were exposed. With this limited information, we are hard pressed to make good risk estimates, too.

Lester Breslow in a recent issue of *JAMA* cautioned against describing diseases as necessary consequences of old age. I'm with him.

W. C. Clark *(Int'l. Inst. for Applied System Anal.)*

I am an outsider in the risk field so I am trying to get a little bit of a feel for relative sizes of problems and efforts as we go through some of the presentations from different disciplines and different perspectives. Would you be prepared to comment at all on the order of magnitude of the number of chemicals that would come under the class of which you gave us some representatives here, that is in which exposure rates, production rates would be of interest to you and compare that to the amount of research effort, epidemiological data collection effort, that you judged would be necessary before your pure fuzz of your dose response curves became something on which we could begin to base our rational, scientific, defensible, call it what you will, arguments about regulation based on responsiveness? That is not an adversarial question but I'd be curious if you could wrap it up in some orders of magnitude.

M. A. Schneiderman

EPA has listed somewhere about 45,000 to 50,000 commercial chemicals which are in use and to which people are exposed. Of the small proportion of these materials that have been tested so far, there are about 250 or so that have been shown, generally in animal systems, to be carcinogens. Some of these are strictly laboratory curiosities and hardly anyone is exposed to them except the people working with them in the lab.

I think we have to struggle with the second part of your question. I don't know what the level of effort would have to be for the epidemiologic approach in order to get substantially better information. I do know now that there are relatively few epidemiologists in this country working on these problems and that they all seem to be very busy — working full time and turning out lots and lots of papers. They are all going to get promoted to full professors pretty soon if one counts the number of papers. They may even get tenure, which is very difficult to do these days. No, I really can't answer your second question, but I think it's a meaningful one. I think it's important, I think we have to look at it.

T. Page *(Environmental Protection Agency)*

When you showed those four extrapolation techniques, you made the comment that essentially we should worry about orders of magnitude in relative rankings rather than absolute levels.

M. A. Schneiderman

Rather than the absolute levels, Toby, yes.

T. Page

If you take this toxic dose rate that gives a 1% response rate and you use the four extrapolation techniques in ranking numbers, will you get the same ranking for these 300 chemicals more or less, no matter which extrapolation technique you use?

M. A. Schneiderman

More or less. These curves do cross each other at certain dose levels. But for most things they don't. So that's why I'd be willing to accept almost any one of the techniques just so long as people didn't think that I was making an absolute estimate of risk but making a relative estimate of risk.

T. Page

But crossing each other is not so important as variations from chemical to chemical.

M. A. Schneiderman

Well, if they cross each other you may get different rankings then for different materials at different toxicity levels.

C. Starr *(Electric Power Research Institute)*

Well, I just want to add a couple of comments just to gain a perspective. My recollection is that if you do extract lung cancers, that the cancer incidence has been fairly flat for quite a long time.

M. A. Schneiderman

For most cancer that statement is nearly correct. But most cancers that we are concerned with that may be related to chemical exposure are probably lung cancers. That means if we extract the lung cancers we would take out much of the things that are related to what we are concerned with and we might then say "See, we've taken out all those things that were related with what we're concerned with and therefore we don't have anything left." And that's what ought to happen. I don't want to play little Jack Horner, and put in my thumb and pull out a plum and say what a good boy am I. I would not take out all of lung cancers. Today, I wouldn't take out as much as 85% because that number was computed some time ago when a higher proportion of males were smoking than are now smoking, and when we had more toxic cigarettes. In addition, I did not include any consideration of the interactions of cigarette smoking with other exposures as the cause of cancer. As I said, I've gone through this kind of arithmetic and I am left, when I play these games that I talked to Bill Rowe about, I'm left with about 7 or 8% increase in total from '70 to '77. That's an increase of about 1% a year.

Something that James Enstrom published in the *Journal of the National Cancer Institute* in April 1979 frightens me. He reported changes in lung cancer mortality for non-smokers. From the end of World War I to the beginning of this decade he finds a 15 fold increase in lung cancer in non-smokers in males and somewhat smaller increase in females.

Unknown

Is that age corrected?

M. A. Schneiderman

That's age corrected, yes. I find this rather disturbing. His results say to me that there must be other things going on in addition to cigarette smoking. Since I was one of the people who urged others to "get off the cigarette bandwagon" trying to tell people not to smoke, to say now that I am finding that there are other things involved with this increase in lung cancer means that I have to back track on some of the things I said earlier. I now recant. I now believe that less of lung cancer is due to cigarette smoking than I believed earlier. Cigarette smoking is the most important single factor — but it's not the only one.

THE KNOWN RISKS WE RUN: THE HIGHWAY

Barbara E. Sabey and Harold Taylor

Transport and Road Research Laboratory
Crowthorne, England

ABSTRACT

The paper reviews known risks on the highway in perspective with the options available for reduction of risk, illustrating principles from experience in Great Britain.

Firstly, the overall risks of accident and injury on the highway are enumerated statistically in relation to the individual and to the community. The subjective view of these risks is discussed in terms of public response to different situations and monetary values ascribed.

The relatively objective assessment of risks associated with different factors has been obtained through a multi-disciplinary study of the contribution of aspects of highway design, vehicle condition, and road user behaviour to accident occurrence. Stress is laid on interactions between the road, the vehicle, and the road user. Other studies give quantification of risks for specific road engineering and environmental factors, and show how probability of injury may be related to vehicle design.

The potential for reduction in accident or injury risk is examined in the light of known benefits from well tried countermeasures, and quantified in relation to road engineering and traffic management, vehicle design and use, road user behaviour and road usage.

Finally, risk factors are contrasted with remedial potential. Future directions for application of countermeasures to reduce risk and for research to enhance quantification of risk and identify new effective countermeasures are enumerated.

INTRODUCTION

The desire to travel is strongly rooted in mankind. Long before the advent of powered vehicles people made many remarkable journeys on foot, by boat and with the aid of animals. They frequently suffered great hardships and in some cases lost their lives in the process. With the growth of towns journeys by road became

References pp. 62-63.

increasingly necessary and were often made on foot. Thus travel by road became established both as a means of extending one's horizon and as a necessity of everyday life; neither class of journey could be relied upon to be comfortable, enjoyable or safe. With the growing use of horse-drawn vehicles the traveller on foot was both endangered and demeaned thus establishing the conflict between vehicle users and pedestrians. Not until relatively recently with the development of modern vehicles and highways could a typical journey be completed in comfort and without incident. Unfortunately uncertainty still remains as to whether it will be completed without accident; the "natural hazards" of former days have been replaced by the statistical probability of conflict which stems from the sheer intensity of modern road traffic and from the disparate nature of the foot and wheeled traffic using the highway.

In considering the risks associated with travel by road it is evident that there is much historical background on the subject which influences both people's attitudes and their behaviour. In broad terms travel by road is understood by most of us to involve some risk though the level of this risk is only dimly perceived and rarely called into prominence because it has been with people almost all of their lives. Daily journeys from the home are commonplace and since, as will be seen later, road accidents are relatively rare events there is normally no reason for the question of risk to assume any great prominence. Road accidents are associated with a well-founded activity which almost everyone needs and wants to be involved in; the road accident situation is therefore very different from many other threats to personal safety which are far less well appreciated and sometimes not even recognised until irreparable damage has occurred to the persons affected.

Historically the introduction of mechanically propelled road vehicles seems to have been the signal for government to take a major hand in regulating the road scene. The highway codes and regulations which resulted were designed to instil some order into the operation of the traffic system and to impose certain requirements upon vehicles and their operators in order to protect the rights and safety of other road users and of property owners. The duties which fall upon road users, especially vehicle operators, are nowadays in Britain a complex amalgam of criminal and civil law with the law of negligence central to the question of recovery of damages following a road accident. The law of negligence as it applies to motoring is based on the principle that every road user has a legal obligation to drive carefully at all times, and to exercise proper care to all those whom he could reasonably foresee might be injured by his driving. Thus it is not surprising that until relatively recently, in most well-developed countries, road accidents and their consequences were thought of in terms of breaches of criminal law and civil lawsuits for damages through negligence; or looking to apportion blame. Though road accident casualties were treated by the medical profession in a similar way to other medical cases, road accidents were not thought of as a serious threat in public health terms. As a result the acquisition and processing of road accident data and the sponsorship of programmes to reduce road accidents have until very recently been almost exclusively the province of police and highway transport authorities supported to a small extent by educational authorities. However with the general improvement in public health the significance of road accidents is manifest; they are now the main cause of death for young people between 15 and 25 years old. The excessive consumption of alcohol, which is a serious public health problem in many countries is also a major influence in road accidents. It is to be expected therefore that the social and health aspects of road accidents which are intimately bound up with behaviour will play a much greater part in future road safety activities.

Risks encountered on the highway are of relatively certain magnitude when compared with risks in other fields and in some circumstances are relatively high compared to known risks in some other fields. Even so there is little unanimity internationally about the levels of road safety that should be sought.

This is in part because road users are exposed to a mixture of voluntary and imposed risks which are, as is well known, often regarded quite differently by the people at risk. To date the concept of "acceptable risk" has not been applied globally to road accidents which formally and politically are regarded as unacceptable. As a result formal target setting in road safety has not been widely adopted but is often implied. In Britain guidelines exist to aid decisions about the adoption of various road safety measures and these might be considered to imply that below certain levels of risk there is no strong case for spending public funds on counter-measures. These implied risk assessments are sometimes but not always linked directly to formal cost-benefit assessments. In the road safety field considerable effort has been devoted to the application of cost-benefit and cost-effectiveness techniques in considering remedial measures but there is much scope for further work in order to make the best use of these techniques. The bulk of road safety improvements are made at the instigation of central or local government. In so doing the authorities are the proxy for the public yet there is a relatively weak link between the two perhaps because historically the welfare of the road user had not been emphasised. It is nevertheless surprising that relatively little work has been carried out to explore public attitudes to road safety or the safety expectations that people have of the highway system. Many road safety improvements confer no other benefit on the road user and some impose a definite penalty such as a limitation in freedom or of choice; this is especially true of measures which require a change in road user behaviour such as the use of seat belts or abstinence from drinking alcohol. A sound knowledge of risk perception attitudes, probable behaviour and expectations is most important to the success of some planned road safety improvements as is an understanding of the extent to which road user behaviour is likely to adapt to the changed situation and defeat the intended safety benefit. It is not unusual for other benefits such as savings in journey time to be substituted for it and thus for the planned reduction in risk to fail to materialise.

This paper reviews highway risks in so far as they are known and examines them in relation to the options which have been shown to reduce risk. The principles discussed are illustrated from experience in Great Britain.

OVERALL RISKS ON THE HIGHWAY - PERCEIVED AND ACTUAL

The overall risks of accident and injury on the highway are quantified statistically in simple terms in many countries, the most common procedure being to relate

References pp. 62-63.

numbers of occurrences to some measure of exposure to risk. Such statistics are widely published, and used (inter alia) in attempts to compare relative performance in different countries, to identify problem areas requiring action, and to assess benefits of countermeasures. Risks are however mostly considered in relation to the community as a whole, or to groups of road users, rather than to the individual. Yet it is the risks as seen by individuals, either actual or perceived, which frequently determine the acceptability or success of countermeasures, particularly in the legislative and behavioural fields.

There are conflicts between the risks as seen by the individual and as relevant to the community as a whole, which may at times hinder progress towards improving road safety, and at other times work towards advancing it. At the basis of this dilemma is the low risk of involvement in accident and even lower risk of injury to the individual. From the reported accidents in Great Britain in 1977* [1] the risk of involvement in an injury accident is once in 57 years: in a fatal accident 1 in 2500 years; and in an accident not involving injury (estimated from insurance claims data) about 1 in 9 years. The risks of injury are lower and vary for different road users, as indicated in Table 1, but even with the most vulnerable road user (the motorcyclist) the likelihood of an injury is so infrequent that it is easy to understand why there might well be disinterest in measures to promote road safety, as is often suggested. It is interesting to compare these figures with work by Sheppard [2] which showed that about one-third of drivers in a random sample who had an opinion on likelihood of an accident thought they would be involved in a serious accident in the future.

TABLE 1

Casualty Rates in Great Britain in 1977
for Different Road Users

	Casualties per 100 Million Occupant km			Years per Casualty	
	Killed	Seriously Injured	All Injuries	Killed or Seriously Injured	Injured
Motor Cyclists	15.9	274	966	69	21
Pedal Cyclists	6.8	107	531	not available	
Car Occupants	0.7	9	42	724	166
Commercial Vehicle Drivers					
≤1½/ton u.w.	0.5	7	31	485	115
>1½/ton u.w.	0.4	4	16		
Public Service Vehicles	0.1	2	21	974	105
All Motor Vehicle Users	0.8	12	52	540	132

---

*National statistics in this paper are based on reports of injury accidents made to the police in a standard format for the whole of Great Britain (England, Scotland and Wales).

In contrast with the risk to the individual, the occurrences of accident or injury present a major problem to the community as a whole. In the whole world, it is now estimated that one-quarter million deaths and over 10 million injuries occur as a result of road accidents each year. In Great Britain in 1977 there were 6600 deaths, 81,700 serious injuries (usually requiring detention in hospital) and 259,770 lesser injuries reported in the police statistics. It is known that injuries are underestimated by probably 30 per cent. In addition there are estimated to be at least another million and a half non-injury accidents reported to insurance companies and an unknown number of accidents not appearing in any statistics. The cost of these accidents to the community has been fairly reliably assessed at 946 million in resource costs, i.e. loss of output due to death or injury, medical and ambulance costs, damage to vehicles and property (over half the total), and costs of police and administration of accident insurance (see Table 2). Over and above these costs are the costs of pain, grief, and suffering to the involved person, to relatives and friends. These are very real costs to society but are by their nature not directly quantifiable in monetary terms. In recognition of the relevance of these losses, current practice in Britain is to include what can be regarded only as a notional minimum allowance for subjective costs, which totals 347 million, and averages £25,880 per fatal accident. However a recent appraisal of these figures [3] suggests

TABLE 2

Cost of Road Accidents in Great Britain in 1977

	Total £ Million	£ per Accident		
		Fatal	Serious Injury	Slight Injury
RESOURCE COSTS				
Lost Output	282	37,450	770	20
Police and Administration	75	150	120	90
Medical and Ambulance	44	300	510	30
Damage to Property	545	820	690	480
Sub-total	946	38,720	2090	620
PAIN, GRIEF, AND SUFFERING	347	25,880	2650	50
TOTAL	1293	64,600	4740	670

that they are not in line with general principles of cost-benefit analyses. A survey of studies where researchers have attempted to evaluate how an individual values risk has revealed figures for value of life between two-and-a-half and ten times this average. It is also true that UK accident values are consistently lower than those of other countries. Further work is therefore to be directed towards getting more realistic figures based on public attitudes on how much money people are willing to pay for a reduction of correctly perceived risk. One of the paradoxes of this assessment or perhaps a limitation of the method of calculation is that property damage is by far the largest element of the accident costs. Yet to most people property damage is undoubtedly secondary to loss of life and all but the most minor of injuries.

In the environmental field, perceived risks are biassed by the drama of the accident. The relatively rare occurrence of a multiple pile-up on a motorway resulting in death (a total of 46 deaths occurred in 33 such accidents involving 3 or more vehicles in Great Britain in 1977), or a multiple fatality in a coach crash (less than 17 cases involving 34 fatalities in 1977) attracts more column-inches of press publicity than the 1940 pedestrians and 1740 other road users killed in towns in the same year. A balanced view must be maintained to ensure that the limited resources available for remedial action are applied in areas where returns are likely to be greatest. Objective risks for these situations can readily be evaluated as in Table 3. The largest numbers of casualties and the highest casualty rates occur in urban areas.

TABLE 3

Casualty Rates in Great Britain in 1977:
Different Road Types

	No. of Casualties		Casualties per 100 Million Vehicle km	
	Fatal	All Injuries	Fatal	All Injuries
Motorways	208	6323	0.84	26
Rural Roads (50-70 mile/h Speed Limit)				
'A' Class	1957	52 717	2.9	79
Others	766	35 304	1.9	89
Urban Roads (30, 40 mile/h Speed Limit)				
'A' Class	1981	121 064	3.1	191
Others	1701	132 601	2.5	192

The consequences of different concepts of risk have been felt in various areas of legislation introduced or proposed for promoting road safety. But some evidence suggests that the risk of apprehension and its consequences are of more import than the risk of accident. Perhaps the most dramatic illustration of this lies in the effect of the British Road Safety Act of 1967 [4] which introduced a legal limit of 80 mg/100 ml alcohol in the blood of drivers. Undoubtedly the public expectation of a high risk of apprehension (quite false and unrealistic in relation to police manpower available) played a large part in the impact of the legislation which overnight produced a reduction of 11 per cent in the national casualty toll. In contrast, an equally false assumption of high risk of accident involvement — the risk of being trapped, caught in fire, or overturned in a car — has been a major argument amongst some opponents of wearing of seat belts in Britain. There are of course some opponents of legislation to compel wearing of belts (but not necessarily of wearing) who regard infringement of liberty of the individual as paramount and in their case the risk of injury and road safety issues in general are irrelevant to their argument.

Overall there is a confused view of risk amongst the population of road users, which can have unpredictable effects of the acceptability or success of road safety measures. Despite this, or perhaps because of it, it is important to get a relatively objective assessment of risk associated with different aspects of highway design, vehicle condition and road user behaviour. The next section of the paper deals with this question.

CONTRIBUTORY FACTORS IN HIGHWAY ACCIDENTS

Most accidents on the highway have a multiplicity of causes (or contributory factors), which may be a combination of human errors and failings, poor road design or adverse conditions, and vehicle defects. Two major studies were conducted over the period 1970 to 1974 by multi-disciplinary teams in the United States (Indiana University) [5] and Britain (Transport and Road Research Laboratory) [6], [7] to try to identify the main contributory factors and their interactions. The planning and execution of the studies were carried out independently, without the teams being aware of each other's activities at the time. The outcome shows remarkable similarity in the findings, which is encouraging in a field of work which is inevitably subjective in judgment to some extent. Illustration of the findings will be restricted to the British study with which the authors are more familiar.

The Role of Road, Vehicle, and Road User in Accident Risk — The study covered 2130 accidents attended by the investigation team which was on call 24 hours a day. At-the-scene evidence was supplemented by later vehicle examination and interviewing of drivers, riders, and pedestrians involved. After all the data, covering up to 400 items of information, were collated, a team discussion aimed at allotting blameworthiness and contributory factors. This was achieved in 2042 accidents.

Of the 3665 drivers assessed, 41 percent were judged to be primarily at fault, 19 percent partially at fault, and 40 percent victims: in contrast, blameworthiness of pedestrians was assessed as 65 percent primarily at fault, 14 percent partially at fault and 21 percent victims.

Driver and pedestrian error, and impairment were main contributory factors* in 1942 accidents (95 percent); road and environmental factors were contributory in 569 accidents (28 percent); vehicle features were contributory in 173 accidents (8-

---

*In this context a main contributory factor means something without which the accident would have been less likely to happen or at least would have been less serious.

References pp. 62-63.

1/2 percent). There were many interacting factors, which can be briefly summarised in Table 4.

TABLE 4

Contribution to Road Accidents

Percentage Contributions

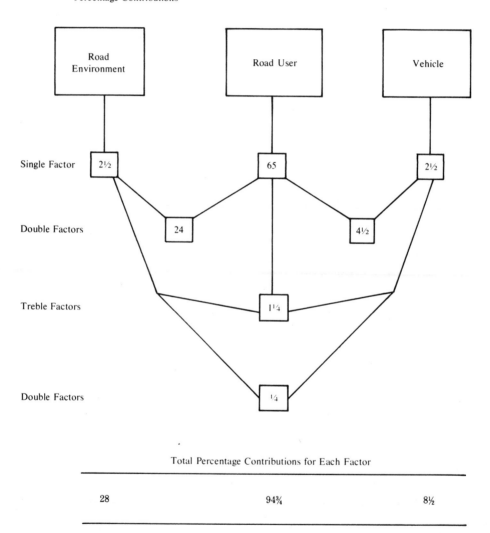

The road user was the sole contributor in 65 percent of accidents; road and vehicle factors were usually linked with a road user factor.

The detailed analyses identified the kinds of error or deficiency most prevalent in accident occurrences. Categories of human error were chosen to some extent arbitrarily, but on the basis of past experience to cover all likely situations. The boundaries between some classifications could not always be clearly defined and in many cases it was necessary to use several factors to describe different aspects of behaviour. In the original analysis one classification used was "lack of care" but this is so general a description that it has now been deleted in favour of more specific aspects. The individual errors can be grouped into areas (Table 5). One important general finding is that the majority of driver errors constitute poor behaviour related to some deficiency in the driver's action rather than deliberate aggressiveness or irresponsibility.

TABLE 5

Road User Factors Contributing to Accidents

Factors	Drivers	Pedestrians
PERCEPTUAL ERRORS — Looked but Failed to See — Distraction or Lack of Attention — Misjudgement of Speed or Distance	1090	53
LACK OF SKILL — Inexperience — Lack of Judgement — Wrong Action or Decision	462	—
MANNER OF EXECUTION (a) Deficiency in Actions: Too Fast, Improper Overtaking, Failed to Look, Following Too Close, Wrong Path	1153	107
(b) Deficiency in Behaviour: Irresponsible or Reckless, Frustrated, Aggressive	94	—
IMPAIRMENT — Alcohol — Fatigue — Drugs — Illness — Emotional Distress	632	7
TOTAL FACTORS	3431	167

Total accidents in which a road user factor was a main
  contributor   *– 1942*
Total accidents assessed – 2042

As road environment deficiencies were usually associated with driver error, features can conveniently be grouped which show similarity in the difficulties they present to the driver (Table 6). Once again there are links between the different factors, which are not mutually exclusive.

The features of "adverse road design" are derived mainly from accidents prior to which a driver did not appreciate a misleading visual situation. "Adverse environment" features are those which contributed to accidents by increasing the

TABLE 6

Road Environment Factors Contributing to Accidents

ADVERSE ROAD DESIGN — Unsuitable Layout, Junction Design — Poor Visibility Due to Layout	316
ADVERSE ENVIRONMENT — Slippery Road, Flooded Surface — Lack of Maintenance — Weather Conditions, Dazzle	281
INADEQUATE FURNITURE OR MARKINGS — Road Signs, Markings — Street Lighting	157
OBSTRUCTIONS — Road Works — Parked Vehicle, Other Objects	129
TOTAL FACTORS	883

Total accidents in which a road environment
*   factor was a main contributor*   − *569*
Total accidents assessed   −*2042*

difficulty of maneuvering a vehicle safely. Features of "inadequate furniture or markings" represent insufficient and unclear information being presented to the driver who is pre-occupied with maneuvering his vehicle. "Obstacles" are unexpected hazards. Although human factors play the largest part in contributing to accidents they are difficult to identify and costly to remedy. In the accident situations described here, remedial measures of an engineering nature, which are easier to identify and often readily and cheaply effected, can be applied to counter human failings.

Vehicle defects making a major contribution to accidents were mainly of the kind which can develop in a relatively short space of time due to lack of regular maintenance by the user of the vehicle. Defective tyres and brakes featured most frequently, making up one-third each of all the main contributory defects.

Such multi-disciplinary studies as these give measures of risk of accident occurrence related to specific features. In relation to application of countermeasures they indicate the areas needing remedial action, though not necessarily capable of remedy. More importantly they indicate interactions which may suggest alternative remedies more likely to be effective. When considering remedial measures, as opposed to blame or risk, the most effective remedy is not necessarily related directly to the main cause of the accident and may even lie in a different area i.e. of road environment, vehicle, or road user. This is particularly true of accidents in which the road user fails to cope with the road environment. Further, even in circumstances in which human error or impairment has been judged to be the sole contributor, it may be possible to influence human behaviour more readily by

TABLE 7

Vehicle Factors Contribution to Accidents

Factors — Defects	Number
Tyres	67
Brakes	65
Steering	7
Lights	10
Mechanical Failure	22
Electrical Failure	4
Defective Load	10
Windscreen	4
Poor Visibility	4
Overall Poor Condition	5
Unsuitable Design	9
TOTAL FACTORS	207

Total accidents in which a vehicle factor
was a main contributor   — 173
Total accidents assessed   — 2042

engineering means than by education, training, or enforcement of legislation.

Before discussing this aspect further, it would be an omission not to recognise many other accident investigations, particularly in the road engineering field, which have quantified risk of accident occurrence. The subject of identification of hazardous road locations, safety principles related to road design — in terms of geometry, surfaces, markings, signs and furniture, and traffic management — and countermeasures applicable has been discussed in detail by an OECD Road Research Study Group, whose report [8] references much of the relevant work worldwide. Table 8 summarises some of the more important aspects quantified.

All the foregoing risks enumerated are relatively objective measures. A recent study, [9] prompted by the multi-disciplinary at-the-scene investigation, has aimed at relating the objective and subjective measures by examining aspects of road layout that affect drivers' perception and risk taking. A total of 45 road locations were investigated by requiring 60 volunteer drivers to make assessment of risk on a 16 mile route, which covered a wide range of road types and hazards, including rural divided carriageways, narrow suburban roads, steep hill crests and level crossings. There was found to be a significant agreement between drivers in ranking the subjective risk at the different locations. The locations were ordered on these subjective evaluations (on a scale of 0 to 10) and compared with the ranking on objective risk, which was obtained from accident and traffic flow data. Although there was a significant rank correlation (Spearman rho 0.37) between subjective and objective risk, at some locations there were wide discrepancies. The five most over- and under-rated location types are described in Table 9.

At 16 of the locations where vehicles were generally freely moving it was also possible to calculate a safety margin based on the difference between the forward

TABLE 8

Road Engineering Measures Quantified in Specific Studies

GEOMETRIC DESIGN Junctions: Layout, Alignment, Sight Distances, Channelisation, Access Control Horizontal and Vertical Alignment: Curvature, Superelevation Cross Sections: Number of Lanes, Shoulder Design, Medians

SURFACES
 	
Micro — and Macro — Texture: Wet Road Performance
    — Skidding Resistance and Speed
    — Visibility by Day and Night
Evenness and Profile

MARKINGS AND DELINEATION Indicators of Prohibitions or Appropriate Manoeuvres Channelisation Guidance
ROAD SIGNS AND FURNITURE Lighting Traffic Islands Anti-dazzle Screens Safety Fences and Guard Rails Warning Signs
TRAFFIC MANAGEMENT Speed Control and Limits Junction Control One-Way Systems Parking

visibility distance and the total stopping distance estimated from recorded speeds. The average safety margin correlated well with average subjective risk ratings. As expected, the margins were found to be smallest at vertical and horizontal curves. There was also variation in performance between different drivers, some having negative safety margins at some locations. At one location however, a left-hand bend, the average safety margin adopted was negative. The results suggest some possible countermeasures, for example, increasing sight-line distances on bends. Further work of this kind could greatly aid the understanding of the interaction between the road user and his environment and identify more effective engineering measures in particular circumstances.

Risk of Injury — Consideration of objective measure of risk should also include assessment of risk of injury once the collision has occurred. The most relevant

TABLE 9

Under- and Over- Rated Hazards

Description of Hazard	Difference in Rank
UNDER-RATED	
Suburban Dual Carriageway Near a Pedestrian Bridge	− 26.0
A Rural Crest on a Single Carriageway	− 25.5
A Left Turn Off a Rural Road	− 25.0
A Rural Dual Carriageway Near a Picnic Area	− 24.5
Rural Cross-roads Controlled by Traffic Lights	− 21.0
OVER-RATED	
Hump Bridge on Rural Road	+ 35.0
Level Crossing on Rural Road	+ 32.0
Surburban Shopping Centre	+ 31.0
Right Turn onto Rural Dual Carriageway	+ 24.5
Right Bend at the End of Rural Dual Carriageway	+ 22.0

issues here are vehicle occupant protection, protection of the vulnerable road user (pedestrian and cyclist), and protection against impact (rigidity of roadside obstacles). A few examples of risk of injury may be cited.

In relation to occupant protection, the simplest concept of risk of injury has been indicated in studies of the effectiveness of wearing seat belts. It is widely accepted and stated that, overall, risk of death or injury in an accident is halved when a seat belt is worn. Such statements related to this or other vehicle safety features tend to imply that there is a clear-cut threshold below which no one will be injured and above which all will be injured i.e. an absolute measure of safety. This is not so, as is recognised by some of the current research into vehicle secondary safety which is directed towards quantifying probability of injury in relation to impact forces and accelerations experienced by human bodies in accidents. Wall, Lowne and Harris [10] have outlined a procedure for estimating how the likelihood of being injured in accidents may be related to different injury tolerance levels, so that with a knowledge of the severity distribution of accident impacts, the overall effect of changes in the tolerance level can be predicted to compare the extra cost of providing cars to this new level with the benefits from injury reduction. Figs. 1 and 2 illustrate the principle by two examples:

1. The incidence of chest injury as a function of shoulder belt tensions, and
2. The incidence of hip dislocation as a function of impact forces on the knee

In the first example a reduction in the tolerance level from 7 kN to 5kN predicts a reduction in 40 percent of the population suffering injury. In the second example a reduction in tolerable impact force to 4kN predicts a reduction in 80 percent of hip injuries.

References pp. 62-63.

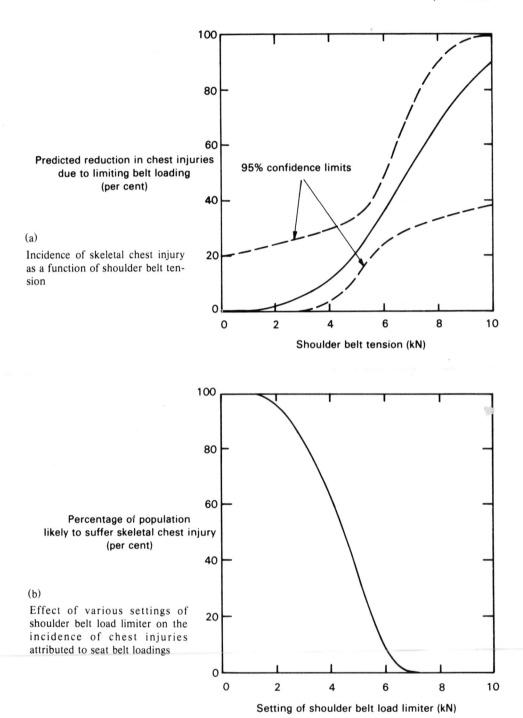

Predicted reduction in chest injuries
due to limiting belt loading
(per cent)

95% confidence limits

(a)

Incidence of skeletal chest injury
as a function of shoulder belt tension

Shoulder belt tension (kN)

Percentage of population
likely to suffer skeletal chest injury
(per cent)

(b)

Effect of various settings of
shoulder belt load limiter on the
incidence of chest injuries
attributed to seat belt loadings

Setting of shoulder belt load limiter (kN)

Fig. 1

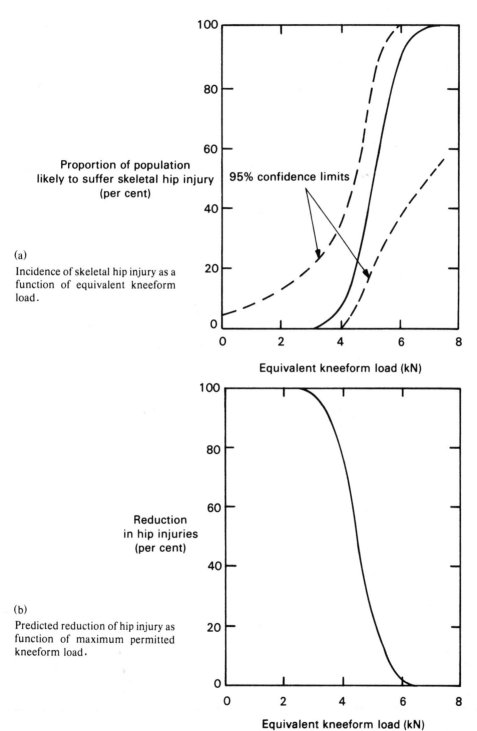

Proportion of population
likely to suffer skeletal hip injury
(per cent)

95% confidence limits

(a)

Incidence of skeletal hip injury as a
function of equivalent kneeform
load.

Equivalent kneeform load (kN)

Reduction
in hip injuries
(per cent)

(b)

Predicted reduction of hip injury as
function of maximum permitted
kneeform load.

Equivalent kneeform load (kN)

Fig. 2

References pp. 62-63.

POTENTIAL FOR REDUCTION IN RISK OF ACCIDENT OR INJURY

As has already been suggested, identifying the risk of accident or injury may not be directly indicative of the potential for reduction in risk. Although studies of accident "causation" indicate that the greatest potential for accident reduction lies in influencing human behaviour, it would be impossible at present to predict future trends on the basis of causes or errors alone: in many cases there is no known effective remedy. In a recent assessment [11] of the potential for accident and injury savings in Great Britain, the approach made therefore was to consider proven remedies, or measures for which there was strong evidence of potential benefits, regardless of blame. The principle applied was: firstly, to quantify savings from individual measures; secondly, to identify the target group of accident or casualty types amenable to change; thirdly, to apply the estimated savings to the target group; fourthly, to relate these savings to the total accident or casualty situation. Thus, if a particular measure suggests a return of A percent in accidents or injuries, and is applicable to B percent of the total accidents or injuries, the potential saving is A x B percent. The various options for countermeasures were identified under the broad headings of road environment, vehicle, and road user. The options are not mutually exclusive and the benefits are not cumulative: some of the interactions were considered.

Over the four years since the assessment was made there have been some changes in the factors A as new research findings have established more reliable assessments of benefits. Factors B (originally based on the year 1973) have also changed in some respects, largely as a result of changes in vehicle usage associated with the environmental changes consequent on the fluctuating fuel supplies. The opportunity has therefore been taken to update forecasts of potential savings based on 1977 (the latest year for which full accident data are available in Great Britain), and to present them here.

The revised benefits from individual road engineering and vehicle safety measures and the reasons behind the revisions are detailed in the Appendix. Overall there is no change in the estimates. A summary of the options and their potential savings is given in Table 10. For the road traffic situation obtaining in Britain potential savings in terms of injury accidents from remedies applied in the three main areas are of the order of:

Road Environment	-	One-fifth (E)
Vehicle	-	One-quarter (V)
(Primary and Secondary Safety) Road User and Road Usage	-	One-third (U)

These are the ultimate which could be achieved at some future date (when all known options in each area were implemented) in relation to the predicted base level accidents at that date. No interactions between benefits have been considered in these estimates, so the total potential is less than the sum of these fractions.

One method of estimating the combined effect is to assume that the fractional reductions E, V, U result from three different and mutually exclusive methods. Then the resultant reduction will be $(1 - E) (1 - V) (1 - U)$ of its former value. The

TABLE 10

Potential for Accident and Injury Reduction in Road Accidents
(Based on 1977 data)

Options	Potential — Percent Savings
ROAD ENVIRONMENT (Low Cost Remedies)	
— Geometrical Design, Especially Junction Design and Control	10½ (11½)
— Road Surfaces in Relation to Inclement Weather and Poor Visibility	5½
— Road Lighting	3 (1½)
— Changes in Land Use, Road Design, and Traffic Management in Urban Areas	5-10 (7½-16½)
OVERALL	ONE-FIFTH of Accidents
VEHICLE SAFETY MEASURES	
Primary — Vehicle Maintenance, Especially Tyres and Brakes	2
— Anti-lock Brakes and Safety Tyres	7 (6)
— Conspicuity of Motorcycles	3½ (3)
Secondary — Seat Belt Wearing	7 (10)
— Other Vehicle Occupant Protection Measures	5-10
OVERALL	ONE-QUARTER of Casualties
ROAD USER AND ROAD USAGE	
— Restrictions on Drinking and Driving	10
— More Appropriate Use of Speed Limits	5
— Propaganda and Information	up to 5
— Enforcement and Police Presence	up to 5
— Education and Training	up to 5
— Other Legislation (eg. Restrictions on Parking)	up to 5
OVERALL	ONE-THIRD of Accidents

Figures in ( ) indicate earlier values based on 1973 data – where different from latest estimates

outcome of this simple calculation is to indicate a potential overall saving of THREE-FIFTHS — a substantial impact on the road accident toll.

RISK RELATED TO REMEDY

From studies of countermeasures to the known risks on the highway it has become increasingly clear in recent years that the priorities for action are not determined by levels of risk alone. Apart from the obvious need to know what remedy suits what risk, attitudes to risk play a vital part in the success of any proposed countermeasure. In the former area there is still considerable research to be done to seek remedies to identified problems, despite much effort already applied. However in comparison with this, the understanding of attitudes to risk is in its infancy.

In this context it is interesting to compare the contribution to accident causation with potential savings as described in the previous two sections of this paper.

Contribution to Accident Causation		Potential Savings in Injury Accidents
28%	Road Environment	1/5
8-1/2%	Vehicle	1/4
95%	Road User	1/3

The potential has been based on remedies proven by past experience, but it has been assumed that these remedies will be equally effective in the future. There may be some doubt about this unless attitudes to risk and understanding of perceived risk are adequately taken into account.

The most clear illustration of this lies in the manner of dealing with the drink/driving problem. The impact and decline of the 1967 British legislation, [12] already mentioned, has had to be carefully analysed to enable recommendations for new legislation to be put forward which will have a chance of making the same level of impact. In particular, since the myth of high perceived risk of apprehension has been exploded, future proposals must lead to the *actual* risk being seen to be increased. This must be backed by a reversion of attitudes to recognise drinking and driving as a social stigma.

The exploitation of the biorhythm theory gives an interesting illustration of the positive benefit of perceived risk. Examples of reduction in accident involvement has been cited by some workers, [13] consequent on warning drivers to take care on days in the biorhythm cycles which are designated critical and said to present a high risk situation. However a recent objective study [14] of a sample of over 110,000 accident involvements in Britain has shown that there is no convincing evidence that the biorhythm theory has any basis for accident involvement. It remains to be seen whether disclosure of this negative result might affect the success of campaigns based on the biorhythm theory.

The success of speed limits in reducing speeds depends critically on attitudes, as has been shown by many studies [15]. Indeed, since the discrepancy between the contribution of road user factors in accidents and the potential for savings hinges on the paucity of proven remedies to change human behaviour, it may well be that a better understanding of attitudes to risk is critical to enhancement of remedial action in this area.

The application of vehicle safety factors and road engineering factors are not immune from the consequences of public attitudes to risk. As already seen, the wearing of seat belts is a critical issue in this respect and illustrates yet another difficulty associated with attitudes of risk; experience in different countries [16] highlights how attitudes depend so strongly on national characteristics and on the value attached to freedom of the individual in relation to the perceived seriousness of the prevailing road accident situation. In general terms vehicle safety features aimed at improving secondary safety are less likely to be influenced by attitudes than are measures related to primary safety. As regards road engineering, public participation in road schemes is assuming greater importance in Britain. The need to explain not only the risk but also the remedy, and to identify associated side-effects, whether beneficial or otherwise, is now recognised as a major factor in many safety schemes.

CONCLUSION

For the future, it will be necessary to pay much more attention to what is likely to be publicly acceptable in terms of countermeasures to risk, and at what cost. One of the difficulties here is that in dealing with accidents on the highway there can be no absolute measure of risk nor overall level of safety which is tolerable. "How safe is safe enough?" is a very real question in this situation. Road safety measures already cost about 1000 million pounds in Great Britain — a sum comparable in scale to the costs of accidents themselves. To achieve further advances it is necessary to balance resource costs, attitudes to risk, and potential benefits, but how this can be done to set targets if indeed targets are desirable is not yet resolved. It is the intermediate factor of attitudes to risk, which is so subjective and unpredicatable, that can sway the balance in decision taking one way or the other.

In terms of application of countermeasures, future directions for road engineers in Britain have to some extent already been defined. Under the Road Safety Act of 1974, a specific responsibility for road safety was placed on local authorities. Effort is now being concentrated on low cost remedial measures, and targets are being set on the basis of potential benefits enumerated in this paper. To encourage and enhance work further a set of guidelines for accident reduction and prevention is being drawn up [17]. The resources required are low compared with the potential benefits or with general road building costs, and road engineering for safety is generally considered acceptable, so the prospects for realisation of the full potential of one-fifth savings in accidents are high.

Future directions for vehicle safety measures and influencing road user behaviour are much more uncertain. Vehicle engineering costs are much higher, in that any safety measure has to be applied to a large population of vehicles. There may also be conflicting requirements for other reasons, particularly to promote energy savings, to improve environmental conditions, and to prevent barriers to trade. Not least of the difficulties is the attitude to protective devices. In the human engineering field attitudes are paramount: influencing the road user is the most difficult safety measure to effect, but when it can be achieved it can also be the most dramatic. A major priority here is the revision of the drink/driving legislation, and the scene will be set for action by a consultative document based on the recommendations of the Blennerhassett Committee of Inquiry [18] which is being published in the Autumn of 1979.

The needs for research to enhance quantification of risk on the highway are still quite extensive, and they are inextricably linked with the identification of more effective countermeasures. The major areas include:

- Risks related to features of road design and traffic flow;
- Demonstration of application of low cost remedial measures;
- Probability of injury related to vehicle design;
- Alcohol countermeasures and medical factors;
- Aspects that affect drivers' perception and risk taking;
- Attitudes to risk;
- Evaluation of engineering and behavioural changes.

In the short term the lowering of risks can best be achieved by application of low cost road engineering measures and some legislation. In the long term it must rely on behavioural changes brought about by education and dissemination of information in the broadest sense.

ACKNOWLEDGEMENTS

The work described in this paper forms part of the programme of the Transport and Road Research Laboratory and the paper is published by permission of the Director.

REFERENCES

1. *Department of Transport, "Road Accidents Great Britain 1977," Her Majesty's Stationery Office, London, 1978.*
2. *D. Sheppard, "The Driving Situations Which Worry Motorists," Department of the Environment, Transport and Road Research Laboratory, Supplementary Report 129UC, Crowthorne, 1975.*
3. *G. Leitch, et al, "Report of the Advisory Committee on Trunk Road Assessment," Her Majesty's Stationery Office, London, 103-105, 1977.*
4. *B. E. Sabey and P. J. Codling, "Alcohol and Road Accidents in Great Britain," presented to the 6th International Conference on Alcohol, Drugs and Traffic Safety, Toronto, 1974.*
5. *Anon, "Tri-level Study of the Causes of Traffic Accidents: Final Report," DOT-HS-034-3-535-77-TAC, Washington, 1977.*
6. *B. E. Sabey and G. C. Staughton, "Interacting Roles of Road Environment, Vehicle and Road User in Accidents," presented to the 5th International Conference of the International Association for Accident and Traffic Medicine, London, 1975.*
7. *G. C. Staughton and V. J. Storie, "Methodology of an In-depth Accident Investigation Survey," Transport and Road Research Laboratory, Laboratory Report 762, Crowthorne, 1977.*

8. *Organisation for Economic Co-operation and Development, "Hazardous Road Locations: Identification and Countermeasures," OECD, Paris, 1976.*

9. *G. R. Watts and A. R. Quimby, "Aspects of Road Layout that Affect Drivers' Perception and Risk Taking," Transport and Road Research Laboratory, Laboratory Report 920, Crowthorne, 1980.*

10. *J. Wall, R. W. Lowne, and J. Harris, "The Determination of Tolerable Loadings for Car Occupants in Impacts," presented to the 6th International Technical Conference on Experimental Safety Vehicles, Washington, 1976.*

11. *B. E. Sabey, "Potential for Accident and Injury Reduction in Road Accidents," presented to the Traffic Safety Research Seminar organized by the Road Traffic Safety Research Council of New Zealand, Wellington, New Zealand, 1976.*

12. *B. E. Sabey, "A Review of Drinking and Drug-taking in Great Britain," presented to the 7th International Association for Accidents and Traffic Medicine, Ann Arbor, Michigan, 1978. Transport and Road Research Laboratory, Supplementary Report 441, 1978.*

13. *Anon, "Biorhythm Theory Makes Safe Drivers and Reduces Accidents," Inviting sound from Japan, Automobile News, 1969.*

14. *P. O. Palmer, "The Effect of Biorhythms on Road Accidents," Transport and Road Research Laboratory, Supplementary Report 535, Crowthorne, 1980.*

15. *E. L. Nordentoft, et al, "Traffic Speed and Casualties," Proceedings of an International Interdisciplinary Symposium at Funen, Denmark, 1975. Odense University Press, 1975.*

16. *G. Grime, "The Protection Afforded by Seat Belts," Transport and Road Research Laboratory, Supplementary Report 449, Crowthorne, 1979.*

17. *Anon, "New Attitudes to Road Safety," Journal of the Institution of Highway Engineers, 26: No. 4, April 1979.*

18. *Anon, "Drinking and Driving," Report of the Departmental Committee of the Department of the Environment, Her Majesty's Stationery Office, London, 1976.*

19. *R. A. Hargroves and P. P. Scott, "Measurements of Road Lighting and Accidents — Some Results," presented to the Annual Conference of the Association of Public Lighting Engineers, Ayr, 1979.*

20. *E. Dalby, "The Use of Area-wide Measures in Urban Road Safety," presented to Traffex '79 Conference on Traffic Engineering and Road Safety, Brighton, 1979.*

21. *C. A. Hobbs, "The Effectiveness of Seat Belts in Reducing Injuries to Car Occupants," Transport and Road Research Laboratory, Laboratory Report 811, Crowthorne, 1978.*

APPENDIX — TABLE 11

Potential Accident Savings from Improvements in Road Environment by Low Cost Remedial Measures

Feature	Improvement	Target Group of Accidents B	Benefit A	Potential Savings A x B (at 1977 Levels)	
				No of Accidents	% of Total
Geometrical Design: Junctions	Control and Design: Mini-roundabouts, Traffic Islands, Speed Control, Visibility	Uncontrolled Junction Accidents on Class A & B Roads in Urban Areas: 63,000	40%	25,000	10½%
		Rural Junctions on Class A & B Roads: 15,100	20%	3,000	
Surface	Rougher Textures	Excess of Accidents Due to Wet Weather: Slippery Roads — 13,000 Impaired Visibility in the Dark — 5,000; Splash & Spray — 7,000	75% 40% 33%	10,000 2,000 2,000	5½%
Lighting	Installation of New Lighting and Guardrails	A Proportion of Dark Accidents on Unlit Roads: Urban — 1,600 Rural — 8,000	30% 50%	500 4,000	3%
	Improvements on Lit Roads	Half of Accidents on Lit Roads — 25,000	20%	5,000	
Urban Areas: Arterial Roads and Residential Areas	Area Application of Low Cost Measures	Accidents on Arterial Roads and Residential Areas: Two Thirds of Total in Urban Areas — 138,000	10-20%	14,000-28,000	5-10%

APPENDIX — TABLE 12

Potential Savings from Improvements in Vehicle Safety

Feature	Improvement	Target Group of Accidents or Casualties B	Benefit A	Potential savings A x B (at 1977 Levels) No of Accidents Casualties	% of Total
PRIMARY SAFETY — ACCIDENTS					
Defects	Tyre and Brake Maintenance	All Accidents: 266,000	2%	5,000	2%
Handling	Antilock Brakes	Car Accidents: 202,500	4%	8,000	6%
		Motorcycle Accidents: 72,000	10%	7,500	
	Elimination of Worn Tyres	Wet Road Accidents: 89,000	3%	2,500	1%
Conspicuity	Wearing of Conspicuous Clothing etc.	Vehicle/Vehicle Accidents at Junctions: 98,500	10%	10,000	3½%
SECONDARY SAFETY — CASUALTIES					
Occupant Restraint	Use of Seat Belts	Front Seat Occupants of Cars Not Wearing Belts: Fatal & Serious — 21,000 Slight — 61,000	60% Fatal & 45% Serious 15% Slight Injury	10,000 9,000	7%
		Rear Seat Occupants: 21,000	25%	5,000	
Other Protective Measures					5% - 10%

DISCUSSION

G. Patton *(American Petroleum Institute)*

I noticed that one of your slides gave the rate of injury per million kilometers with three classes of road: the limited access highway, the rural and the urban. And the rates are quite significantly different though I don't remember the numbers. It seems to go against your statistic that said something like 2-1/2 percent of the accidents were attributed to the road and the other 95% were attributed to the driver. I wonder if you would comment on that.

B. E. Sabey

B. E. Sabey

On the face of it this could seem so. Without going into all interactions, it's difficult to explain. But for example on the high speed roads the factor of speed would come in, which, in many cases, we would class as a human error; i.e., somebody going too fast for the situation, rather than a failure in design speed of the road itself. In fact, it was quite common that one of the individual errors was going too fast for the situation; not necessarily going too fast in absolute terms or exceeding the design speeds of the road. That's the sort of example to help to answer your question.

E. V. Anderson *(Johnson and Higgins)*

Can you clarify the difference between the 95% and another figure you had of 77% as human error?

B. E. Sabey

You need to see the detailed tables in the paper. In the one case the 95% is the number of accidents in which the human is a contributing factor. The 77% is the number of factors which are human errors; the number of errors is larger as there is more than one factor per accident.

H. Joksch *(Center for the Environment & Man)*

I would like to make a few comments on the assigning of human factors, vehicle factors, and highway factors. If I could comment only on the Indiana study. We have very carefully reviewed the methodology of the Indiana study as well as others and, very simplified, the situtation is as follows. The assignment of causes, to a large extent, follows established folklore in the accident investigation field. If you take a completely unpreoccupied look at the data, you will notice the following. The manner of dealing with various factors has been taken as given and the failure of the driver to accommodate the situation is what is determined by these factors, taking the drivers out.

Now, there is the possibility to argue the other way around. Since a very large percentage of the population is being licensed and we cannot reduce this matter very much for political reasons, one could also take the behavior of the vast majority of drivers as given and then the failure of the vehicles and the highway to accommodate the behavior of the vast majority of drivers could be considered the highway and vehicle failure. Now, this is a methodological question. There are some points on the detail. What are "causes" identified as contributing factors? To say "contributing factor" in the Indiana study is an improper outlook and this is not operationally defined. Here, essentially, the entire usefulness of such a study breaks down because if we cannot define, before a crash has occurred, what constituted improper outlook and what constitutes proper outlook. So identifying a contribution of this precrash behavior to that fact cannot be qualified. Essentially, percentages found in the Indiana study were arbitrarily assigned to causes after the fact although they cannot be identified before hand. For example, for a case in which neither the closing speed nor the distance between vehicle could be established, the cause "following too close" was assigned. "Following too close" was really the rephrasing of the observation that the rear end collision has occurred.

B. E. Sabey

These are exactly the kind of things that worried us before and during our study. We were aware that some of our assessments were going to be subjective, which is why we were so encouraged to see similar results from the Indiana study conducted independently. I will read to you what is given in the paper as the definition of contributing factor: "a main contributing factor means something without which the accident would have been less likely to happen, or at least would have been less serious." We assess this after we have gathered together all our information from different data sources; at the scene of the accident, the examination of the vehicle subsequently, the detailed police report, the reports of witnesses and our own interviews of drivers and other persons involved. Also our team is essentially one which has a multidisciplianry experience in road safety, including mechanical engineering, highway engineering, physics, mathematics, statistics, and police operations. So we think we have the view of every kind of person involved in road accidents. We tried to list all kinds of errors in the road, vehicle and human factors and then assign these "contributing factors." In some cases if we were in doubt we might assign more than one human factor and then of course, as you say in the paper we brought them all together. Now, I agree with you entirely about the importance of the interaction between road engineering and human factors in particular; rather less so for the vehicle. But there are a lot of errors identified in our study, in about 1/4 of the cases, where there was a road deficiency coupled with a human factor. If you take your suggested new methodology, that would merely be translated into something which just had a road factor. But either way, when one is looking for a remedy now, you would tend to say it is going to be cheaper and easier to effect a road engineering measure than it is to change human behavior directly.

Over and above that, even amongst the group of 65% of accidents where human error alone has been assigned we would say that there are ways of influencing behavior by road engineering features even though we don't, at this time, see any deficiency in the network. In particular you may have seen some examples of physical means that we are now using for influencing speed: for example, transverse yellow bar lines across the road at approaches to junctions, and humps in residential areas to keep vehicle speeds down to very low values — in the range 10 to 15 miles an hour.

D. B. Straus *(American Arbitration Association)*

I think everybody here perceives what they are listening to from their own particular viewpoint and mine has been in dispute settlement. And I hear from underlying thesis, if I hear it correctly, that one of the problems is the wide gap between the perceptions of risks and the quantitative analysis of risk that comes out from the experts that have been working on it. Trying to bring perceptions and a sence of what the actual data shows is an essential part of the dispute resolution process. It occurs to me however that as groups like this become more expert in analysis, the wider will be the gap between perceptions and the actual analysis. This will continue unless some process is developed to bring the thought process of decision makers who are seldom the technicians in line with the thought processes of the technicians. I address this question to you and to others who may speak after you. Has this been given any thought in this whole area?

B. E. Sabey

This is something that has been increasingly our concern in recent years. Perhaps I should explain how we relate in the Research Laboratory to the policy and decision makers in the Department of Transport, before I answer your question. We have a different situation in the UK from the situation in this country. In the Department of Transport we have a Road Safety Directorate which is responsible for the policy on road safety, for legislation and rule making. Also, in the Department of Transport, we have our own Transport and Road Research Laboratory, but we are regarded as the technical advisors and our Director has the absolute authority to decide what is or is not published. So there is a link there between the scientific and technical appreciation of the problem and the actual requirement to make decisions. In fact, we work very, very closely together, and we have regular discussions with our opposite numbers in the policy directorate, which helps give mutual understanding. More and more we are realizing the difficulties between the perceived and the actual risks, and certainly one of our tasks is to keep things in perspective. One of the facts to which I constantly refer when I talk to anyone concerned with road safety is that 3/4 of our accidents occur in urban areas. Just a simple fact like that can keep things in perspective when people want to put effort into making the motor ways even safer than they are.

S. Pollock *(University of Michigan)*

I wonder if you care to comment on something that may be mythology, but I believe that I read that in most Scandinavian countries they have an extremely stiff drunken driving ordinance and apparently has solved that problem. And that commercial aviation and general aviation people have extremely stiff qualifications for getting into an airplane and flying it. That seems to reduce tremendously the operator error problem. Aren't these rather cheap solutions for these problems?

B. E. Sabey

You are talking about what has become known universally as the Scandinavian myth. In fact, we have a government committee of inquiry on drinking and driving which was setup about four years ago and I was a member of that committee. One of the things that we looked at was the Scandinavian problem, and in fact I think people get the wrong impression of what the law is in Sweden. It is not specifically a no drinking law. There are basically three levels. There is a level of 50 milligrams, below which you will only be penalized if you have an accident or commit any other offense. Between 50 and 150 milligrams the law is very much like the British law where you get disqualification and a fine. It's only above 150 milligrams that you would actually get a jail sentence. Though it isn't actually jail, it's something like a labor weekend camp. So that the legislation and its consequences are not quite as tough as is sometimes made out. When one compares internationally the levels of blood alcohol in drivers killed, which is the only thing that is consistently reported in different countries you find, in fact, that the distribution of alcohol levels in Sweden is just about the same as over here and rather higher than in Britain. So I don't think we have very much to learn from the Scandinavian legislation.

S. Pollock

I don't understand that statistic on the levels of blood alcohol in drivers killed.

B. E. Sabey

If you want to do a comparison of the level of drinking which has consequences in road accidents in different countries, the only statistic available which can give any reasonably consistent comparison is the level of blood alcohol in drivers killed in accidents. And that has been compared for the United States, for Britain, for Sweden and other countries as well; Sweden has about as high a distribution as the United States.

S. Pollock

Should we normalize that by the number of accidents per passenger mile?

B. E. Sabey

We should, but it has not been done because of lack of exposure data.

C. Starr *(Electric Power Research Institute)*

Miss Sabey, thank you very much. I just want to say parenthetically that the Swedes start from a higher alcoholic base.

PERCEPTIONS OF RISK AND THEIR EFFECTS ON DECISION MAKING

Raphael G. Kasper

National Academy of Sciences, Washington, D.C.

ABSTRACT

The evaluation of risks is a part of any rational decision-making process. But what risks? And whose evaluation? Measures of risk tend to fall in two broad categories: those that purport to observe or calculate the actual risk of a process or project and those that rely upon the judgments of those assessing the risk. A nomenclature problem arises here. Some characterize the categories as "objective" and "subjective"; others, perhaps somewhat arrogantly, call them "real" and "imagined." Technical experts tend to consider the use of measures of the first sort the only legitimate way of describing risk, yet measures of the second sort tend to dominate the thinking and actions of most individuals. Whatever they are called, the two measures of risk seem seldom, if ever, to agree. The social and psychological reasons (or explanations) for the disparity are interesting but, beyond them, there are important implications of the disparity for decision making in an increasingly technological society. The distinction between the two measures of risk creates difficulties for decision makers and regulators among which are: an increase in the use of propaganda and indoctrination by government, industry, and technical experts in attempts to convince the public that technical estimates of risk are valid; a continued erosion of trust and understanding between experts and the rest of the public; further complication in the already involved process of setting priorities for government or corporate programs; and a challenge to decision makers to explain uncertainties about the effects of their actions.

INTRODUCTION

Now is a particularly appropriate time to discuss risk and its assessment. A public already confused and unsure about the impacts of technology has recently been forced to consider issues of risk more than ever before. In rapid succession new incidents, like the accident at Three Mile Island, the crash of the DC-10, and the return of Skylab, have taken their place in the public consciousness alongside continued safety recalls of automobiles, disclosures of dangers of chemicals in the air, water, and foods, and any number of other indications of the pervasiveness of risk in our society.

It has been recognized for some time that the increasing complexity of modern technological society leads to consequences that are often unknown or uncertain. But as we learn — or are forced to learn — more about the possible extent of those consequences on our well being, our way of life, and our surroundings, it becomes more and more necessary to improve our ability to identify and evaluate the impacts of technology on our lives. The tools are imperfect, at best, and an awareness of the inadequacy of our understanding of risk has led to the creation of a burgeoning, vigorous, and sometimes illuminating field of research known as risk assessment.

The evaluation of risks is, of course, only a part of any decision-making process. A rational decision maker must also determine a set of benefits and then weigh (often intuitively, on occasion explicitly) benefits against costs and risks. (Some have defined risk assessment as the entire process of identifying and evaluating risks, identifying and evaluating benefits, and performing trade-offs of risks and benefits. Here risk assessment refers only to the first of these activities; "decision-making" is used to indicate the entire process.)

These remarks are concerned largely with decision making in government, although, not surprisingly, much of what is said has implications for corporate or individual decision making.

Measures of risk tend to fall into two broad categories: those that purport to observe or calculate the risk of a process or project and those that rely upon the perceptions of those assessing the risk. Measures of the first type employ experimental evidence, long-term experience, or sophisticated analytical calculations to describe actual risks or project potential risks. Often framed in probabilities, these measures seek to tell us, for example, what the chances are of an explosion at a liquified natural gas terminal, or what the probability is of an oil spill from a large tanker, or what effects on health can be expected from various levels of atmospheric pollutants. Measures of the second type tell us what people think the risks of a particular activity are. A nomenclature problem arises here. Some characterize the categories as "objective" and "subjective"; others, somewhat arrogantly, call them "real" and "imagined." There may be no solution to the problem. In any case, whatever they are called, the two measures seem seldom, if ever, to agree. The social and psychological reasons (or explanations) for the disparity are interesting and can teach us much about the ways in which people react to technological undertakings. But aside from their intrinsic interest, the differences between the two measures of risk have important implications for decision making in an increasingly technological society.

CALCULATIONS AND PERCEPTIONS OF RISK

Quantitative Methods — *Experience* —There have been a number of attempts to observe or calculate the risk of particular activities. Some involve collection of data to describe the risk; highway fatality statistics fall in this class. The generation of such data allows a fairly accurate picture of certain kinds of risks, most notably those that involve easily recognized and countable risks (such as death) for which the cause is clear and for which there exists a body of available statistics. If there is a relatively non-controversial and easily applied aspect of risk assessment, it is in the collection and use of this kind of data.

Unfortunately, few risks fall into this class. For some the relationship between cause and effect is less clear; this may be so because the effect follows the cause by some period of time (perhaps by as much as decades in cancer caused by certain chemicals), or because the immediate cause cannot be distinguished from other possible causes (as in the case of the effects of air pollutants on the functioning of the lungs), or because the effect is not a dramatic one that is observable and amenable to simple measures (as in the effects of trace metals on behavioral responses in people or in the effects of toxic substances on ecosystem dynamics). The difficulties are apparent in one example of a risk (or possible risk) that has generated considerable controversy: the effects of low levels of lead (say, on the order of 30 μg per 100 ml in blood) on learning ability and other neurobehavioral characteristics of urban children. There, even the best of epidemiological studies is confounded by the myriad explanations for low level neurobehavioral effects; the same effects attributed to lead may be caused by exposure to low levels of many other trace metals and, indeed, by exposure to the pace and stress of urban life itself. The result is that careful studies yield not proof but only suggestions (which become quite strong as additional, confirming evidence is gathered) that exposure to lead from automobile exhausts, from paint, and from foods at doses below those that produce clinically diagnosed symptoms leads to impaired neurobehavioral functioning of urban children. Detailed quantitative estimates of the effects are not currently available.

Calculations — For other risks, experience may be only a very partial guide and assessors are forced to use elaborate methods to project or predict highly uncertain effects. The potential long-term consequences of carbon dioxide accumulation in the atmosphere due to the combustion of fossil fuels and the possible results of experimentation in the field of genetic manipulation are but two of a large (and apparently growing) list of such risks.

But if these risks are less tractable than those that are the subject of simple accident statistics, they are nonetheless the object of extensive activity by risk assessors. Risk assessment exercises vary widely in type and scope, although if one such effort could be thought of as capturing the essence of what has been referred to by some as objective measurement of risk it would be the *Reactor Safety Study: An Assessment of Accident Risks in U.S. Commercial Nuclear Power Plants* [1]. (This is so despite the study's deficiencies, which have been examined in detail

References p. 80.

elsewhere [2].) That study applied techniques developed in the U.S. space program and elsewhere to an estimation of the probability of various occurrences in nuclear power plants and was presented, with great fanfare, as a demonstration of the safety of the nuclear power program. Other, less ambitions, examples abound.

Uncertainties — An attraction of objective measures of risk is the apparent elimination of subjective elements in at least one part of the decision-making process. But the attraction is more apparent than real. Even objective measurement of risk, like any intellectual endeavor, involves some elements of uncertainty. The very choice of questions to be asked, issues to be considered, and methods to be used involves judgment. Thus, in examining the effects of lead on urban children, why does the analyst choose to consider some measures of learning ability (such as IQ) and not others? Why isolate urban children and not all children? Or in examining the risk of nuclear power plant accidents, why consider particular chains of events and not others? The choice is clearly subjective, based at least in part on the analyst's judgment and conception of the problem under investigation. Because it is impossible to conceive of any analysis that is purely objective, any attempt to reduce risk assessment to a rote, mechanistic application of technique is doomed to failure. (This has not dampened the spirit and audacity of those who propose schemes that purport to take judgmental elements out of analytical techniques or to completely separate facts and values [3].)

I do not propose to discuss in detail the mechanics of risk assessment; it is however, worth bearing in mind that even the most coldly analytical of methods is likely to be inaccurate because of deficiencies in the data base or in the method itself. This is not to say that quantitative risk assessments are not useful or illuminating, only that it pays to be somewhat skeptical of the quantitative results of risk assessments and to recognize that the appearance of great accuracy that precise numbers in such analyses carry with them is spurious.

Discrepancies — While technical experts hone their craft and generate more detailed quantitative estimates of risk, subjective measures continue to dominate the thinking and actions of most individuals. However, a growing body of research, most notably that of Paul Slovic and his colleagues, demonstrates that perceptions of people about risk do not always (and, in fact, rarely) coincide with what we know to have been the actual risk of certain activities or with the best available expert projections of other risks [4].

Between "Real" and Perceived — Many, though not all, of Slovic's examples [5] illustrate discrepancies between "real" and perceived risks in situations where the nature and size of the risk (often death or injury) are well understood and documented. Thus a sample of members of the League of Women Voters estimated that motor vehicles cause 28,000 deaths per year in the United States; a sample of students estimated 10,500 deaths. In fact, there are about 50,000 deaths each year caused by automobile accidents. Similar differences exist in estimates of other accident fatalities. For example, the two groups mentioned above estimated 930 and 370 drowning deaths annually; accident statistics show that the actual number is about 3000. The discrepancies in these cases can probably be attributed, at least in part, to a lack of complete or accurate information about the frequency of occurrence of certain events or to the inability of individuals to comprehend and use statistical evidence.

Between Calculated and Perceived — But differences in estimates also exist in instances in which adequate numerical evidence is not available and in which projections or calculations of risk are involved as well as in the simple accident cases mentioned above. Thus, it seems clear that inadequate information or data handling is not a complete explanation for the differences between calculations or projections of risk and perceptions of risk. Slovic discusses a number of possible determinants of individual perceptions of risk including the extent to which the risk is faced voluntarily, the length of delay between cause and effect, the degree of control over the risk, the novelty of the risk, and the potential for large or catastrophic effects [5]. Another kind of disparity, more difficult to articulate, also seems to play a role. The difference between experts and the rest of the public may lie in part in a difference in the very conception of the risk or, one might say, in a cultural difference. The difference may be best illustrated by some examples. Knowledgeable individuals active in the anti-nuclear movement, even those with a sound appreciation of the technology, seem to retain a deep and pervasive fear of nuclear power plants exploding like nuclear weapons. It seems clear that no number of Rasmussen reports demonstrating the safety of power plants could ever change the way these people, and quite likely many others, feel about nuclear power. Similarly, expert assurances of the safety and benign aspects of a liquified natural gas terminal near Santa Barbara, California, cannot still the fears of a small group of Indians that the terminal, located at the site of the passageway through which the Indians believe the souls of their people pass to join the spirits of their ancestors, is a threat to the very core of their beliefs and existence.

Context — Even the discrepancy (or at least a part of it) noted earlier between perceptions of highway safety and accident statistics may have a root in the conception of the risk. After all, the figure of 50,000 deaths has been widely publicized and it is hard to imagine that educated laymen, who constitute the samples used in the study, have not come across the number many times. Consider what possible current conceptions of the automobile might be if the introduction of petroleum products to the public had been in the form of napalm rather than in the form of oil for light and heat. (Nuclear power, it should be recalled, was introduced to the public through the awesome destructive power of the atomic bomb and it is almost certainly true that current public perceptions of nuclear power are shaped largely by the first impression.) If people's visual images of petroleum were the hideous scenes of Vietnamese cloaked in flames (as their visual image of uranium is a mushroom shaped cloud) rather than the cozy warmth of a fire on a cold day, it might be more difficult to cruise down the highway with equanimity atop a twenty-gallon tank of gasoline.

Myths — Some prevalent myths about perceptions and calculations of risks deserve debunking. It is popularly thought (at least among technical experts) that

References p. 80.

people always tend to judge risks as being greater than they are in fact (or that people act in ways that indicate that this is how they judge risks). That this is not invariably the case may be seen from the accident figures quoted earlier or from the reaction of people to information about the relationship between smoking and cancer or to the proposed ban on saccharin in foods. (It is worth noting, however, that the cases in which individuals underestimate risks tend to be those in which the risk is assumed voluntarily, although Slovic shows similar results for fatalities due to electric power as well, a case in which much of the risk is imposed on the public [5]). In a similar way, the lay public tends to think that experts underestimate the risks of scientific or technological enterprises. There are notable exceptions to this as well although, it must be admitted, it is somewhat harder to think of them. Thus, it was scientists who first sounded the alarm over the possible dire consequences of experimentation with DNA, an alarm that now appears to have been overstated. (It may be in reaction to their initial, perhaps excessive concern, that some scientists are now changing their position to the other extreme and claiming that there is no danger at all.) And it was the automotive experts who, in the mid-1970's, expressed grave concern about potential effects of sulfuric acid mist emissions from automobiles with catalytic converters. A flurry of interest led to experimental studies and field measurements that showed the problem was not of major import.

Another popular misconception holds that the public, despite its vocal concern about risks, will inevitably change its tone when its comfort or convenience is compromised. The widespread practice of driving without seat belts is often cited as support for this view, but there is evidence that the public does not always react in this way. A recent article in the *Wall Street Journal* quoted Albert Casey, Chairman of American Airlines as saying:

"During the early days of August our DC10s had far lower load factors than they carried during the same period a year ago — a clear indication that some of our customers are avoiding the aircraft [6]."

The article noted that some people are going out of their way to avoid flying on DC10s, even when travelling on another aircraft entails considerable inconvenience. The public's willingness to inconvenience itself persists even in the face of assurances, such as that of an aviation accident investigator at the National Transportation Safety Board, that the DC10 is a "safe plane" [6].

IMPLICATIONS FOR DECISION MAKING

Various aspects of calculated and perceived risk provide fuel for psychologists who seek to explain or predict the reactions of people to risks and for technical experts improving and refining the techniques of quantitative risk assessment. But differences between the two views of risk also have important implications for policy makers and decision makers. It is unfortunate that most of the implications that come to mind create problems rather than improvements in the decision-making process. It is true that one could by a stretch of the imagination conceive of benefits arising from the differences. One might argue that the differences serve to slow the inevitable technological determinism of modern society (though empirical support for this point is sparse), but this is only a benefit to those who believe that there is such a trend and are disturbed by it. Or it could be held that since public perceptions of risk tend to differ from "actual" risk, only the existence of calculations or projections of risk protects a gullible public from danger and assures that limited resources are used as effectively as possible. This too is difficult to prove and is only an advantage to those who feel that there is no benefit in relieving public apprehension, even if that apprehension seems unsupported. In short, arguments for beneficial effects of the difference between calculated and perceived risk are unconvincing.

The root of the problem lies in the tendency of technical experts (and many government and industry decision makers) to view objective characterizations of risk, illuminated by the experts' calculations, as somehow more real or more valid than the perceptions of the rest of the public. There is, as we have already noted, a kind of arrogance in such a view.

One might have expected the Three Mile Island accident and the DC10 crash to have tempered this arrogance, but unhappily, this does not seem to be the case. Just two weeks ago, on the day of an exceptionally large rally and concert protesting the continued use of nuclear power, a group of pro-nuclear scientists issued a statement (on a television newscast) to the effect that rock singers should stick to singing and leave technical issues to the experts.

That people's fear (even if that fear is of a risk that has been characterized as having an extremely low probability, say 10^{-7} per year) should be viewed as any less real than the results of abstract, complex, arcane calculations has always seemed difficult to explain. Yet that is the view of many experts. In fact neither perception nor calculation is more or less real; they are just different.

One implication of the disparity between calculated and perceived risk has been an exaggerated emphasis on propaganda and indoctrination, often at the expense of real progress toward solutions to existing or potential problems. The experts' approach to their differences with the rest of the public has been to somehow persuade the public to alter its perceptions so that they more nearly approximate the calculated or projected results. (I know of no serious effort directed toward shifting expert's views so that they will coincide with those of the rest of the public.) The history of the nuclear power program, from its inception through the explanations of the Three Mile Island accident, is replete with examples of this tendency. The same history demonstrates the futility of the approach.

But efforts at propaganda would be of less concern were it not for the fact that they materially detract from programs designed to protect the public. Nuclear power is a conspicuous example of this because of its great public visibility. Nuclear safety programs and nuclear waste management programs were allowed to lag through the 1960's and early 1970's as though the experts were swayed by their own public relations. One cannot help but think that some recent problems in nuclear power plants, like the Brown's Ferry fire whose effects were exacerbated by poor placement of electric cables or the discovery of potential difficulties in the emergency coolant systems of certain boiling water reactors, might have been avoided if more effort had been put into solving problems and less into manipulating the public. This may be one implication of a recent statement by Chauncey Starr, former president of the Electric Power Research Institute, who was quoted in the *New York Times* as saying:

> "One of the major results of the Three Mile Island accident has been the recognition by the utilities that they must undertake a national and leading responsibility to ensure the reliable and safe operation of nuclear power stations [7]."

What, one feels compelled to ask, was the industry doing before Three Mile Island?

Nuclear power is not the only culprit in this regard. Examples of attempts to use propaganda to persuade a supposedly gullible public, and of the failure of these attempts are ripe for picking. While pollution control programs languished through the 1960's (and even into the 1970's), automobile companies invested huge sums in advertising campaigns to convince people that automobile emissions were not harmful to public health and welfare. (Only the force of new laws, laws plainly designed to stress the technology of the automobile manufacturers, induced the automobile companies to place a real effort on pollution control.) The manufacturers of the Concorde SST and the airlines that use the plane have tried to convince airport neighbors that the Concorde is not a noisy plane while every overflight provides damning evidence to the contrary. It is not surprising that public indoctrination programs do not bring experts and the rest of the public together; psychologists have known for years that once strong and divergent opinions are formed, only clear, incontrovertible evidence (which rarely, if ever, exists or can be obtained in the controversial issues of risk being discussed here) can lead to consensus.

Another distressing result of the difference in views of risk between experts and the rest of the public is the continued erosion of trust and respect between the two groups. The era of blind trust for technical experts has ended; the public now often views experts as insensitive and, sometimes, dangerous. The image is not totally unwarranted; significant problems have been caused or exacerbated by lack of concern of experts. The disastrous collapse of the Teton Dam can be attributed in large measure to neglect of common-sense rules of dam safety. (The dam, it may be recalled, was constructed quickly and the reservoir filled at a rate four times as rapidly as ordinary practice dictates.) The failure might have been thought of as a fluke if subsequent investigation had not revealed that numerous dams in the United States presented potentially similar problems. Yet there is no evidence on the part of the dam building fraternity, including the government agencies that build and regulate dams, that anything will change to assure that similar accidents are not repeated. The usual interagency study was commissioned, but the participants appeared more interested in protecting their respective turfs than in protecting the public.

Programs for the disposal of hazardous wastes seem to be following a similar path, although here unfounded optimism like that of the dam builders is not a factor. The problems at Love Canal and elsewhere were purely and simply the result of a lack of concern.

Problems of distrust between experts and the rest of the public are exaggerated in the case of industrial scientists and engineers. Here the problems are exacerbated because people tend, not always without good reason, to look with suspicion on the motivations of industry. Reluctant recalls of automobiles for safety adjustments cannot have helped the public image of automobile companies and their technical experts. Nor can the almost unseemly display of impatience on the part of McDonnell-Douglas officials in trying to get the DC10 back into the air have strengthened the public image on the airline industry.

The difficulty is not one-sided. When, as is the case in many technological enterprises, the public sees or imagines greater risks than the experts do, the experts begin to doubt the rationality of the public and decry the emotionalism that slows technological progress. The result is confusion, disaffection, and sometimes confrontation.

Differences between perceptions and calculations of risk add another dimension to the already vexing problem of setting priorities for government action. There are, for example, about 35,000 pesticides sold and used in the United States for the control of agricultural and other pests. Every pesticide must be examined and, if necessary, regulated. But it takes a long time and expenditure of considerable resources to regulate pesticides; which should be controlled first? The answer would not be simple even if everyone agreed about the nature and extent of the risk. But which problems should take precedence, those for which the objective measures of risk are greatest or those, usually different, about which public fears or concerns are greatest.

Clearly, the degree of accountability of the decision maker to the public is important in this regard. As a general rule, the more directly accountable a decision maker is to the public, the more likely it is that public perceptions will receive consideration in priority setting. Thus legislators, popularly elected every several years, tend to show more concern for perceptions of risk than do regulators in the executive branch who are often shielded from public accountability. It was the Congress, not the regulatory agencies, that set stringent limits on automobile emissions and that passed the law that mandates perhaps the strictest of risk-reducing decisions, the Delaney Amendment.

CONCLUSION

There is no reason to expect that the two measures of risk, calculated and perceived, will ever coincide; the challenge is thus one of making decisions in the face of the continuing disparity.

One approach to decision making — often taken in the recent past — is to ignore public opinion and to insist on using only objective or calculated estimates of risk. But the result of such an approach is continued dissatisfaction on the part of the public with government and with the process of making decisions. Yet the other extreme is no more palatable: an approach that seeks always to quell public apprehension could result in excessive conservatism in the development of technology. There may be no solution to the problems posed by differing views of risk, but there are at least two changes in the manner in which we make decisions about

References p. 80.

technological enterprises that would make things easier. One involves a change in attitude, the other a change in process.

It must first be recognized that the gap between experts and the public will not be narrowed by forcing change only on the part of the public alone. The respect that people once held for experts will be restored only if experts work at reversing an image that makes them appear self-serving, dangerous enemies. One approach to this reversal would be for the experts to develop an appreciation and an understanding of public perceptions of the risks of technology and to realize and accept the fact that these perceptions must play an important role in decision making.

The process change alluded to above would be one that assures the early and real involvement of all affected parties in decision making. What the precise nature of the change should be has thus far eluded us; although early involvement sounds eminently sensible, it rarely occurs. Yet if no workable process is found, many people affected by decisions with far-reaching consequences will continue to feel that they have had little say in their destiny.

REFERENCES

1. U.S. Nuclear Regulatory Commission, Reactor Safety Study: An Assessment of Accident Risks in U.S. Commercial Nuclear Power Plants, WASH-1400, 1975.
2. H. W. Lewis, et al., Risk Assessment Review Group Report to the U.S. Nuclear Regulatory Commission, NUREG/CR-0400, Nuclear Regulatory Commission, 1978.
3. K. R. Hammond and L. Adelman, "Science, Values, and Human Judgment," Science, 194: 389-396, 1976; K. R. Hammond and J. Mumpower, "Risks and Safeguards in the Formation of Social Policy," presented at the Beijer Institute International Review Seminar on "Impacts and Risks of Energy Strategies: Their Analysis and Role in Management," Stockholm, Sweden, September 1978.
4. P. Slovic, B. Fischhoff, and S. Lichtenstein, Cognitive Processes and Societal Risk Taking, Oregon Research Institute Monograph, Eugene, Oregon, 1976.
5. P. Slovic, B. Fischhoff, and S. Lichtenstein, "Rating the Risks," Environment, 21: no. 3, 14-20, 36-39, April 1979.
6. A. Casey, "Some Travellers Still Go to Great Lengths to Avoid the DC10s," Wall Street Journal, 1, August 30, 1979.
7. C. Starr, "Three Mile Island Accident: A Cloud Over Atom Power," New York Times, B5, September 24, 1979.

DISCUSSION

C. Starr (Electric Power Research Institute)

Let me just drop the other shoe; we have to distinguish between cosmetic propaganda and announcements. The utility industry has undertaken three major post-Three Mile Island efforts. One is a Nuclear Safety Analysis Center (NSAC), located at the Electric Power Research Institute, to analyze in great detail what all the existing power stations have in way of technological content relative to risks and accidents, a new responsibility for EPRI. The utilities have undertaken the formation of an Institute Of Nuclear Power Operations (INPO) which will monitor the operation of all the nuclear power plants in the country. They have also undertaken the formation of a mutual fund, Nuclear Electric Insurance, Ltd., to cover the cost of any accident. These actions represent a philosophic revolution within the utility industry. In effect all utilities are looking upon each nuclear utility as "his brother's keeper" in operations. This relationship has never existed before. The answer to your question is that the utilities had depended on the manufacturers and the government agencies, particularly the NRC, to establish the basis for good operation. That obviously wasn't adequate and the industry was naive in not recognizing that. So the utility actions are not just a propaganda piece. They are a substantial change in operation. That doesn't mean that your point wasn't well taken. We're talking perceptions of these matters which are a dominant issue and we haven't handled public perceptions well.

R. G. Kasper

D. Okrent *(University of California)*

First a comment. I guess I myself would be a little careful applying the word arrogant only to those who make risk estimations. I think that those who are proposing alternative energies and not providing risk estimations in various areas, are putting them forward as almost panaceas when, in fact, there may be problems. The question I have is related to how much the public perception of risk should be used in setting priorities for risk management, and I would like to use the example question of dams. As you indicated, the public, including the League of Women Voters, perceives dams as not very hazardous. If we took public perception, we would not have a program to reduce risk from dams. In fact, its been a fairly slow process for the Federal Government to do anything. It's been a rather slow process for most states to do anything and I wonder whether we could count on public perception to adequately set priorities for where we should be putting our efforts?

R. G. Kasper

Well if I used the word arrogant to describe all risk assessors, I didn't mean it. I said only that I thought that the view that public perception of risk is somehow not real is arrogant.

Now, as for whether one can count on the public's perceptions of risk to drive priorities, I think that you do not count on those perceptions alone nor do you count on the experts' calculations alone. I think we tend more toward dependence on experts' calculations alone. One result has been a dissatisfaction on the part of people who feel that decisions are being imposed upon them — decisions that affect their lives and have perhaps far-ranging consequences for them — and they feel they've been taken out of the decision process. I don't know the answer to the question of how to set priorities, but I do know that neither sole reliance on the calculations nor sole reliance on perceptions is likely to be appropriate.

B. Altshuler *(New York University)*

I find that no matter how one defines it, risk gets used in two different ways. Lowrance, at the beginning, spoke of it as a compound measure of the probability of occurrence and magnitude of results. The probability of occurrence is something that is a matter of scientific objective estimation. I think the differences come in when there is a value judgement about the degree of magnitude or severity of the damage that's done. Now *that* the scientist has no patent on. He's a member of the public like other people. He may use experts to tell him of the consequence, but the value is a matter that's anybody's judgement. When it comes to the question of what most people think of an objective fact, which is the frequency, that's something that must be done by technical experts. It seems to me that, as you put it, you run that second issue together with the first.

R. G. Kasper

I do not agree that determination of probability of occurrence is a totally scientific, objective matter. As I pointed out, in the lead example, figuring the probability of a learning deficiency in a child with a low level of lead in his or her blood depends a lot on the way you design the experiment, on the kind of assumptions you make in designing the experiment, on the groups you choose to examine, and on the measures of effect you use — and all these are subjective.

B. Altshuler

Science doesn't get into that. That's just a question of differences among people who decide.

R. G. Kasper

I am afraid I don't see why it's just a matter of differences among people in the sciences. If my question is: "Why look at urban children, why not look at all children?" or "Why look at children and not adults?", it seems to me that I could go out and ask just about anybody on the street what he or she thought was important and that would have the same value as the opinions of scientists.

C. Starr *(Electric Power Research Institute)*

You're saying that the question was developed out of public concern. I think the question was posed through professional examination in a special area. Now the interpretation of the results will also be of public concern, but there's a period when the public stays out of it simply because of a lack of professional involvement.

S. M. Swanson *(American Petroleum Institute)*

We must be somewhat careful in talking about this difference between objective and perceived risk. For example, if you are a healthy person who has no bad habits living next to Three Mile Island it may be a most significant risk you face even though you may overestimate the actual magnitude of that risk. So, from an individual point of view, they look at what society ought to remedy. That's a perfectly reasonable decision. So you begin at the objective measures in a more individualistic fashion. That is, as you divide the population into a finer grain, you begin to see that the difference between the objective and perceiver diminishes. It may not go away because healthy people have different perceived risks than unhealthy people, let's say. Vis-a-vis nuclear power.

C. Starr

You're dealing with a definition of a question that really has to do with the differences between the calculation of the probability that something happens, its consequence, and the value measure. And the individual value measures were not discussed.

J. H. Wiggins *(J. H. Wiggins Company)*

I take exception to the implications of your talk, mainly because I think I am qualified. My firm did the Liquid Natural Gas study for the terminal near Santa Barbara, before the fact. My firm also did the Supersonic Transport noise analysis, before the fact, and my firm did the DC-10 analysis after the fact. The implication is that people like us are not professional, and I think that is the key that I would like to see come out of this meeting; getting more professionalism, professional attitudes, into this business. When you have a professional design your structure, you have one form of a structure. If you have your next door neighbor who happens to sell Kentucky Fried Chicken, design your structure, you get a different kind of structure. When you have your next door neighbor analyze the pain in your gut as opposed to going to a reputable physician who has had training in looking at these things, I think you're going to get two different kinds of answers. I have four different licenses. We have professional people doing these kinds of things and we are concerned. I just take exception.

R. G. Kasper

I didn't want to imply that people who perform risk assessments aren't professionals. I want to say two other things. One is that, for my part, I would just as soon have the Kentucky Fried Chicken salesman structure the issues if he does it in a way that is logical to him, as to have a professional do it. The other is a question: what was the percentage of the souls that passed through the holy place?

J. H. Wiggins

We have to look at Vandenberg Air Force Base which was there before and since they didn't protest that, there were no souls prevented — i.e., zero percent.

G. Patton *(American Petroleum Institute)*

I wonder if you care to comment on the recent set of projected calculations that shows that mobile homes cause tornados.

SESSION II
"ACCEPTABILITY" WITH FIXED RESOURCES

Session Chairman

Allen Kneese

Resources for the Future
Washington, D. C.

INTRODUCTORY REMARKS

SESSION II

Allen Kneese

Resources for the Future
Washington, D.C.

It's not going to be necessary to make extensive introductory remarks because the things that were said this morning by Chauncey in opening the conference and also the papers of the morning really set out a very good context for what we're going to try to be doing. The focus this morning, while rather broad, was still mainly on defining what the risks are that we face as a society. In this afternoon's session we are going to begin to focus on what we could possibly do about them. In one sense we've already done a lot. There is legislation on the books, much of which was mentioned this morning. There are hazardous substance provisions of the clean air and water acts. There's OSHA, there's TOSCA, FIFRA, and there are several others. However, we are just really getting into the opening phases of trying to implement this legislation and the enormous difficulties involved with it are becoming clearer as we then try to come to grips with the specifics of instituting regulations in the area. Furthermore, while there is a clear social desire in the United States to reduce risks, there is at the same time an increasing concern about the possible economic affects of regulations on the economy, especially at a time when there are economic difficulties for other reasons, when productivity has slowed, and when we have a high rate of inflation. These possible economic impacts are perhaps the chief reason why there is now an extremely strong interest in doing economic analyses of these laws and of these regulations. It's a reason for the intense interest in what's called benefit cost analysis and in other kinds of decision methodologies. Indeed, as someone pointed out this morning some of the acts require that benefit cost analyses be done for the regulations. Others have pointed out the enormous numbers of chemicals that are introduced into the economic system every year. It was pointed out this morning that someone has estimated that there are 35,000 pesticides to regulate. Consequently, just in terms of sheer volume, the problem of trying to apply quantitative rational decision making processes is an enormously large one.

In addition, the technique of benefit cost analysis was developed in the United States many years ago for the analysis of public investment projects. Many of us have worked with the technique for a long time. It was especially applied to the analysis of investments in water resource development projects over the years, I would say for 40 years anyway. As time has passed, it has become applied to more and more difficult kinds of problems, including the kind that concern us today. For example, the old AEC made a benefit cost analysis of its fast breeder reactor program. There have been several analyses of automotive emissions programs. There have been analyses of emissions from stationary sources. There have been some analyses of toxic materials. There have been analyses of the storage of nuclear wastes.

Well, these applications to increasingly difficult problems stretch the technique to it's limits and possibly beyond. There are a couple of reasons why these are so much harder than the traditional applications. One is the sheer lack of data. We simply don't know very much about the cause and affect relationships in many instances. Another is that we are talking about situations that more often than not involve risks to health and possibly to life. This raises the very severe question of evaluation of risk. What are the proper concepts? What are the kinds of ethical considerations that enter into this evaluation?

Another aspect is that the distributional questions, which have always been difficult for benefit cost analysis, are exacerbated because we're not only speaking in many cases about distribution among members of the existing generation but we are speaking about distributional questions having to do with future generations. How we should view their rights in making decisions for the present raises tremendous problems about some of the traditional techniques that are used, particularly discounting of the future. At some finite rate of interest, 5 - 10%, nothing matters after a relatively few years. Which raises ethical questions about whether this is a proper basis for making decisions. These are things being explored.

One may have great qualms about some of the applications that are being attempted and yet we are still not sure what other kinds of more suitable decision criteria we might be able to develop and apply. These are things that will be addressed not only to some extent in the present session but in continuing through following sessions.

In addition to the evaluation of these programs, there is an increasing interest in the efficiency of our efforts to control environmental degradation and a search for alternatives to the direct regulation which has been the traditional approach. People have raised the possibility of applying economic incentive measures of one sort or another in the field. There is interest in the effect that different kinds of liability rules may have on the introduction of hazardous materials and on the innovative process. There is renewed interest in insurance as was pointed out, I guess, by Chauncey this morning with respect to the electric utilities.

Well, I won't speak anymore. I'm simply underlining things that have already been said in one context or another. I would like then to turn to the first speaker.

ON MAKING LIFE AND DEATH DECISIONS

Ronald A. Howard

Stanford University, Stanford, California

ABSTRACT

Recent research has provided us with methods by which an individual can make decisions that involve risk to his life in a way that is consistent with his total preferences and with his current risk environment. These methods may ethically be used only by the individual himself or by an agent designated by the individual. In the absence of such delegation, anyone who imposes a risk on another is guilty of assault if the risk is large enough. Just as society has found ways to distinguish a "pat on the back" from physical battery, so must it now determine what risk may be placed upon another without his consent.

The research on hazardous decision making creates a framework for this exploration. The basic concept of this approach is that no one may impose on another a risk-of-death loss greater than a specified criterion value established by the experience of society. If anyone attempted to do so, he could be forbidden by injunction. The only way that an injunction could be avoided would be by showing evidence of insurance that would cover the damages to be paid by the imposer of the risk if the unfortunate outcome should occur. The methodological framework is used both to estimate the risk-of-death loss and the amount to be paid if death occurs, an amount that is likely to be much larger than present "economic" values of life. Evidence would be required both on the preferences of the individual-at-risk as revealed and corroborated by his behavior and on the magnitude of the risk as assessed by experts.

Such a system is likely to require revisions in the present legal codes. It is to be expected that when a logically and ethically based risk system is functioning, there will be an increased interest in purchasing the consent of people to imposed risk. Problems of securing the consent of contiguous property owners, for example, could be handled by interlocking options. People will also be more likely to be informed of the risk implied by using products or services. Thus risk would become an explicit part of purchasing decisions. The joining of logic and ethics in these new procedures offers hope for a more effective and humane treatment of risk issues in society.

References p. 106.

NOTATION

p:	probability of death
x:	required payment to undertake specified death risk
W:	present level of wealth
c:	constant annual consumption
ℓ:	remaining length of life
$\bar{\ell}$:	expected remaining length of life
w:	worth numeraire
η:	consumption-lifetime trade-off
u:	risk preference function on worth
γ:	risk aversion coefficient
ρ:	risk tolerance; $1/\gamma$
ζ:	annuital factor; amount of annuity that \$1 will buy
i:	interest rate
p_{max}:	maximum acceptable probability of death
$v(p)$:	life value in expected value sense when facing death with probability p
v_s:	small-risk life value
v_e:	economic life value; c/ζ
p_n:	probability of death in year n of life
q_n:	probability of death in year n of life given that individual was alive at beginning of year n.

INTRODUCTION

What risk may one impose on another? This question has achieved increasing importance as the sources of harm in our environment have increased. The spectrum of risk that one person imposes on another ranges from the relatively minor risks posed simply by existence up to the very serious risks represented by assault or attempted murder. Some of these risks society has chosen to ignore, while others have been treated as very serious matters requiring extensive social action. We shall examine both the ethical and practical questions of risk in society, propose measures for risk, suggest procedures for evaluating risk, and indicate how these procedures could be used in practice.

EFFICACY AND ETHICS

Social arrangements for any purpose may be judged in terms of both efficacy and ethics. Efficacy refers to what works in pursuing specific human goals; ethics refers to what actions are morally desirable in achieving those goals. For example, killing babies with genetic defects might be a very efficacious way of achieving the human goal of physical perfection, but it would be ethically unacceptable to most people. When we wish to judge any action or arrangement, we can think of examining it against standards of physical knowledge, ethics, and efficacy. For example, if someone threatened to bring the wrath of God against another, that threat would not be actionable in a court of law today because it is the present belief of a majority of our society that no one has such power. However, in the 12th century in Europe, such a threat may have been taken very seriously: the one who threatened might be condemned as a witch. Actions that seem physically feasible can then be subjected to the further tests of ethical acceptability and practical efficacy. Since there is much more discussion of efficacy than of ethics, our primary concern here will be the ethical one.

The ethical basis we shall use in our discussion is that every individual has a right to his own person. Or to put it in negative form, no one may initiate force against another without his consent. Of course, this allows for the use of force against the initiator of force in the sense of self-defense. Imposing a large risk of the use of force upon another is enjoined by the same principle. If the imposition is intended to be coercive, then the imposition is a threat. The robber who says "your money or your life" is thus violating the ethical principle even though you may avoid the use of force by surrendering your money.

Even when there is no intention to harm, the principle prohibits the imposition of a large risk on another. Thus, someone who is firing a gun in random directions may be restrained even though he has no intention of hurting anyone simply because he poses too great a threat to others.

While there might seem to be a wide variety of ethical principles from which to choose, the choice is not so large as one might think. In fact, the only other system with a claim to consistency (although a faulty claim, in my opinion) is that the king, czar, party, government, or church can do to any person whatever it likes. In such a system, of course, we don't have to worry about risk management; we simply ask the king-equivalent what to do.

Therefore, the ethic that shall guide us in this paper is that no one may impose a large risk on another without his consent. The remaining question, then, is how to measure risk and how to determine how large a risk may be imposed involuntarily.

It is important to distinguish this discussion of ethics from the usual discussion in terms of political and economic systems. The political system in many countries does incorporate ethical elements, such as the U.S. Constitution's Bill of Rights. However, it may also allow actions that many individuals consider unethical. Thus the political system technically contains both ethical judgments and other features based on the power possessed by various groups. As long as there exist two systems, political and economic, in the same society, then there exists the possibility of arbitrage, of people using political power to achieve what they cannot achieve economically or using economic power to achieve what they cannot achieve politically. For example, rent control is an action to transfer property ownership at least partially from owners to tenants. Environmentalists' objections against development can be attempts by some to raise their standards of living by political means at the expense of the standard of living of those not so well economically situated.

The main point is that unless political and economic systems have a common ethical basis, ethical conflict is bound to arise. The approach we take here is to follow ethical principles that preclude political and economic contradictions.

References p. 106.

MEASURING RISK: AN INDIVIDUAL DECISION MODEL

Recent research has shown one way life and death decisions can be made consistent with the non-coercive ethical principle [1], [2]. Naturally, then, this is a way for people to make their own risky decisions, *not* a way for other people to impose risky situations upon them. However, by seeing how an individual would view such an imposition by his own lights, we obtain a starting point for constructing a legal position regarding the imposition of risk.

The Black Pill — As a useful thought experiment, we imagine an individual faced with what we call the black pill question. He is offered the chance to take a pill that will kill him instantly and painlessly with a probability he assigns as p. If he takes the pill, he will receive x dollars. Should he accept? For example, should he accept a $p = 1/10,000$ incremental chance of death for a payment of x = \$1000? The choice is diagrammed in Fig. 1.

If the individual rejects the offer, he will continue his life with wealth W and face whatever future life lottery he presently faces. His future life lottery is the uncertain, dynamic set of prospects he foresees beginning with today. If, on the other hand, he accepts the proposition, his wealth will increase to W + x. If he lives after

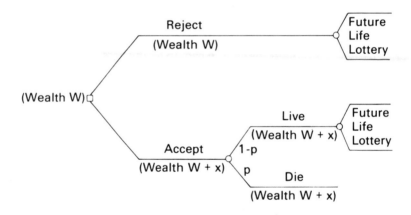

Fig. 1. The black pill decision tree.

taking the pill, he will begin his future life lottery with wealth W + x, presumably a more desirable situation. If he dies, he will leave W + x in his estate, and, of course, have no opportunity to enjoy it. Clearly the value of this benefit might be different for different people, and could be included. But let us say, for the moment, that it has no value to him. Naturally, there would also be tax effects, but these too we shall ignore.

We have analyzed this question in quite general form [2], but here we present the simplest model we have used to answer it. We assume that everyone has a fundamental preference on both level of consumption and length of life. We begin by asking the individual how much consumption (measured in today's dollars) he expects to have at each year in the future. We then ask what constant level of consumption beyond bare survival over his lifetime would make him indifferent between this level and his present prospects. We call this the constant annual consumption for that individual. Now we give him choices between the different futures described by different constant annual consumptions c and different lifetimes ℓ, and find to what combinations he is indifferent. For the simple example, we shall assume that the indifference curves have the form

$$w(c, \ell) = c\left(\frac{\ell}{\bar{\ell}}\right)^{\eta} \quad \eta > 0 , \tag{1}$$

where $w(c,\ell)$ is the worth numeraire associated with each indifference curve. The numeraire equals c when ℓ equals $\bar{\ell}$, the expected lifetime remaining.

Now we measure the risk preference on worth of the individual. For the example, we shall use the exponential form

$$u(w) = -e^{-\gamma w} = -e^{-w/\rho} , \tag{2}$$

where γ is the risk tolerance. With this structure and the assessment of the individual's joint probability distribution of c and ℓ, we can compute the utility of the individual for the case when he does not accept the black pill.

When he does take the pill, the probability p of dying immediately will transform his probability distribution on remaining life. The payment he receives, x , will increase his wealth. We assume that the individual will use the amount x to purchase an annuity over his remaining life at the prevailing interest rate i. In the calculation of annuity cost, we assume further that the seller of the annuity assigns the same probabilities on remaining life as those assigned by the individual. If we let ζ be the amount of annuity that one dollar will buy, then

$$\zeta = \frac{i}{1 + i} \frac{1}{1 - \left\langle \left(\frac{1}{1 + i}\right)^{\ell} \right\rangle} , \tag{3}$$

where $< >$ denotes expectation.

When we set the utility of taking the pill equal to the utility of not taking it, we determine that at the point of indifference p and x must satisfy the equation.

$$p = \frac{\left\langle e^{-\gamma c}\left(\frac{\ell}{\bar{\ell}}\right)^{\eta} \right\rangle - \left\langle e^{-\gamma(c + \zeta x)}\left(\frac{\ell}{\bar{\ell}}\right)^{\eta} \right\rangle}{\left\langle 1 - e^{-\gamma(c + \zeta x)}\left(\frac{\ell}{\bar{\ell}}\right)^{\eta} \right\rangle} \tag{4}$$

By inverting this relationship computationally, we find for a given value of p what value of x will make the individual indifferent between taking the pill and not taking it.

References p. 106.

If we let x grow without limit in this equation, we find that p approaches a value p_{max} given by

$$p_{max} = \left\langle e^{-\gamma c \left(\frac{\ell}{\bar{\ell}}\right)^\eta} \right\rangle \tag{5}$$

Thus no amount of money could induce the individual to accept a death probability as large as p_{max}.

To derive a measure of "life value", suppose that a risk-neutral observer examines the relationship between x and p. He could interpret $x = x(p)$ as the expected loss that the individual would incur from the risk if the individual valued his life at a number $v(p)$,

$$x(p) = pv(p) , \tag{6}$$

and he could then determine this number from

$$v(p) = \frac{x(p)}{p} . \tag{7}$$

Thus $v(p)$ is the value that the person is placing on his life in an expected value sense when he confronts a risk of magnitude p.

Of special interest in a safety context is the magnitude of this life value when the death risk is small. From a limiting analysis of the equation relating p and x, we find that as p approaches zero (and, of course, x also approaches zero), the ratio $v(p)$ approaches a value v_s given by

$$v_s = \frac{1 - \left\langle e^{-\gamma c \left(\frac{\ell}{\bar{\ell}}\right)^\eta} \right\rangle}{\zeta\gamma \left\langle \left(\frac{\ell}{\bar{\ell}}\right)^\eta e^{-\gamma c \left(\frac{\ell}{\bar{\ell}}\right)^\eta} \right\rangle} \tag{8}$$

We call this value the small-risk life value. It is the one number that an individual would need to keep in mind to make his safety decisions.

We shall be interested in comparing this small-risk value of life with an economic value of life comparable to that produced by other analyses. We shall define the economic life value, v_e, as the amount of money required to purchase an annuity paying the constant annual consumption c. Thus v_e is given by

$$v_e = \frac{c}{\zeta} \tag{9}$$

Illustrative Results — To illustrate the calculations implied by the model, let us consider a base case individual who is a 25-year-old male with a constant annual consumption of \$20,000 per year and a lifetime probability distribution given by a standard mortality table, Table 1. He chooses $\eta = 2$, which means that if he is sure to live his expected life (46.2 years), then a 1% decrease in his life would require a 2% increase in consumption for him to remain indifferent. From further questioning, we find that his risk tolerance is $\rho = \$6000$, which means roughly that he is indifferent between his present situation and equal chances of constant annual consumption of \$17,000 or \$26,000 for the remainder of his life. We also find that he faces a prevailing interest rate of 5% per year.

The results of the calculation appear in the upper part of Fig. 2 where we show the amount x that he would have to be paid corresponding to each probability of death p. We observe that this amount increases proportionally to p until about

TABLE 1

Life Table for White Males, U.S.
of 100,000 Born Alive, Number Dying During Age Interval

Age	Number Dying During Age Interval	Age	Number Dying During Age Interval	Age	Number Dying During Age Interval
0	2592	37	229	73	2775
1	149	38	251	74	2815
2	99	39	278	75	2841
3	78	40	306	76	2853
4	67	41	339	77	2855
5	60	42	376	78	2844
6	55	43	415	79	2821
7	52	44	458	80	2789
8	47	45	505	81	2738
9	43	46	556	82	2639
10	40	47	613	83	2482
11	40	48	681	84	2280
12	46	49	754	85	2096
13	56	50	835	86	1898
14	73	51	916	87	1693
15	90	52	995	88	1490
16	107	53	1071	89	1288
17	121	54	1144	90	1086
18	134	55	1216	91	888
19	143	56	1295	92	709
20	153	57	1383	93	548
21	162	58	1486	94	413
22	167	59	1598	95	300
23	163	60	1714	96	216
24	157	61	1827	97	152
25	149	62	1935	98	103
26	141	63	2039	99	70
27	137	64	2136	100	45
28	137	65	2231	101	29
29	141	66	2323	102	17
30	147	67	2409	103	11
31	154	68	2487	104	6
32	161	69	2559	105	3
33	170	70	2621	106	2
34	180	71	2678	107	1
35	194	72	2729	108	1
36	210				

References p. 106.

$p = 10^{-2}$, when it increases more rapidly and finally becomes infinite at $p_{max} = 0.103$. No amount of money could induce this individual to play Russian roulette $(p = 1/6)$.

The lower portion of Fig. 2 shows how the life value $v = v(p)$ depends on p. We observe that for small values of x, v is approximately equal to $v_s = \$2.43$ million, the small-risk life value of the individual. This means that for small probabilities of death (here less than 10^{-2}) the individual is acting as if his life were worth $2.43 million in an expected value sense. Thus, if the individual faced the black pill problem with $p = 1/10,000$, the required compensation would be $v_s p = \$243$. He would accept any payment x greater than $243 as an inducement to take the pill.

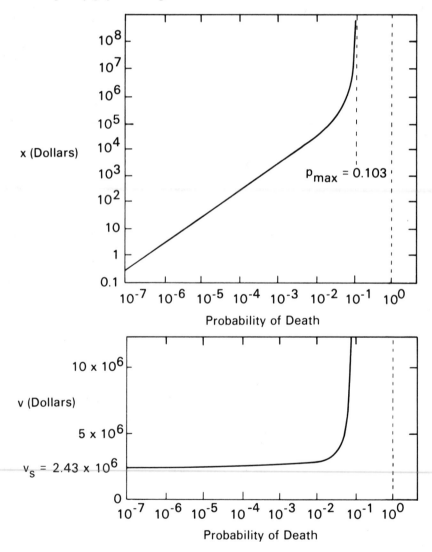

Fig. 2. Black pill results.

The economic life value v_e for this individual is \$363,000. Such a number has sometimes been used for decision purposes. We observe that the small-risk value is about 6.7 times the economic value. If this model and the numbers used in it are representative, the economic values that have been used in the past considerably underestimate the individual's own value. This discrepancy has implications for both the efficacy and the ethics of risk decision-making in our society.

The White Pill — Our analysis up to this point has emphasized the question of what we must pay an individual to undertake an additional risk. However, more often we face the problem of spending resources to avoid risk or in other words increase safety. The same theoretical model serves to illuminate this problem with only a few small twists.

Suppose that an individual faces a hazard that will kill him with probability p; for example, an operation. If he survives, he will live his normal life with whatever wealth he possesses. However, now someone arrives with a white pill that if taken will surely eliminate the death risk from this hazard. How much, x, would the individual be willing to pay for the white pill? Fig. 3 shows the relevant decision tree.

The unusual feature of the white pill question is that, of course, the amount x that he is willing to pay cannot exceed his wealth, no matter what death risk he faces. We assume that the individual can sell an annuity based on his lifetime distribution to pay the amount x for the purchase of the white pill. Since the most he can give up is his consumption beyond survival c, this means that in the white pill case the x versus p curve terminates on the economic life value of the individual

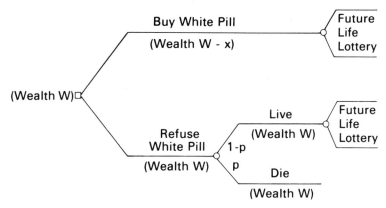

Fig. 3. The white pill decision tree.

when $p = 1$. The equation relating p and x,

$$p = \frac{\left\langle e^{-\gamma(c - \zeta x)}\left(\frac{\ell}{\bar{\ell}}\right)^{\eta}\right\rangle - \left\langle e^{-\gamma c}\left(\frac{\ell}{\bar{\ell}}\right)^{\eta}\right\rangle}{1 - \left\langle e^{-\gamma c}\left(\frac{\ell}{\bar{\ell}}\right)^{\eta}\right\rangle}, \quad (10)$$

confirms this observation, since at $x = \dfrac{c}{\zeta} = v_e$, $p = 1$.

Fig. 4 show the results for the base case individual. When $p = 1$, $x = \$363{,}000$, the economic value of his life. However, as p decreases, the x versus p curve becomes coincident with that of Fig. 2, and in particular implies the same small-risk value of $2.43 million derived for the black pill case.

Table 2 shows how the small-risk value depends on the model variables. The first row shows the effect of changing annual consumption level from $10,000 to $30,000 while fixing the risk tolerance at 30% of consumption. We observe that the small-risk value is then proportional to consumption level. The second row shows that the effect of varying the interest rate i from 10% to 2.5% is to change the small-risk value from $1.421 million to $3.622 million, because the individual needs a higher cash payment to obtain the same increase in consumption. The third row shows the relative insensitivity to the consumption-lifetime trade-off ratio η, whereas the last row illustrates how the small-risk life value falls with age.

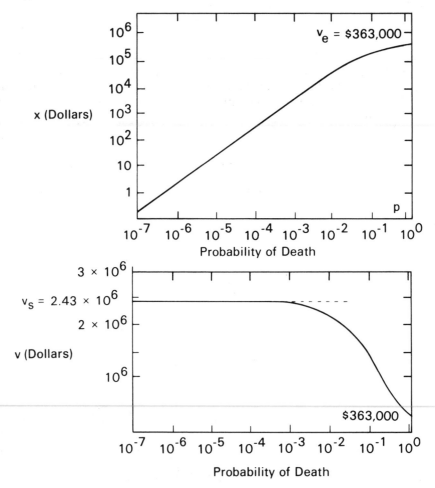

Fig. 4. White pill results.

TABLE 2

Sensitivity Analysis

Variable				Small-Risk Value, v_s ($ million)		
c:	10,000	20,000	30,000	1.215	2.430	3.645
ρ:	3,000	6,000	9,000			
i:	0.10	0.05	0.025	1.421	2.430	3.622
n:	3	2	1	2.418	2.430	2.541
Age:	35	25	15	2.157	2.430	2.671

Buying and Selling Hazards — Now that we have both the black pill and white pill results before us, we are in a position to make a few general observations. First, we see that the disparate results of the black and white pill cases for $p = 1$ show that we have answered a continual objection to analyses that place a finite value on life without regard to the distinction between accepting an additional risk and removing an existing risk. Since few people, if any, will sell their lives for any finite sum, all such analyses are doomed to failure. *However, the present model shows that it is perfectly consistent to refuse any finite offer for your life and yet be limited in what you can spend to save it.*

Of greater practical importance, however, is the result that for the wide range of hazardous decisions where we are buying and selling small hazards in our lives, the small-risk life value offers a simple and practical procedure to assure consistency.

To simplify the use of the small-risk life value and to emphasize the necessity that it be used only when the risk to life is small, we find it useful to define a unit for small risks to life. We shall use the term "micromort" to mean a one in one million chance of death, with symbol μmt. Then the small-risk life value can be conveniently expressed in dollars per micromort, or $2.43 for the base case individual. With this terminology, it is easy to explain why an individual can set a value for a micromort that is valid up to, say, 1000 micromorts, but also why that price is inappropriate for larger risks.

The Value of Reducing Risk - We can use the base case individual's value of $2.43/$\mu$mt to see what he would be willing to pay annually to remove various hazards in his life. The first column of Table 3 shows U.S. accident statistics for 1966. The second column shows the number of micromorts/year each risk poses to the base-case individual if, as we now assume, he uses these statistics as his probability assignment to death from each risk. The final column shows what the base case individual would be willing to pay each year to eliminate each hazard, an amount obtained by multiplying the number of micromorts by the individual's value of a micromort. Note that he would be willing to pay $900 just to eliminate the dangers of motor vehicles and falls. All other sources of accidents contribute collectively to an expected loss of less than $500. This calculation is an important

TABLE 3

U.S. Accident Death Statistics for 1966

Type of Accident	Total Annual Deaths*	Probability of Death in Micromorts/Year (1 Micromort [μmt] $= 10^{-6}$)	Payment of Base Case Individual to Avoid Hazard @$2.43/$\mu$mt
Motor vehicle	53,041	270	$ 656.00
Falls	20,066	100	243.00
Fire and explosion	8,084	40	97.20
Drowning	5,687	28	68.00
Firearms	2,558	13	31.60
Poisoning (solids and liquids)	2,283	11	26.70
Machinery	2,070	10	24.30
Poisoning (gases and vapors)	1,648	8.2	19.90
Water transport	1,630	8.1	19.70
Aircraft	1,510	7.5	18.30
Inhalation and ingestion of food	1,464	7.3	17.70
Blow from falling or projected object or missle	1,459	7.3	17.70
Mechanical suffocation	1,263	6.3	15.30
Foreign body entering orifice other than mouth	1,131	5.7	13.90
Accident in therapeutic procedures	1,087	5.5	13.40
Railway accident (except motor vehicles)	1,027	5.1	12.40
Electric current	1,026	5.1	12.40
Other and unspecified	6,163	31.0	76.50
Total	113,563	580.	$1,384.00

*U.S. Accident Statistics for 1966

starting point for determining whether feasible safety expenditures to modify these hazards would be worthwhile. It is clear that spending $1000 to be free of motor vehicle accidents would not be a wise choice for the base case individual. There is a limit to the value of safety.

Continuing Risks - Hazard Modification — Many of the risks to life occur not at a single instant, as does the black pill, but rather over several years or even a lifetime. The risks of living with automobiles, of smoking, or of living near a power plant are of this type. We can use the previous formulation to analyze this situation after we deal with the concept of hazard as follows.

The lifetime mass function is defined by p_n, n = 1,2,3, . . . where p_n is the probability that an individual will die in the nth year of his life. Let q_n be the probability that the individual will die in the nth year of his life given that he was alive at the beginning of that year. Then q_n for $n = 1,2,3, \ldots$ is the hazard distribution or force of mortality. The lifetime mass function and the hazard distribution are related by the equations:

$$p_1 = q_1$$

$$p_n = q_n \left[1 - \sum_{j=1}^{n-1} p_j \right] \quad n = 2, 3, 4, \ldots \tag{11}$$

and either may be constructed from the other.

We can now ask what present payment x would be required to induce an individual to accept a given modification of his hazard distribution, with the previous assumption that the payment will be converted into an annuity.

Consider first increasing the hazard in every remaining year of a person's life by adding 250 micromorts, a risk about equal to that posed by automobiles in American society. To induce the base case individual to accept such a hazard modification, which would decrease his life expectancy by 0.3 years, we would have to give him a lump sum of $13,000, or an annuity paying $700 per year.

If we doubled his hazard in every year, a risk considered by some the equivalent of heavy smoking, life expectancy would fall by 7.8 years and he would require a present payment of $212,000 or an annuity of $12,400.

If all benefits and costs associated with a general pattern of hazard modification are reduced to dollar terms, then the model can be used to determine the additional payment that the individual would demand or offer to be just indifferent to the modification.

Summary — We can use this model to evaluate how much an individual would have to be paid in money or its benefit equivalent in order to accept any given level of risk. Of special interest are those situations where the additional risk is small, for in this case the payment that the individual would require is equal to the probability of his death multiplied by a small-risk life value in dollars. This small-risk life value is likely to be constant over the range of risks involved in safety situations, for example, 1000 or less micromorts per year. The small-risk life value is typically several times larger than the economic value of life and is of the order of a few million dollars.

RISK ISSUES IN SOCIETY

Now that we have discussed both the ethical basis of imposing risk and a procedure by which an individual can make or delegate decisions that affect his chances of dying, we can proceed to an examination of the implications of those observations for various situations involving risk in our society. These situations include the treatment of risk in the marketplace and on the job, the imposition of excessive risk, and the creation of risky projects. We shall also discuss how these observations bear on the question of corporate liability.

Risk in the Marketplace and on the Job — Risks involved with the purchase and use of products or with holding a job are likely to represent 1000 or less micromorts per year, and hence, for practically all individuals, they will fall in the region of life value characterized by the small-risk value of life. This means that the individual who has assessed his chance of death can quickly calculate the expected death loss from the situation and balance it against other benefits and costs associated with the situation to make his decision.

However, we should also note that the designer of the product or the safety engineer of the job has already been making decisions that balance death against other considerations. In fact, if he is to be consistent, he should be using some small-risk life value in his design [3]. A logical next step would be to reveal this value to the purchaser of the product or the job applicant in a statement like, "We used a $3 million small-risk value of life in designing this car (or coal mine)." Naturally, the individual will hope that the value used is at least his small-risk life value. Otherwise, he would be rightly concerned that the situation will not be safe enough for him. The small-risk value of life used in design could then become one of the features of the product or job that is advertised to the public. Companies that used too low a value would experience competitive pressure, based on safety concerns, to raise it; whereas, those who used too high a value would find their products overpriced relative to competition. A similar result would apply to jobs. Thus companies would be encouraged by the marketplace to balance safety and economics. As the standard of living increased, so would the level of safety.

A further step in this development would be for the companies to buy insurance that would pay the small-risk life value used in design to anyone killed as a result of the design. Since this amount would be listed on the product or in the job description, the product liability or safety liability of the company would be specified in advance. The estate of the person who bought a cheap hammer designed with a small-risk life value of $10,000 would be able to collect only $10,000 if the head came off and killed him. Of course, someone who wanted the hammer for use as a paperweight might still buy it.

The idea of describing products (or jobs) by the small-risk life value used in their design is only useful if the number can be believed. It would be fraud to post a number higher than was actually used, or, of course, to say that insurance paying this value in the event of death through product design is in force when it is not. To be fair, the insurance would pay off only if the product failed while being reasonably employed in its intended use. (If the purchaser of the cheap hammer commits suicide by hitting himself over the head, his estate has no claim.)

Recently one of America's largest automobile companies lost a multi-million dollar suit involving product design. Evidence presented at the trial showed that the value of life used in the design was an order of magnitude below those we have discussed. How many of the purchasers of the car would have bought it if they had known the design basis? Placing this number in view may be the most important single step that can be taken to insure the proper balance of safety with other considerations.

Risk Imposition — The question of risk imposition can be addressed in terms of the legal procedures used when one person claims that another is imposing an unacceptable risk on him. Since nearly everyone is a potential danger to everyone else at some level, no absolute standard is possible. The question is at what level of risk different legal remedies may be imposed. We propose to measure the risk to an individual using the model described above. This model provides a monetary value of the risk in terms of the probability of death imposed and the preferences of the individual. To make the model operational, we must specify the source of the probabilities and the preferences.

Since there may be differences of opinion as to the probabilities of death imposed on one person by the actions of another, one function of the legal process would be to assign this probability in as objective and impartial a manner as possible. This may mean reviewing historical evidence, examining experimental findings, and ultimately considering the statements of experts. This procedure will not be easy, but it is necessary if serious concerns are to be separated from paranoia.

The individual's preferences, of course, are his alone. But to establish them in court, the individual will have to show that he acts consistently with his stated preferences. For example, it would be difficult for a circus performer who took large risks for money as part of his profession to then claim that no amount of money could compensaste him for much smaller risks. The past practices and decisions of the individual would in most cases provide good evidence of his preferences.

In the majority of situations where this procedure will be implemented, the risk faced by the individual will be small enough that his preferences can be summarized by his small-risk value of life. In these cases, the procedure will reduce to the court's determining the probability of death and the small-risk life value. The product would then measure the extent of the risk imposed on the individual, a number we shall call his risk evaluation.

The risk evaluation would in turn indicate the kind of relief to which the individual is entitled. If the risk evaluation were very small, say less than 10¢ or perhaps 10^{-5} times the average annual income, then no relief would be provided under the principle that the law does not concern itself with trifles.

On the other hand, if the risk evaluation were greater than a serious level, say, $10 or 10^{-3} times the average income, then the individual might be entitled to injunctive relief. That would mean that the imposer of the risk would be prohibited from imposing it. At this point, the risk imposer would either have to cease his activity, buy the right to impose the risk from the individual (for whatever he demands), or reduce the level of the risk evaluation below the serious level by making his activity considerably safer.

For risk evaluations in the intermediate region between trifling and serious, the medium range, a different remedy could be applied. This could be allowing the activity to continue only if the risk imposer buys insurance sufficient to pay damages if the activity actually kills the individual at risk. Moreover, the damages would not be the economic loss to the dead person's estate but rather his small-risk value of life, typically many times higher. This would mean that the risk imposer would always find it at least as desirable to pay the individual the risk evaluation in exchange for the right to impose the risk as he would to buy insurance. However, the insurance option does allow people to impose relatively small risks on others who may have unreasonable fears of certain kinds of activity.

Regardless of the level of risk involved, if the imposer buys the right to impose that risk on another individual from that individual, there would be no cause for litigation. This would include even cases of high risk, such as paying someone to play Russian roulette. However, in cases of high risk, the principle that every person has an inalienable right to his body would mean that the court would not allow the risk to be imposed in a situation where the seller of the right to impose risk changed his mind. The seller would, of course, remain liable for any damages he had promised to pay in the contract should he change his mind. Contracts involving the selling of rights to impose risk in the domain of safety, say, 10^{-3} chance of death or less, would not be subject to the alienability criterion, but would be considered as transfers of property.

On Creating Risky Projects — Much of the modern concern with risk arises from the building of what we might call risk-creating installations. These are installations that cause increased risks to the public, that is, to people who have not made any agreement to accept the increased risk. Such installations might be oil storage facilities, airports, or nuclear power plants. According to our preceding discussion, those creating such risks could proceed unencumbered only if the risk evaluations they created for those affected fell in the trifling range. If the risk evaluations fell into the medium range, then insurance would have to be bought that would pay the estate of anyone killed his small-risk life value. Of course, if no insurance company were willing to sell such a policy, the project could not proceed. Finally, if the risk evaluations fell in the serious level, the project could be prohibited regardless of insurance.

The entrepreneurs wishing to build risk-creating installations would be strongly encouraged by such a system to purchase in advance the risk rights from all individuals involved. But one immediately thinks of the problem of the holdout — someone who refuses to sell. When the risk evaluation is at the serious level, no one, including the government, could compel him to do so. For it is a violation of our basic principle regarding the initiation of force to use ideas like eminent domain to justify the initiation of force. How, then, can the practical entrepreneur proceed?

The basic idea that can solve this problem is the idea of risk options. The entrepreneur can buy from an individual an option to purchase his risk rights under certain conditions and at a specified price. For example, the entrepreneur might pay $10 for the option to buy at some time in the next year the right to impose 100 micromorts per year at a price of $100 per year. Then if the entrepreneur decides to build a 100 micromorts per year installation in the individual's area, he pays him $100 per year. If he decides to build some place else, then the individual has received only the $10 for the option. The entrepreneur can then buy options in several different areas knowing that he will in fact build in only one area. If the entrepreneur encounters holdouts in one area, he can move on to another.

The option and the rights could both be negotiable to create a market in risk. Communities of people who were relatively more willing to accept risk for money would then be more likely to be those places where the risk-creating installation was built. Since risk-rights could be bought more cheaply in remote locations, such locations would also be favored for the building of the installations. Thus the risk market, like any market, would encourage the more efficient use of resources. However, we should remember that the efficiency in this case is not being achieved at the expense of ethical principle.

Liability — We must still consider the case where someone imposes a risk in the medium or serious range without having at least purchased insurance. Logically in this case if someone is killed as a result, the liability of the risk imposer should be at least the victim's small-risk life value, and more if the risk extended beyond the safety range. There would be an additional heavy penalty if the insurance was not bought after a finding that the risk was in the medium range. Deaths resulting from serious range risky activities would incur criminal penalties.

The lower portion of Fig. 2 shows how the minimum liability might depend on the prior death probability. The region where the value becomes infinite we might call the "murder" region.

However, to put this principle into operation when the risk imposer is a corporation will apparently require changes in corporate law. The reason is that today corporations have the same limited liability to third parties (like the victim of the risk) that they have to second parties, their knowing creditors. This means that if a corporation is so structured that its assets are insufficient to satisfy a claim, the victim's estate cannot reach beyond those assets to the stockholders for the settlement of the claim. While I have no objections to the limited liability to creditors because they entered into the credit arrangement with knowledge of the limited liability, I see no reason why this limit should extend to third parties. For example, if a group of individuals organized themselves into a corporation and then the actions of the corporation resulted in someone's death, the personal assets of those individuals would ordinarily not come into play, whereas if the individuals had organized as a partnership, their personal assets would be available to satisfy the judgment.

The limited liability to third parties is not a recent feature of corporate law. The corporate form is, after all, a human invention. At the time corporations were first allowed legal status, there was debate on this issue of third party liability. Unfortunately, from my point of view, limited liability to third parties was instituted as a feature of corporations. But there is no reason why this decision could not be reversed.

Suppose that corporations were treated like partnerships in matters of third party liability. This would mean that every stockholder would be liable for damages done by the corporation to third parties. The result would be increased care among corporations in controlling their effects on third parties, effects on their property as well as on their lives. A corporation that was careless in this regard would soon find that it had lost the favor of investors.

Consider a company that is in the business of constructing or operating nuclear reactors. At the moment its stockholders are protected not only by the Price-Anderson Act, but by the limitation of liability to third parties. If both of these limitations were removed, anyone investing in such a company would have to be quite sure that he was protected from a calamitous loss. This would mean, most probably, that such companies would have to buy insurance against all such losses.

If such insurance were unobtainable or prohibitively expensive, there would be good reason to question the eonomic viability of the industry. Furthermore, even if the insurance were available, it is likely that the insurance companies in their own self-interest would require that independent agencies certify the safety of production and operation. Thus, the ultimate effect of unlimited liability to third parties would be either prohibition of unsafe industries or considerable improvement in their safety.

CONCLUSION

Our present apparent impasse on many safety issues stems mainly from a reluctance to re-examine the ethical basis for risk management in our society. As long as these issues are stuck between the economic and the political system, we can expect little progress. Only by returning to more fundamental ethical considerations can the issues be clarified and ultimately resolved.

REFERENCES

1. R. A. Howard, "Life and Death Decision Analysis," Second Lawrence Symposium on Systems and Decision Sciences, Berkeley, California, October 1978.
2. R. A. Howard, "Life and Death Decision Analysis," Research Report EES DA-79-2, Department of Engineering-Economic Systems, Stanford University, December 1979.
3. R. A. Howard, J. E. Matheson and D. L. Owen, "The Value of Life and Nuclear Design," Proceedings of the American Nuclear Society Topical Meeting on Probabilistic Analysis of Nuclear Reactor Safety, American Nuclear Society, May 8-10, 1978.

DISCUSSION

F. E. Burke (University of Waterloo)

I have one question of clarification. You had a slide in which you had computations where you compared various fairly modest risks. In one case an 8 year life expectancy change would require a present payment of $212,000 or an annuity of $12,400. That I could understand because my mental arithmetic was fast enough. But then you have a life expectancy change of 0.3 years which corresponds to a lump sum payment of $13,000 or an annuity paying $700/year. My mental arithmetic left me there and I wonder if you could help me out?

R. A. Howard

The payment is not proportional to the change in life expectancy because the effect is nonlinear. The changes in life expectancy and payments are computed from the same data but they are not obviously related to each other by any constants.

R. A. Howard

F. E. Burke

One is instant and the other is at the end of the expected life.

R. A. Howard

Unfortunately, there is a common belief that if I lose some expected life, the decrease always comes off the end. If I believed this, I could think there's no problem with smoking because it's going to hurt me after I'm too old to enjoy anything else. Of course, that's not true. The increased hazard is a change in the whole probability distribution of life.

E. V. Anderson *(Johnson and Higgins)*

One possible error is the use of the "fall" statistic to apply to a 20-year-old. About 90% of your fall deaths are of people aged 65 and over and you probably should take the deaths between 20 and 50 and use that as a basis.

R. A. Howard

Right, I think that's a very good observation. Everything in here should be interpreted from the individual's point of view. A 20-year-old should use a probability of falls that he believes describes his own risk. I am sure he would like your information in assigning his probability. I believe such modifications of general experience are proper. But the individual should realize that he may be biased. As we know, everybody thinks he's a better driver than everybody else.

J. H. Wiggins *(J. H. Wiggins and Company)*

Have you done anything that says, I'm not talking about gambling on me, but say on my wife, my children, my next door neighbor, the fellow down the street, or finally some person in Miami, Florida who I don't even know? In other words, how would this same kind of a thing deal with the case when it is a person other than myself?

R. A. Howard

That's what we call the value of a friend; it is discussed in the report referenced in the paper. The model shows that when you value your friend as yourself, you are willing to pay for him as yourself. As the degree of friendship goes down and down, of course, you will logically spend less for him than you will for yourself. But, only in the white pill case do I find this an ethically acceptable idea. You may not impose serious risks on others, but you can save people's lives so long as you don't affect them in any way in terms of coercing them. If you want to contribute to someone's medical plan, that's terrific.

J. H. Wiggins

We're doing Black Pill things all the time. All society is imposing serious risks on others.

R. A. Howard

Many people are, but I am not intentionally going to do anything like that, or to encourage it.

J. H. Wiggins

If you vote for a man who votes for certain legislation, then you are responsible?

R. A. Howard

I vote not because I support government coercion, but because I think it's o.k. for a slave to use any means at his disposal to secure his freedom. I am not responsible for the government any more than a slave is responsible for slavery.

M. E. Pate *(Massachusetts Institute of Technology)*

I have a practical question about the evaluation of public policy. I would like to know what kind of figure you would recommend for a group of people, with different ages, incomes, and risks. How would you aggregate their individual figures?

R. A. Howard

That's precisely what I would not do. If we start a flying club of three or four people and we want to decide what kind of airplane to buy, then I can provide a small-risk life value to be used in a decision process I have agreed to. But I think it's very dangerous for people to say what other people's lives are worth. Because once you allow them that possibility, of course, you are open to their making the value as small as they like. So I don't like the idea of people setting a value on other people's lives. Ethically I think that each person may want to set a value for his own life; that's up to him. It's all right for him to make such decisions, but not for other people.

W. D. Rowe *(The American University)*

My question, Ron, is how many times a day do you have to make a calculation? For example, I get into my car and I'm driving along having made one calculation, and suddenly I see an accident and I decide I want to remake my calculations because of the imminence or the reminder of the reality factor. So there's a dynamic aspect here, isn't there?

R. A. Howard

Life value should change with age and changing circumstances, as we have seen. However, sudden changes would be unusual. You could spend your whole life making life and death decisions, but that's not what I am recommending. What I am saying is if you feel these issues are important this is a way to make such decisions.

A. S. Curran *(Dept. of Health, Westchester County)*

Getting back to the last question about putting a value on others' lives. In my position as Commissioner of Health, I frequently have to do that in a flip-flop way in that I am asking the taxpayer to pay a certain number of dollars so that he won't have trichloroethylene in his well or something. I am faced constantly with this type of analysis that has to be done, and then peoples' perceptions of what the risk really is. I think what you're talking about today can be very beneficial, but I think we do have to sometimes assume that responsibility.

R. A. Howard

I don't like the imposition of that responsibility on people who don't want it. There was a discussion earlier today about the freedom of the individual. It seems to me the ultimate freedom of the individual is to own and control his own life. So, I don't like the idea of health commissions making decisions about my life, but I guess a lot of other people must because we have a lot of such activity in our society.

B. Bruce-Briggs *Slave. (Laughter) (New Class Study)*

How would you handle a Typhoid Mary problem?

R. A. Howard

Typhoid Mary problem? I wouldn't eat in a restaurant that didn't inspect its workers and require that they be healthy. Typhoid Mary worked in a restaurant. Right? Look at the companies that have had problems with botulism in their canned goods. What has happened to them? Well, they have been in real business trouble. You can ruin your reputation when you take risks with what you put in the can. No restaurant is going to risk its reputation by not having health examinations for its employees. If you're in the restaurant business, you don't need a government regulator to tell you that having such examinations is a pretty smart idea, particularly when there is no corporate limited liability to third parties.

A. Curran

But then you're doing what you said you didn't want to do because if I'm taking the responsibility of saying I'm going to protect your health by sending in sanitarians to inspect that restaurant, I'm making some kind of decision for you. I'm assuming you want to be protected.

R. A. Howard

You heard me wrong. I don't want you to protect me. I want the owner to do it in his own self-interest. For one thing, he won't ask the taxpayers for money to do it.

M. G. Morgan *(Carnegie-Mellon University)*

What sort of plans do you have to use the technique to examine a significant sample of people so that we have some notions of how it would apply to different individuals in different walks of life?

R. A. Howard

That's a very good question. We have done, not what I would call experimentation, but rather class exercises with people of different ages and situations. There is quite a bit of divergence; some people have $50,000,000 small-risk values and others have $1,000,000 values. I have not done, nor am I likely to do, a sort of demographic study based on this model. Some other people, I understand, are interested in doing that. I wish them luck. I just hope that it won't be used for public policy decisions.

E. A. C. Crouch *(Harvard University)*

Then how do you evaluate the small-risk life value to apply to products if each individual has his own?

R. A. Howard

Good point. You just display it on the product. In other words, if General Motors is going to use a $1,000,000 value on life (I could have used other companies, of course), they just stamp it on the bumper of the car. A Mercedes Benz could have another number. As I said earlier, it becomes a product characteristic, just like color or how soft the seats are or anything else. Stating that a particular life value was used when it was not would be fraud. I would like the small-risk life value stamped on the product to be the indemnity paid by the manufacturer if someone is killed using it as a result of its design.

E. A. C. Crouch

Then you leave it up to the individual whether he should get that car or not?

R. A. Howard

Who else?

J. Huntsman *(Applied Decision Analysis)*

Ron, how can we be sure that the information that companies state is truthful? A sufficiently large company can lie about their information and there's no other information to go on.

R. A. Howard

Well, that would be fraud. Presumably at the time of trial all this would come out. The papers would be subpoenaed and so forth and so on.

J. H. Wiggins

We have found that when you ask people what they would do and then see what they really do, it's different. Can you use this method for real decisions?

R. A. Howard

I use this method but I can't tell you whether you ought to use it. I know I make my safety decisions this way.

D. McLean *(University of Maryland)*

I'm pursuing this small-risk life value number that's been put on cars. Do you propose that your analysis is limited to such personal product choices like cars or that you put the same sort of number on a highway?

R. A. Howard

Yes, highways, too. Of course the small-risk life value applies only to small risks.

I didn't have a chance to go into it, but where do you find death probabilities outside the safety range, that is, probabilities of death from one in one thousand up to one? They do exist in our society, mainly in medical problems. When you go to the doctor and he recommends an operation, you're often dealing with probabilities in that range and then you might use the more detailed model rather than simply the small-risk life value. But in the safety region, with death probabilities less than one in one thousand, then I'm quite content to use the small-risk life value.

A. Kneese *(Resources for the Future)*

I'm sorry to cut off this very interesting discussion. If somebody has a question that is just burning, I''ll let him take one more question. Yes, sir.

M. Thompson *(Insitute for Policy & Management Research)*

You say you use this method yourself.

R. A. Howard

That's correct.

M. Thompson

Well, honestly, I could not understand a word of it, so I can't use it for myself.

R. A. Howard

From this short presentation, I wouldn't expect anybody who hadn't heard it before to understand it. The paper will be clearer, and the report that is referenced in the paper even more explicit. But perhaps you're saying not that you didn't understand it, but that you disagreed with it.

M. Thompson

Yes, I did disagree . . .

R. A. Howard

There is hope on that issue, too.

ECONOMIC TOOLS FOR RISK REDUCTION

Lester B. Lave

Brookings Institution, Washington, D.C.

ABSTRACT

Risks are inherent in human activity. Society has devised a series of tools (or institutions or modes of behavior) to cope with risk; some of the coping takes the form of risk reduction behavior and some takes the form of rationalizing the acceptance of risk. Much of the nonrational behavior can be seen as suboptimizing, given the difficulty and expense of obtaining data, the difficulty and expense of analysis, and the cost of changing behavior. Three contradictions exemplify the difficulty of regulating risk: we are safer than ever before but more worried about risk, consumerists have achieved virtually all of their objectives but there is general dissatisfaction with risk regulation, and regulations never seem to achieve their objective. These contradictions show the need for explicit estimation of risks, costs, and benefits and recognition of the difficulty of changing individual's behavior. Only by recognizing the difficulties of regulating risks and developing thoughtful approaches can we hope to achieve progress.

INTRODUCTION

Three contradictions dominate our thinking about risk and constrain our ability to take actions to mitigate it. Without an understanding of why these statements aren't contraditions, our thinking about risks and attempts at formulating policy are doomed to confusion.

1. People are more concerned about risks, particularly risks of catastrophic events such as nuclear reactor accidents and destruction of atmospheric ozone, than ever before; but in overall terms, there is less risk to people living in the developed countries than ever before.
2. Regulatory agencies have promulgated hosts of rules to lower risks, many of which are well designed in the sense that they would be effective in decreas-

ing risk if properly used; but regulations never seem to succeed in their objectives since people persistently seem to engage in perverse behavior that increases their risk in the face of the new regulations.

3. Public interest groups have achieved virtually all their goals in securing the passage of new legislation, staffing regulatory agencies with their nominees, in the promulgation of myriad regulations, and in garnering vast increases in government budgets and staff, and private expenditures on risk reduction; but many people desiring reduction of risks consider our efforts to date to be a failure, subverted by business, a government that knuckles under to special interest groups, and consumers who won't buckle their seat belts, who put leaded gasoline in their cars, and who won't maintain their emission control system.

In the course of showing why these sets of statements aren't contradictory, I will attempt to sketch some alternatives that might allow society to respond better to the desires for risk reduction. Unfortunately, there are no magic wands. Returning to a laissez faire market economy would probably increase economic efficiency, but only at the cost of increasing risk. Rather, the solution is to do a better job on those activities that have already been undertaken.

WHY ARE PEOPLE MORE CONCERNED ABOUT RISKS?

Americans enjoy greater economic prosperity today and better health than ever before. Our real income per capita (after accounting for inflation) is several times the level at the turn of the century and more than double the level in the 1940's. Life expectancy at birth has about doubled compared to the beginning of this century, and has continued to increase, although slowly, in the last two decades.

Life expectancy, or rather health status more generally, is the best overall indicator of risks to health. Any increase in health risks that is not offset by other decreases must eventually manifest itself in a decrease in health status. Thus, the increase in life expectancy is a general indication of lower risk.

Whether one measures economic prosperity by the wealth or income of Americans, it is greater than ever before. This is true not only of mean income and wealth, but also of the level of income enjoyed by the poorest segment of the population. Furthermore, unemployment, disability, and retirement have lost much of their economic sting due to extensive social programs.

Thus, there is good evidence that Americans have never faced lower overall levels of physical and economic risk. Why then should they be so concerned with the levels they do face?

People who have nothing, have nothing to lose. Someone in good health has a great deal to lose in an accident or from a chronic disease. People with high incomes and with wealth stand to lose them. A necessary adjunct to the vast gains made in increasing economic prosperity and improving health is the fear of risks that would take away these gains. There is a one-to-one relationship between the level of prosperity and how much one has to lose.

Thus the first contradiction is resolved by pointing to our prosperity: It is our success in achieving affluence and in solving health problems that has given us so much to lose. With so much more to lose than our parents or grandparents, it is only natural that we should be more concerned with protecting what we have.

WHY DO PEOPLE ACT TO MITIGATE THE GAINS OF NEW REGULATIONS?

Well run regulatory agencies gather data and analyze the risks faced by individuals. To alleviate economic risk agencies require full disclosure of financial information, audit and regulate corporations (especially banks and insurance companies). To reduce injuries from accidents, NHTSA has changed the design of the automobile and added safety features.

Yet people appear to react by investing in highly risky companies and in changing their driving behavior so as to give up at least part of the gain in risk reduction. A number of interpretations of this phenomenon have been given. Economists such as Peltzman argue that the individual is attempting to maximize some objective function, given various constraints. When the risk associated with a set of behavior is lowered, the individual is able to achieve a still higher utility by adjusting behavior toward some increase in risk that allows improvement in other aspects. For example, an individual can get to his destination more quickly by slightly increasing his speed; this adjustment becomes attractive after the car has been redesigned to enhance safety, since the individual can trade-off some of the increase in safety for getting to his destination more quickly.

A psychologist might say that the individual desires a certain level of risk and so some individuals will elect to increase their risk levels after the environment has been made safer. A consumerist might argue that individuals have been seduced by Madison Avenue to engage in irrational behavior because of some masculinity image.

I am partial to the economic explanation since it identifies the behavior adjustment as a natural process that would be unaffected by regulating advertising or stopping appeals to the less appealing human traits. Indeed, it is possible that the trade-off between, for example, speed and safety might be shifted so much by redesigning the automobile that people might elect to increase risks to higher levels than those prevailing before the redesign because of the huge increases in speed or other desirable attributes that could result.

The point of this observation is that the regulatory agency is not to be judged by overall safety statistics and that people are not necessarily irrational if they modify their behavior toward increasing risk. According to economic theory, such a modification indicates a greater gain in utility to the individual than that associated with the increase in safety. Thus, although regulators may be frustrated by what they perceived to be perverse behavior, consumers have benefitted. Nor can regulators afford to place a much higher value on safety than do consumers, since the resulting behavioral adaptations will make it impossible to achieve the desired level of safety and what is gained is likely to be only at an exorbitant cost.

WHY HAVE THE NEW REGULATORS NOT SUCCEEDED?

Many people do not differentiate between process and outcome. To them, securing the passage of legislation, the creation of regulatory agencies, and the hiring of staff was evidence of success. When these people review the history of this campaign, they find that the amount of new legislation, number of new agencies, people in the agencies, and new regulations indicate success. But popular desire is not for such process measures, but for the reduction of risk. People looking at outcomes find the lowering of risks to be disappointing, particularly in view of the scope of resources going into the process.

Furthermore, even zealots who are appointed to these regulatory agencies find many competing goals, instead of the single minded search for risk reduction. Unless an agency secures its appropriation and general Congressional support, it can't accomplish its goals, e.g., the Federal Trade Commissioner, the Consumer Product Safety Commission. Unless the agency decisions are sustained in court, the whole process is wasted. Unless it can get some degree of business and consumer cooperation, it can achieve little; going to the wall with each violator requires so much time and resources that general success is impossible. For a host of reasons, an agency finds that it must compromise its ideals if it is to achieve any of its goals. Furthermore, if the agency isn't managed with skill, these confrontations will appear late in the process when the agency has staked out a position which it discovers is untenable. This "knuckling under" to business or buckling under consumer pressure (as with the interlock device or with saccharin) is no more than the expression of any agency that has learned later than it should that its goals are unattainable. It is a sign of learning, although one hopes they would begin to learn earlier in the process, before the need for public confrontation.

More importantly, the world doesn't share the single minded concern, or even the focus, of the regulatory agency. It is natural that an agency regulating risk should attract individuals who tend to be concerned with risk and then socialize these individuals to be even more concerned. The result is almost certain to be that the agency will be more of a zealot than general public opinion. Thus, unless the agency is aware of this tendency and corrects for it, its proposed regulations are unlikely to be acceptable to the vast majority of people. The agency can even find itself in a position where it is railing at the vast majority of consumers for refusing to act in the way it desires, e.g., the interlock device.

Thus, much of the lack of success of the new agencies stems from setting goals not shared by most Americans. At the same time these "idealistic" agencies have been naive about what is required to implement new regulations. Finally the process of lowering risks is terribly complicated, with ramifications throughout the economic system and for civil liberties. Confronted with these realities, it is inevitable that progress will be slow.

INDIVIDUAL ADJUSTMENTS TO RISK

Before intelligent goals can be set for regulating risk, we need a good picture of what consumers desire. This requires recognition of the twin concepts of consumer control of risks, and consumer actions to mitigate the damage resulting from untoward events. The first notion recognizes that I face a wide range of control levels for risk, ranging from my ability to turn off the electricity when I repair wiring to my inability to affect the probability of thermonuclear war. Where individuals have control over risk levels, there can be little role for regulatory activity. After consumers become informed of the level of risks and consequences of their actions, regulation can do little to prevent them from taking risks; they will choose the level of risk they deem appropriate. Perhaps something can be gained by the difference between having an individual take action to increase safety and increasing safety to the point where individual action is required in order to undo safety. But there is a real danger of using this subtle effect in inappropriate ways.

The second notion recognizes that, while I can't stop nuclear war, I can take steps to minimize the damage to me and my possessions should it occur. The first device for doing so is insurance. Fire, flood, and earthquake insurance have the effect of minimizing the *economic* loss should these events occur. Minimizing the loss from nuclear war or general warming due to carbon dioxide is more difficult. Yet, even here, by migrating, rearranging my portfolio of assets, and engaging in various kinds of training, I can lessen the loss.

The individual who desires regulation to lessen the risk of an untoward event over which he has almost complete control is asking the government to take on the role of his keeper. Most people do not desire to have anyone assume this role. Regulations taking on this role are necessarily controversial. But, individuals do look to government to take on roles that they cannot accomplish individually, or at least those that are extraordinarily expensive to take on. For example, if no company offers insurance against flood, earthquake, or nuclear reactor accidents, individuals have an unmet desire. Some could elect to organize an insurance company to provide this service, but that is hardly a casual enterprise. It is probably easier to take collective action via an already organized collective, the government, than to attempt to create a new institution.

For another set of risks, the government is in control of the level of risk. For example, government operation of flood control systems put the government in a better position than any private company to assess the risk and to minimize damage should heavy rains create a dangerous situation. Thus, the government should be able to offer the lowest cost insurance. Probably only the government would be in a position to offer insurance against nuclear war.

Thus, consumers will tend to desire government action in situations where individual action is not possible or is very costly; they will also desire government action where existing institutions don't appear to be lowering the risk or protecting them against the loss as cheaply as they believe it can be done.

WHAT IS THE GOVERNMENT'S ROLE?

To discover the proper role for the government in regulating risk, it is important to begin with the lesson from the previous two sections: The government cannot "do the right thing" i.e., it can't unilaterally lower risk. For example, the accident rate for small, private aircraft is so high that the government desires to lower risk; but it has been able to do little since the risk is generally within the control of individuals who assume it with general understanding of the consequences.

Changing Behavior — Legislating behavioral changes is generally unsuccessful, e.g., protecting people from alcoholic beverages. Government can provide information about the risks and consequences of actions, and even make it slightly more difficult for individuals to "misbehave," but it cannot force changes in behavior or lower risks more than is desired.

Some of the attempts to regulate the automobile illustrate the difficulties with attempting to get consumers to do something they don't want to do. NHTSA did have the power to require manufacturers to design and install an interlock device. But to prevent individuals, or even small repair shops from deactivating the devices proved impossible. The lesson is even more forcefully made with respect to emission control devices and the inability to:

1. Prevent consumers and repair shops from modifying the engine and control system,
2. Prevent drivers from using unleaded gasoline, and
3. Getting owners to keep the emission control system in good repair.

As most parents can verify, attempting to get children to modify their behavior because of its long term consequences is nearly impossible; they are myopic with respect to the benefits to be gained from brushing their teeth or even getting to bed on time. Some adults are similarly myopic in that they wish they were not given certain choices or say they would have made a different decision had they known at the time what they later learned. Some myopia is inevitable; certainly, people will always want to have their lives to live over again and decisions to be made the right way. The adult who moans his failure to brush his teeth in his youth finds it no easier to convince his children than does the lung cancer victim to get people to stop smoking. Most of the time one should doubt whether the individual really would behave differently if given the decision again. In any case, the government's role here is extraordinarily limited; beyond providing information of the consequences of alternative decisions, restricting the range of alternatives is the sort of paternalism that nearly everyone rejects.

The most one can do is to publicize untoward outcomes and make the risks more salient. After an auto accident killing several people we all drive a bit better. Thus, agencies can act to keep the greatest excesses from occurring by dramatizing risks. One success of such attempts is, for example, the lower rate of fatalities per passenger mile over the Labor Day weekend than at other times. However, it is much more difficult to change behavior permanently.

Correcting Externalities — Correcting externalities is the largest role for government seen by economists. When my actions interfere with you (except through the marketplace as with auctions or competing for the same job), laissez faire cannot produce an efficient outcome.

Three sorts of externalities are most important for risk. The first consists of preventing accidents which would injure people other than the person in control.

Someone falling off a ladder which he is using incorrectly hurts himself and there is little role for government other than providing information about how to use a ladder properly and the consequences of not doing so. But if the fall injures someone nearby, an externality has occurred. Thus, NHTSA ought to give particular attention to "first collision" devices, those that lower the risk of an accident (which generally involve innocent second parties).

The second sort of externality revolves around the financial support given an individual who becomes disabled or the dependents of an individual who is killed. The cost to society of assuming these financial burdens can be extremely large. Obviously, society has a stake in making sure that motorcyclists do not become paraplegics and that wage earners with many dependents are not killed. Thus, many states require motorcyclists to wear helmets and NHTSA regulates "second collision" safety features. Perhaps a simpler solution would be to require people taking risks to have sufficient insurance to pay for the costs that society now assumes. However, attempting to enforce the motorcyclists' purchase of medical insurance policies of this magnitude or the purchase of life insurance policies by people with many dependents seems a hopeless task.

The third sort of externality is the loss faced by family, friends, employer, and community when someone is killed. As Schelling points out, this loss can be substantial. However, there seems to be no way of using this loss to get individuals to change their behavior or accept government paternalism.

Externalities provide a valid basis for government action. However, many of the situations must be accepted as simply unfortunate, since there is no feasible action the government can take to alleviate it.

Technology Forcing — Some of the congressmen associated with passing the Clean Air Act Amendments of 1970 remark that they knew that the technology did not exist to satisfy their emission limitations on the automobile. The point of the stringent standards was to force industry to be innovative in developing new technology. The "technology forcing" role is often cited as a reason for stringent standards. However, it can lead to settling on an inferior technology because of time pressures and act to delay technological change. Certainly, there is something to be said for getting industry's attention and providing significant incentives to innovate. But this process is much more complicated than the simple statement above suggests.

In summary, government cannot simply do the right thing; it can only attempt to influence industry and consumers. The greater is the behavior change required, the less is the chance of it taking place. Providing information about the consequences of alternatives seems much less powerful than mandating action, but the latter is paternalism and generally unsuccessful. Externalities are sufficiently important that government has a potentially major role to play in certain types of risk reduction. However, the effectiveness of government actions, and limitations on their scope are constrained by resentment of paternalism and the inability to get people to change their behavior. Speeding innovation is not nearly so simple as holding someone's feet to the fire. Less heroism and more good sense should produce greater progress.

ALTERNATIVE REGULATORY FRAMEWORKS

A series of alternative regulatory frameworks might be used by an agency to direct its actions in lowering risk: no risk, general balancing of benefits and costs, regulatory budget, and formal benefit-cost analysis. The no risk framework is an attempt to see that the consumer need never bear any needless risks. Perhaps the most dramatic expression is in the Delaney Clause in the Food, Drug and Cosmetic Act, which forbids the introduction of any cancer causing substance in food. "No unnecessary risk" has an attractive ring. However, the results of an attempt to implement such a policy are far from pleasing. No unnecessary risk would prohibit many of our activities, change virtually all others, increase the prices of goods and services dramatically, and generally lower our standard of living. When confronted with the detailed implications of such a policy, few people would espouse it.

The framework receiving most attention in Washington is a general balancing of costs and benefits. The Congress and Administration recognize that such a framework commits neither them nor the agencies to do anything. However, this framework appears to satisfy the desire for reform. The most formal expression of this framework is in the various executive orders from Presidents Ford and Carter mandating that agencies consider costs and benefits in setting new regulations.

I include in this category risk-benefit analysis, which is similarly vague in specifying precisely what is to be considered and how decisions are to be made. In addition to imprecision, this framework suffers from ignoring the costs of a project. While there is intuitive appeal to some general balancing of costs and benefits or risks and benefits, this framework does not provide concrete guides to action.

The fourth framework, the regulatory budget, provides each agency with an implementation budget as well as its traditional internal budget. In considering regulations, each agency would have to choose among the many possible ones; it would be constrained in that the costs of implementing all new regulations could not exceed its implementation budget. Thus, the agency would be forced to plan and set priorities, to estimate the detailed costs of implementation, and to estimate the benefits of each regulation. The net effect is likely to be salutary, especially the setting of agency priorities and feedback mechanisms whereby OMB and the Congress review prior budgets and the resulting experience in assessing the new budget requests.

Formal benefit-cost analysis has been applied to government projects since the 1930's when the Army Corps of Engineers was divided to apply it to waterway projects. This detailed quantification of costs and benefits has set economists off merrily chasing the social rate of discount and values of untraded goods and services, such as the dollar value of a premature death. Formal benefit-cost analysis has not fared well; economists vastly overstated its value, it has been extraordinarily difficult to implement, and the resulting applications have been far from definitive. It seems doubtful that formal benefit-cost analysis could ever become the standard procedure relied upon to provide guidance in risk reduction.

CRITERIA FOR CHOOSING AMONG FRAMEWORKS

An ideal new regulation would:
1. Promote economic efficiency,
2. Enhance social equity,
3. Be easy to implement and enforce, and
4. Require no major organizational changes (including new legislation).

For example, correcting an important externality would enhance economic efficiency. Most regulations trade-off economic efficiency against attaining the other goals; the question becomes how much efficiency should be given up in order to achieve these other goals.

Social equity is not a well-defined concept. To the utilitarians, it meant simply equal incomes. Now it has more of a connotation of aiding the disadvantaged and rewarding those who are worthy (and not permitting those who are not to receive benefits). Perhaps the only well defined notion is that of not tolerating a reduction in real income to any single group; this seems to be the equity criterion used in practice.

Economists and engineers are inclined to think of terribly complicated schemes that have good theoretical properties, but which are nightmares to implement. Administrative costs, administrative ineptitude, and delays all loom larger now than they did in 1965 at the beginning of this era of health, safety, and environmental regulation. This criterion has grown in importance, and may be of predominate importance.

Getting a regulatory agency to act is difficult; when an action requires concurrent action by the Congress or other parts of the executive branch, it becomes nearly impossible to achieve. Thus, there is a premium on decisions that can be taken without reorganization, new legislation, or the concurrence of other agencies.

These four criteria provide not so much a check list, as a set of important tests for a proposed regulation. Rarely will a proposal rate well on all four criteria. Thus, policy making consists of deciding the relative importance of each criterion generally, and of making trade-offs among the criteria in deciding among alternative regulations.

FUTURE DIRECTIONS IN RISK REDUCTION REGULATION

The first of the major legislation on risk regulation, the Highway Transportation Safety Act, was passed in 1966. A great deal has been learned by regulators, industry, and the public in the past decade and a half. The no risk framework, which was the hero of the early legislation, is now generally recognized to be unworkable, and generally undesirable. Agencies have begun to accept the notion that they must estimate the costs of implementing a proposed regulation, and these estimates must be sufficiently good that they can be defended. Agencies are slowly beginning to accept the notion that they will have to estimate the benefits of a proposed regulation. The executive orders that started off as merely red tape have begun to loom larger in importance. This upwelling of good sense is due in large part to the slower rate of growth of real income and constant budget pressure that government and the private sector have had to face. Rather than the estimation of costs and benefits being a nuisance that could be neglected because the costs were small relative to what the public felt could be gained, we have seen the costs finally being borne by the public and the resulting disenchantment.

"Regulatory reform" in one form or another seems inevitable. The issue is whether form will continue to be more important than substance and whether the Congress will pass a bill that seems plausible, but will be impossible to implement. If so, the new effect will probably be that no new regulations will be issued for some time. While that course might seem to have appeal to people who are disenchanted with the current state of health, safety, and environmental regulation, the pleasure will be short lived. The public demand for risk reduction grows with each improvement in economics and health; bottling up new regulations is rather like closing the safety valve on a boiler: the resulting explosion is unlikely to be attractive to anyone.

The time has passed for arguing about whether we should have regulation of risk. The question now is how to do that constructively and what areas deserve first priority. Public pressure for risk reduction will grow and sweep away any opposing force (short of disastrous economic conditions or war). With 15 years of experience, this is surely the time to think about some better ways of reducing risk than the tired and ineffective legal mechanisms used to date.

DISCUSSION

L. B. Lave

C. Starr *(Electric Power Research Institute)*

Would you say something about the internalization of cost, social cost, environmental impact, as an alternative to government regulation.

L. B. Lave

Yes, that's a good point. I would not go as far as Ron Howard did in my right wing economics. It's always amazing for a right wing economist to find somebody, generally a non-economist, who is even more right wing. There is a long way to go in trying to internalize costs; at that point there will be much less for regulatory agencies to do. We are asking too much of the regulatory apparatus. Wisdom would seem to call for vastly reducing the sorts of tasks that we ask of them. We should focus on the really difficult questions that the market place isn't going to handle and let the market place handle those questions which it can.

R. J. Tobin *(State University of New York)*

I find your presentation somewhat simplistic in that you seem to lump regulation and all the ills of regulation together. For example, it might be entirely possible that the regulations are very good, except that you have poor regulators, perhaps you have insufficient support for the regulatory agencies, perhaps there is a difference between the formulation of the regulatory policy and its implementation. I think many of your criticisms directed toward regulation, while appropriate, at the same time are perhaps unfortunate because everything that's wrong apparently is involved with the regulation.

L. B. Lave

I would disagree. You have to expect that people will take their best shot; if it turns out that after ten years the Environmental Protection Agency hasn't done what you want, you could say, "Oh Shucks, we need better people and to try harder." Alternatively, you can say, "Gee, there's something we've learned from the historical experience." I submit that the sort of record across regulatory agencies is such that it's very difficult at this point to say "Well, we just had a bunch of bad administrators," or people who weren't really qualified running these agencies to date. There is enough uniformity so that one can look at what occurred, what are the faults, and what have been the strong points. The principal problem is the emphasis on legal mechanisms, the process orientation of making sure that everybody has their say rather than with trying to make use of individual incentives and market mechanisms. At this point Allen (Kneese) probably would like to interrupt to make his plea for effluent fees, something that I wholly support. That doesn't mean that I think effluent fees don't have any problems, but I think that, compared to the alternative, they look wonderful. Five years ago, effluent fees were rejected as "licenses to pollute". They didn't really get a serious hearing in Congress. At this point there seems to be a great deal more sympathy from Congress for effluent fees, not because the Congress no longer believes they're licenses to pollute but because we've had five more years of bad experience with their alternative.

W. C. Clark *(Int'l. Inst. for Applied System Anal.)*

Yes, I agree with you with the notion that first we shoot all the lawyers, but so help me, if I didn't misunderstand your presentation, you're going to shoot all the lawyers and replace them with economists, who are then going to go through the same set of calculations to devise these cunning externalities, incentives, and effluent charges. You are now accusing the lawyers of bunging things up by trying to go so far beyond what they know, that is, what knowledge is available, in trying to specify outcome of procedures and so forth. I am sympathetic to the problem but please tell me why five years from now we're not going to be meeting here saying, "Gee, the economists have had their spell and they bunged it up as well."

L. B. Lave

Well, these are questions that are best left alone, but my mother never raised a smart child. The first sort of issue that you are talking about is really Ron Howard's issue of what role is there for government. Shouldn't government be getting out of this and encouraging individuals to do more? I agree. I am a right wing economist and I think that individuals are capable and should do much more of their own deciding. However, there are problems that individuals cannot deal with very well, because there are strong externalities associated with them; the government must get involved if anything is to be done.

Actually, you need not worry about lawyers — before you can shoot them, you've got to try them.

If economists set effluent charges anew for each region with new nonlinear schedules, then the economist will be going through the same madness that the lawyers are doing now. If instead, we focus on the basic notion of decentralized incentives, recognizing that the effluent charges will only be approximate, we could do better than the lawyers. No system is going to be perfect, unless we devote tremendous resources — too much — to the system itself.

There is some shift in emphasis from due process, giving everyone a chance to be heard, to worrying more about the outcome. We want a cleaner environment, but we haven't gotten it out of the current mechanisms so far. The kinds of complaints I hear are not that the EPA has cost society too much money, but rather that we should have gotten a lot further by now. This is particularly true in view of the publicity about the cost. Society seems to want more emphasis on efficiency and outcomes and may be willing to give less emphasis to due process.

H. Joksch *(Center for the Environment & Man)*

Your second thesis was that users of automobiles trade off safety versus some other risk, for instance speed. Now that is the sort of argument economists like to make very often. Unfortunately, the majority of car drivers don't make well informed, rational, decisions when they are driving. Even if they had the intelligence as is theoretically required by most economists, they lack the basic data.

To be precise, right now, we don't know with any degree of accuracy what the effect of increasing speed is on accident risk — and much more. We know that the majority of drivers have very wrong perceptions not only in absolute terms but in relative terms of the risk. So your assertion that drivers make a trade between a decreased risk and sacrifice a little bit of that for decreased travel time has no good theoretical basis.

Second, it contradicts our empirical evidence. Namely the studies by Peltzman to which you refer, have no empirical basis at all: they are inferences and contradict with other evidence. So, the situation is from what we know, the regulations have had some effect. Now that may not be a desirable answer.

There are two kinds of regulations. Actually three. The 100 series standards which reduce the risk of crashes and the 200 series reducing the risk of injury. The third is for relatively minor problems such as preventing fires after the crash. Now, the 100 series are supposed to, and those are regulations we do emphasize because they are supposed to prevent the accidents from occurring such as the other guy from running into you. Unfortunately, we know very little about the effectiveness of these standards. There is a good reason to doubt that they have any appreciable effect at all. However, we know that the 200 series has a decided effect. We cannot pinpoint it exactly. Now we have the problem, if anybody is going to regulate anything, you know you can't do much about preventing crashes. Either about changing driver behavior because empirical evidence about what we can do about driver behavior is of a minute order of magnitude. Maybe a risk reduction of about 5%, 10% at the extreme, 15-20% if you are going close to a police state with the regulated drivers. Whereas, we can do quite a lot of the order of 20, 30 maybe 45% mainly by reducing injury and fatalities. And, by the way, the ignition interlock was not invented by NHTSA, it was invented by an auto company but not GM. So would you comment on some of my comments.

L. B. Lave

I will try to make the comment shorter than the question. In college I was a part owner of a Model A. I hate to admit that to General Motors. It was clear to me that I was never going to try to drive that Model A at 70 miles an hour. Indeed, if you go back to the 1930's and look at what speeds people drove Model A's or Chevrolets, it wasn't very fast. In part that was due to the road, but it was due mostly to the automobile itself. Our current automobiles are capable of sustained high speed driving with much greater safety than we had before; people willingly drive these automobiles much faster. If you were to look at how safe it was to drive cars at high speed, you'd see that people drove about as fast as they could safely do so. There is gross evidence that people take the inherent safety of a vehicle into account.

I completely reject the notion that people have neither the data nor the ability to make the calculations. That's an elitist notion that we few experts should make a judgment for everybody else because they're all too dumb or too uninformed to take care of themselves. In the United States we have 200,000,000 experts on the automobile. Go ask anybody on the street whether they're an expert on automobile safety. With the possible exception of real experts; most people will answer that they know about their safety. It may be that I don't know all of the technical data about how my automobile behaves on an icy surface, but when streets are icy, I make some tests and find out. I don't have the data to four decimal points, but I know pretty well by the time that I'm out on the highway how the car handles. Each one of us has these feedback mechanisms that tell us how safe it is to drive the car under these conditions. Did you ever stop to think on a day when the roads were icy just how good all those drivers are?

But there is still room to make them better by improving the vehicle — the first collision devices. Some drivers aren't so good. GM had an ad about an inboard computer that prevented ignition until you passed a sobriety test. It would flash numbers on the screen and you had to repeat them within twenty seconds. Unfortunately, 10 or 15% of the drivers couldn't get it right when they were stone sober. Since we still let those people drive, I want better first collision features on all cars and good second collision features on at least my car.

TRADE-OFFS

Richard C. Schwing

General Motors Research Laboratories, Warren, Michigan

ABSTRACT

Finite resources meeting competing desires require efficient expenditures in all areas of life, even those that are life extending. Efficiency thus becomes synonymous with trade-offs. A survey of longevity for longevity trades introduces the concept of choosing among alternative life extending programs.

The 55 mph speed limit introduces more dimensions into the problem and the trade-offs become more difficult. Finally, the risks inherent in different energy futures present a very complex trade-off problem. The situation is complicated by individual and group perspectives and illustrates why a political consensus is difficult to achieve. The paper graphically illustrates why our institutions are virtually paralyzed by attempting to gain a mutually acceptable policy.

INTRODUCTION

"Who Shall Live?" [1], *"Tragic Choices"* [2], *"Tyranny of Survival"* [3] — these are not only titles of books, but also the phrases that describe the subject matter of this symposium. For all of our discussion revolves around the choices or trade-offs that eventually must be faced, made explicit, and ultimately decided.

Underlying my discussion is the assumption that our supply of time, capital, and resources has finite limits. No country is as healthy or as safe as it could be. Regardless of national culture and regardless of economic system, no country does as much as is theoretically possible. The "right" to health or the "right" to safety is always less than complete because societies have other purposes to pursue. We seldom, if ever, acknowledge that health and safety compete for the same limited pool of resources as education, the arts, shelter and other less noble purposes. And, as Victor Fuchs has pointed out with regard to medical care [1], resource

References pp. 141-142.

limitations are not manifest only in the absence of amenities, delays in receipt of care, and inconvenience; they also result in loss of life.

In comparing the merits of various health programs Grosse [4] has written that "the truly moral problem is not to distinguish between good and evil but rather to select appropriately among alternative goods." Because our lives are part of the calculation, the engineers' label, "trade-offs," is disturbing. We are reluctant to face up to the fact that trade-offs are necessary in a world of finite resources.

Engineers cannot apply their craft without engaging significant trade-offs. It is conventional to swap the operating costs of an air conditioner or heat pump against the initial cost of the heat transfer surface area; or the ride and handling characteristics of a tire for fuel economy and quiet. I would like to extend the engineer's tradition in trade-offs to less traditional dimensions starting with relatively straightforward trades where the units are explicit and comparable, then escalate to emotionally more difficult trade-offs, concluding with trade-offs where our perceptions and biases virtually paralyze the discussion and reduce our collective ability, as a nation, to make *any* decision let alone a "correct" one.

DIRECT COMPARISONS — ONE DIMENSION — LIVES FOR LIVES, LONGEVITY FOR LONGEVITY

In our previous work on life-saving programs [5], we emphasized that so-called "life saving" efforts were not "life-saving" at all but life-extending We are all mortal and although there is risk associated with the timing of our mortality, the event itself is certain. So we have argued for a "longevity gain" metric instead of a "lives saved" metric. While some find the translation from life-saving to longevity change offensive when discussing violent or accidental deaths, especially for the young, it is generally acceptable to relate life-extending terminology and concepts to disease related mortality which effects the elderly. Of course, we can further refine our classification scheme but for the first illustration in this discussion I would like to consider accident and disease reducing programs separately.

Coleman [6], Schwing [5], and Wilson [7] have compiled extensive catalogues of environmental risks. Probably the most complete is the catalogue of Cohen [8].

Lives for Lives — Using the vertical dimension to indicate a measure of efficiency and the horizontal dimension to portray the magnitude of the impact,* Fig. 1 displays the available accident data tabulated by Cohen. Note that the extremes in the efficiency measure, dollars per fatal accident prevented, range over three orders of magnitude. This enormous spread of costs illustrates the need to set priorities when some programs cannot be implemented because of lack of funds.

---

*Though various programs have an impact on either large or small components of the population, each program is located with respect to the horizontal scale as though everyone in the population is a potential beneficiary. For example, though substantial longevity benefits from programs 1 and 6 accrue to miners, the impact is presented as a small gain in the U.S. longevity. Locating program efficiencies and impacts on this graph is more thoroughly presented in reference [5]. An average longevity gain of 40 years per fatality is used for all accident programs as a simplification, Cohen [8] provides specific values for each program.

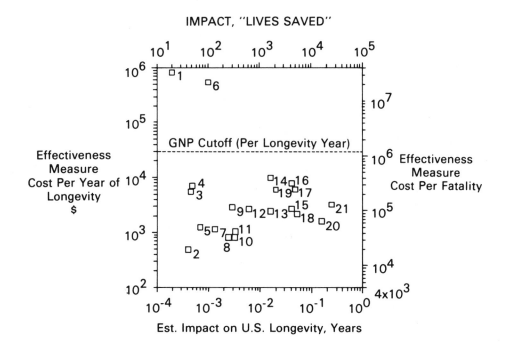

Key

Safety Program

1. Mine Safety (Non-Coal)	11. Skid Resistance
2. Highway Construction	12. Impact Absorbing Roadside Development
— Maintenance Practice	13. Steering Column Improvement
3. Median Barrier Improvement	14. Tire Inspection
4. Clear Roadside Recovery Area	15. Passive Torso Belt-knee Bar
5. Wrong Way Entry Avoidance	16. Air Bags (Driver Only)
6. Coal Mine Safety	17. Passive 3-point Harness
7. Bridge Rails and Parapets	18. Driver Education
8. Guardrail Improvements	19. Smoke Alarms in Homes
9. Breakaway Signs and Lighting Posts	20. Rescue helicopters
10. Regulatory and Warning Signs	(1 Helicopter/100,000 People)
	21. Auto Safety Equipment 66-70

Fig. 1. Cost per fatal accident prevented (Data obtained from Cohen [8]).

For perspective, a "GNP cutoff" line is indicated in this graph. If an economy was willing to expend the total GNP to reduce fatal accidents, this benchmark gives us the amount available to allocate to each accident fatality, ignoring all other forms of mortality and other allocations which typically consume the GNP. Of course, just because this sum is expended does not imply that society will be successful in eliminating the fatality. If society behaves as though the value of life is infinite and if a resource constraint is not imposed, even a partially successful,

References pp. 141-142.

inefficient program can be rationalized.*

The main point of this graph is the vertical dimension of efficiency in a world of fixed resources. When expenditures on inefficient programs consume our ability to implement efficient programs, we are not merely trading dollars for lives, we are trading lives for lives.

The horizontal scale shows that some programs can have a positive impact on either larger or smaller components of the population. Research efforts might be invested where costs are high but potential impact would be great if more efficient technology could be found.

Longevity for Longevity — Life extending programs can be represented on a similar graph. In Fig. 2, the vertical dimension for disease data is dollars per longevity year** as indicated. Again, the remarkable range of values and our

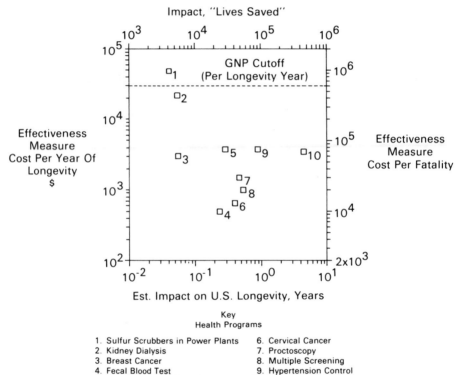

Fig. 2. Cost per year of longevity gain (Data obtained from Cohen [8]).

---

*Calabresi [9, p. 392] has reflected on why we are at times willing to spend enormous sums on an individual, especially an identifiable, victim. He states, "After all, the event is dramatic; the cost, though great is unusual; and the effect in reaffirming our belief in the sanctity of human life is enormous." It allows us to "preserve the essential myth that human life is a pearl beyond price."

**An average longevity gain of 20 years per fatality is used for all disease programs as a simplification. Cohen [8] provides specific values for each program.

inability to implement all programs because of constraints on resources, makes obvious the advantage of setting priorities.

The cutoff line on this figure is an estimate of the amount available for extending each life if the total GNP was available for life extending programs.

"Efficiency" in matters of life and death has often been cast as a symbol of *immorality;* yet here, where efficiency is shown to save more lives, it can actually be considered a criterion for a moral choice; for if information is sufficient to show that within a fixed budget we are actually shortening more lives by implementing an inefficient program, we are confronted with an explicitly immoral situation. The concept that "efficiency" is linked more closely with *morality* than *immorality* is often lost in contemporary debate.

TWO DIMENSIONAL TRADE-OFF — LIVES FOR TIME OR MONEY

In this section I will choose an example that illustrates the need to include trade-offs between two dimensions. Much of the literature concerning the 55-mph speed limit has discussed the fuel-saving and life-saving benefits. The life saving is substantial, while the fuel saving is modest.

Fuel savings attributed to the 55-mph speed limit range between 2.5 and 2.8%. This savings, of the order of 3 billion gallons of fuel per year, is comparable to:

- one-third the gain achievable with proper spark plug maintenance [10],
- one-half the savings gained by the increased use of radial tires [11],
- the savings gained by increased tire pressure [12];
- or the gain achieved by choosing the proper viscosity when changing oil [13].

Yet these alternatives are not aggressively promoted.

The safety implications will be examined here to assess the trade-offs for risk reduction benefits. J. D. Williams was among the first to address the trades imposed by speed limit regulation and pointed to the difficulties in setting wise policy [14]. Only a few authors since Williams, [15] [11] [17] have considered the additional travel time which lower speed limits impose.

If we consider two components of the U.S. highway system, primary rural roads and interstate highways, and use the values for the effect of speed on highway fatalities derived by O'Day [18], we can assemble the information in Fig. 3.

Note that the greater number of lives are saved on the primary rural system whereas the greater time penalty is spent on the interstate system. The cost effectiveness ratios for these two systems are therefore:

$$\frac{5 \times 10^4 \text{ years}}{427 \text{ lives}} = \begin{array}{l} 110 \text{ years of travel time per life saved on the} \\ \text{primary rural road system and} \end{array}$$

$$\frac{9 \times 10^4 \text{ years}}{147 \text{ lives}} = \begin{array}{l} 600 \text{ years of travel time per life saved on the} \\ \text{interstate system.} \end{array}$$

References pp. 141-142.

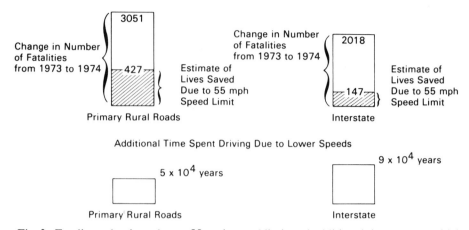

Fig. 3. Fatality reductions due to 55-mph speed limit and additional time spent on highways.

Values calculated from data in references [19] and [20]. Extra travel time = 1.8 persons/

$$\text{car} \times \text{Vehicle Miles Traveled} \times \left(\frac{1}{\substack{\text{Average Speed} \\ \text{in '74}}} - \frac{1}{\substack{\text{Average Speed} \\ \text{in '73}}} \right).$$

There are those who enjoy the extra time on the highway as an increment of solitude or a chance to better observe the landscape. There are also those, specifically truckers and salesmen, whose livelihood depends on maximizing travel within time and safety constraints. This latter group may not view their portion of the 600 years of added travel an appropriate trade for one life. For the purpose of discussion, if the minimum wage rate is used as the value of time, and if the above times are included, the 55-mph limit could be viewed as costing* $13 million per life saved on the interstate system and $2 million per life saved on the primary rural system.

Clearly, disparate groups, having differing motivation and value systems, will view the 55-mph speed limit in drastically different manners; however, an acceptable compromise may be achieved by examining the values of users as they apply to the different components of the highway system.

MULTI-DIMENSIONAL TRADE-OFFS — THE COAL-NUCLEAR-OIL ALTERNATIVES

Probably the most difficult set of trade-offs come into consideration when we choose among alternative energy scenarios. All too frequently, however, the trades enter the debate one at a time to justify a given position or perspective, a form of single issue politics. My purpose here is to make explicit and confront several of the major trades which must be addressed as the nation moves towards an energy policy.

---

*The value of fuel savings offsetting this cost is of the order of 10^6 per life saved.

Economic — From the perspective of traditional engineering and/or economic criteria, the choice between alternative fuels to generate electric power could be made simply on the basis of cost per kilowatt-hour of power. One could debate whether or not work place hazards are internalized in the cost, but if the model of Thaller and Rosen [21] is valid, then the work place hazards are captured by differences in wage rates and it is legitimate to view the problem as pictured in the ternary diagrams* in Fig. 4.

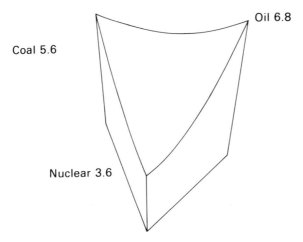

Fig. 4. Power generation costs estimated in 1975 for coal, oil, and nuclear (cents/ kilowatt-hour).

These data, which include capital, fuel, and operation, are from Balzhiser [22] at a time when energy economics were not subject to short term supply disruptions or environmental uncertainties. In the ternary diagram depicting the average price for various mixes of energy, the cheapest available energy supply would be a mix towards the nuclear apex. (The minimum average cost must be on an interior region on the surface if the cost of the first kwh of oil or coal is less expensive than the last kwh of nuclear.)

Externalities - Work Place and Environmental — If, on the other hand, one does not believe the wage rates adequately internalize the risks in the mines, oil fields, power plants and industries supporting the energy sector, the perspective depicting an energy mix in Fig. 5 might be more useful. This uses the disease and accident fatalities estimated for various technologies by Inhaber [23], and the perspective is that of concern for work place hazards. This diagram makes explicit the costs born by the labor force.

However, a wider perspective is achieved if one adds the estimated environmen-

---

Ternary diagrams or composition diagrams for three-component systems allow the presentation of a dependent variable for any mixture of three components. The "pure" systems having 100% coal, nuclear, or oil are represented at the corners of the equilateral triangle. The dependent variable in each graph is depicted in the vertical dimension.

References pp. 141-142.

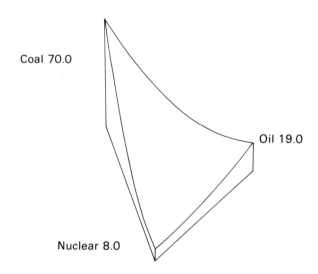

Fig. 5. Occupational man-days lost per megawatt-year estimated for coal, oil, and nuclear.

tal damages in terms of fatalities to people living near the respective generating facilities. The disease and accident estimates of Inhaber [23] show the incremental net mortality cost of each technology or technology mix. This aggregate representation is shown in Fig. 6. It indicates that the nuclear risks in the work place are low relative to coal and oil.

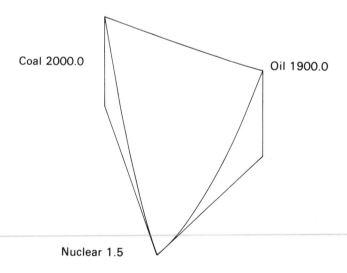

Fig. 6. Public man-days lost per megawatt-year estimated for coal, oil, and nuclear.

Perceived Domestic Risks — The mode of quantification above, using extrapolation and statistical inference, cannot be reconciled with answers obtained when we probe the psyches and elicit the fears of the lay public. As Slovic, Fischhoff, and Lichtenstein have shown [24], Fig. 6 does not adequately represent the "subjective" responses of sophisticated laymen to energy technology risk estimates. Assuming the lay public does not differentiate strongly between coal and oil power plants, the perceived risks of estimated lives lost in a disastrous year are plotted in Fig. 7. These I will label "the perceived domestic risks."

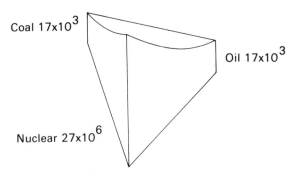

Fig. 7. Potential loss perceived by laymen, estimated fatalities in a disastrous year for coal, oil, and nuclear.

Clearly, a comparison of Fig. 6 with Fig.7 highlights the discrepancy between "expert" estimates (the "objective") and "lay" estimates (the "subjective"). Yet in a democracy, the views represented in Fig. 7 are an equal if not more important consideration to our policy makers.

Perceived National Risk — To some Americans, our nation's vulnerability to the will of other countries is the greatest risk of all. The risk to national security is directly linked to the degree to which we depend on foreign resources and increases with the percentage of that commodity imported [25] [26]. Fig. 8 is a schematic of such a perception. No scale is available to properly measure the level of concern, I have merely indicated the relative future imports forecast for 1985.

Other Complexities — The perspectives above are offered here to compile a list of major concerns which dominate perceptions. Yet we have still not included other trades or political issues which include economic growth, unemployment, and the distribution of impacts across population groups or regions. How should we handle trading the risks between various subgroups defined by income, geography, age, race, or sex, etc.? With difficulties in arriving at agreement in the issues discussed in preceding sections, it is no wonder that this array of decisions affecting so many different groups paralyzes our ability to arrive at a consensus.

One additional parameter must still be considered, moreover, that of the timing of the impacts. Some of the risks being traded are "clear and present" while the timing of others is much more remote.

Dollars expended (Fig. 4) and work place lives (Fig. 5) are with us today and

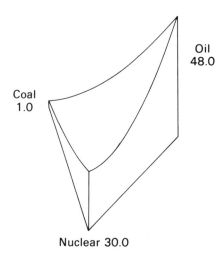

Oil
48.0

Coal
1.0

Nuclear 30.0

Fig. 8. Perceived risk as indicated by dependency; percentage to be imported for coal, oil, and uranium according to 1985 forecast.

represent relatively firm data. Nuclear risks (Fig. 7) are also with us today, yet some of the stated fears relate to the distant future. The trade-offs in Fig. 7, therefore, must be smeared over a long time horizon. Environmental impacts (Fig. 6) are only slightly displaced into the future while risk to the nation's security (Fig. 8) must be spread over a long time horizon, as shown in Fig. 9.

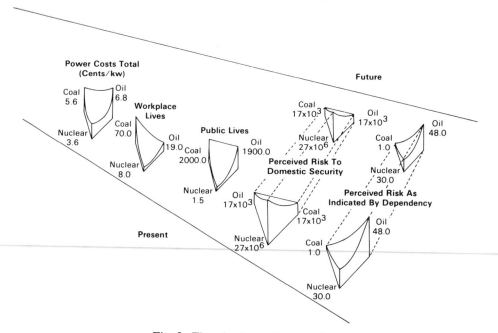

Fig. 9. Time horizon of perceptions.

Relative weightings to these concerns are related to one's discount rate as one proceeds with decision making. If one values future lives as much as current lives and if one accepts forecasts as gospel, then a zero discount rate is appropriate. If, on the other hand, one is skeptical of the forecasts, then a high discount rate should be employed when comparing impacts. This theme has been captured nicely by Linstone [27] in diagrams concerning the "Limits to Growth" debate.

Society is also prone to limit action to incremental adjustments of the current social system because of its inability to cope with uncertainty. As Calabresi has noted, "The costs of moving from an existing social system to a new one ... are unknown and involve a substantial risk that, whatever pattern of life the new system brings, *more* lives will be taken than in the old system."

RECAPITULATION

Does it help to make these trades explicit? I'm not so sure the presentation helps in making a decision, but it may guide us in how society should design a process in making a decision. Let me use a pair of sketches (Fig. 10) and a discussion from Mary Midgley's *Beast and Man—The Roots of Human Nature* [28].

Herring Gull   Oystercatcher

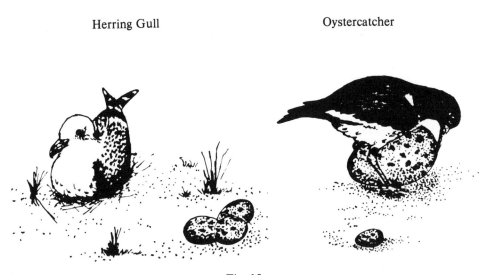

Fig. 10.

One sketch depicts a contented gull, sitting on an empty nest, oblivious to the fact that its eggs remain in the cold, a foot away. A helpful ethologist has removed the eggs to see which the creature would prefer, and it has settled for the nest.

The other sketch is that of an oyster catcher trying to perch on a monstrous egg, larger than itself, ignoring its own egg of manageable size. Ethologists provided the

References pp. 141-142.

dummy to test the powers of discrimination. These are two of many examples which show how easily garbled are the natural cues creatures act on, even in the cases essential to survival. Herring gulls care more about nests than eggs; oyster catchers prefer big eggs.

Mary Midgley goes on to illustrate the ways in which Homo sapiens behave in analogous ways to supranormal stimuli. I would argue that as long as we select just one of the perspectives from Fig. 4 to 9 in order to make a conclusion or take a position, we are responding to a limited set of cues and behaving as an oyster catcher or herring gull.

This illustration demonstrates the need to develop tools for integration and consensus, not just cleverness in looking at parts of the problem. We must be forced to recognize when we are looking at just parts of the problem without falling into the traps of "holism."

Though intended to look at the whole, "holistic" analyses produced thus far tend to have buried within them unstated value systems. All too often the objectives implicit in the models reflect those of the analyst rather than those present in the social system being analyzed or modeled. Specifications of the imposed relationships are not always based on sound empirical evidence and the analyses themselves are undertaken for reasons other than the solution of problems.

The most obvious characteristic of this class of problem is what Daniel Callahan [29] calls *inherent uncertainty*. The uncertainty built into the very nature of the problem is dominated by the difficulty in obtaining good, solid, scientific data and, possibly more important, by the even greater difficulty getting a consensus on what is important. With variation in age, experience, goals, demographics and psychology, various groups are bound to approach a problem with a different set of perspectives and the alleged "facts" brought to the analysis — that is what are chosen to be called facts — may differ radically from those chosen by a different group having a different value system. "Values" and "facts" are inseparable and we must avoid the hazard of assuming that better science could perfectly answer the question of "how safe is safe enough?". We cannot be distracted from the more difficult issue of making decisions in the face of substantial uncertainty in either a technological or non-technological society each carrying a burden of evident and hidden risks. These risks await the individual and even society itself but to avoid all risk would lead to the ultimate risk of stagnation and "rotting in a dungeon of security."

The discussion need not end on such pessimism. Since we are not about to change human nature, we should face up to the conflicts in values and not hide them in a general systems model. There is no magic methodology for analysis or privileged group of analysts (including those in the OTA, GAO, EPA, UAW, or GM) that has a value-free, interest-free perspective on society. The political process, as cumbersome as it is, has evolved to resolve questions and proceed with the necessary decision making. The value of a systematic analysis is to provide tools to examine the necessary trade-offs. The only "holistic" perspective appropriate to issues as complicated as the illustration here is that we have limited time, capital, and resources to build a path for society through the risky (i.e., uncertain) landscape of the future.

ACKNOWLEDGMENTS

The author is particularly grateful to Craig Palmer and James Licholat for computation and computer graphics, and to Brad Southworth, Barry Bruce-Briggs, Gerald Lieberman, and David Martin for comments on earlier versions of this paper. A special thanks to Linda Meyer and Marilyn Southern for typing the manuscript with care and good humor.

REFERENCES

1. V. R. Fuchs, *Who Shall Live?*, Basic Books, New York, 1974.
2. G. Calabresi and P. Bobbitt, *Tragic Choices*, Norton, New York, 1978.
3. D. Callahan, *The Tyranny of Survival: And Other Pathologies of Civilized Life*, McMillan, New York, 1973.
4. R. N. Grosse, "Cost-Benefit Analysis of Health Service," *Annals of the American Academy*, 399: 89-99, 1972.
5. R. C. Schwing, "Longevity Benefits and Costs of Reducing Various Risks,"*Technological Forecasting and Social Change*, 13: 333-345, 1979.
6. W. T. Coleman, Jr., "The National Highway Safety Needs Report" by the Secretary of Transportation to the United States Congress, U.S. Department of Transportation, April 1976.
7. R. Wilson, "Risks Caused by Low Levels of Pollution," *Yale Journal of Biology and Medicine*, 51: 37-51, 1978.
8. B. L. Cohen, "Society's Valuation of Life Saving in Radiation Protection and Other Contexts." (*Health Physics*, in press.)
9. G. Calabresi, "Reflections on Medical Experimentation in Humans," *Daedalus* 98, 2, 387-405, 1969.
10. "Neglect of Maintenance Worldwide Study Finds," *Automotive News*, p. 20, November 20, 1978.
11. A. C. Malliaris and R. L. Strombotne, "Demand for Energy by the Transportation Sector and Opportunities for Energy Conservation" published in *Energy: Demand, Conservation and Institutional Problems*, M. S. Macrakis, ed. MIT Press, Cambridge, Mass., 1974.
12. *Automobile Fuel Economy*, Motor Vehicle Manufacturers Association, September 21, 1973.
13. M. C. Goodwin and M. L. Haviland, "Fuel Economy Improvements in EPA and Road Tests with Engine Oil and Rear Axle Lubricant Viscosity Reduction." Presented to Society of Automotive Engineers Passenger Car Meeting, Troy, Michigan, June 1978.
14. J. D. Williams, "Comments on Automobile Traffic," *Rand Corporation Report*, P-1556, 1958.
15. L. A. Lave, "The Costs of Going 55," *Newsweek*, p. 37, October 23, 1978.
16. G. H. Castle, "The 55 MPH Speed Limit: A Cost/Benefit Analysis," *Traffic Engineering*, 46: 11-14, January 1976.
17. L. Goldmuntz, "The Public Interest in Auto Fuel Efficiency," *Economics and Science Planning, Inc.*, May 11, 1978.
18. J. O'Day, D. J. Minghan, and D. H. Golomb, "The Effects of the Energy Crisis and the 55 MPH Speed Limit in Michigan," University of Michigan Highway Safety Research Institute Report No. UM-HSRI-SA-75-9, 1975.
19. U.S. Department of Transportation, Federal Highway Administration, *Highway Statistics*, 1973.
20. U.S. Department of Transportation, Federal Highway Administration, *Highway Statistics*, 1974.

21. R. Thaller and S. Rosen, "Estimating the Value of Saving a Life: Evidence from the Labor Market." Preliminary report prepared for the Conference on Research in Income and Wealth, New York, National Bureau of Economic Research, 1973.

22. R. E. Balzhiser, "Energy Options to the Year 2000," Chemical Engineering, 74-90, January 3, 1977.

23. H. Inhaber, "Risk with Energy from Conventional and Nonconventional Sources," Science, 203: 718-723, 1979.

24. P. Slovic, S. Lichtenstein, and B. Fischhoff, "Images of Disaster: Perception and Acceptance of Risks from Nuclear Power." Prepared for Beijer Institute International Review Seminar on Impacts and Risks of Energy Strategies: Risk Analysis and Role in Management, Stockholm, Sweden, September 1978.

25. Foreign Uranium Supply, Electric Power Research Institute Report EA-725, Research Project 883, April 1978.

26. National Energy Outlook, Executive Summary, Federal Energy Administration, 1976.

27. H. Linstone, "On Discounting the Future," Technological Forecasting and Social Change, 4: 335-338, 1973.

28. M. Midgley, Beast and Man: The Roots of Human Nature, Harvester Press, London, 1979.

29. D. Callahan, "Morality and Risk-Benefit Analysis," prepared for the Congress/Science Forum, Risk/Benefit Analysis: Its Role in Congressional Science and Technology Policy Decisions; Cosponsored by the Senate Subcommittee on Science, Technology and Space; the House Subcommittee on Science, Research and Technology; and the American Association for the Advancement of Science, in Washington, D.C., July 24-25, 1979.

DISCUSSION

R. C. Schwing

F. E. Burke *(University of Waterloo)*

Dick, first of all, I want to congratulate you and then I want to ask you if the costs that you gave in your single dimensional trade-off diagram are project costs or if they also include costs, for instance, in the 55 mile an hour case, of time and inconvenience to the affected publics.

R. C. Schwing

The costs would be those incurred by the public; Cohen's data did not include the cost of administrating the regulation.

F. E. Burke

So is it hardware costs?

R. C. Schwing

If it were for a pollution device, it would be the hardware cost plus the operating cost, but not the cost of running the regulator operation.

D. B. Straus *(American Arbitration Association)*

Your last words were that the final payoff here would be whether or not you could use this model for reaching some consensus. Have you had an opportunity to use this model of varying perspectives to test their views as to which is a better type of fuel to use?

R. C. Schwing

No. These are taking values out of the literature to show why we have no consensus. Groups have not been exposed to these models to facilitate making a decision. However, I think that the people at the University of Colorado have attempted such an effort. They tried to arrive at an acceptable bullet which was acceptable to both the ACLU and the police department. When they really got down to displaying the value space, they found out there really wasn't as much disagreement as had appeared on that issue.

C. Starr *(Electric Power Research Institute)*

I just want to comment on the longevity use. This is a good start, but I think is very primitive for what we really need. First, you mixed the kinds of things that affect the young children, young adults, and old people, equally as though it affected just the aged alone. So, that in fact, we are not really measuring either the social response or the social attitudes on loss of life. Secondly, there is a subtle matter in the quality of dying which is as important as the quality of living. And one of the great psychological fears of cancer is not that it kills you but it kills you in a very difficult way. You know there's the old joke that at the age 90 I would like to die from a heart attack with a blonde in bed. I don't want to die from cancer. That kind of value has to somehow be included as a quality measure. So the age distribution and the quality measure has to be put in. If that could be refined then the allocation of national resources to improve some weighted longevity number might really be useful.

R. C. Schwing

I agree totally and tried to demonstrate that by including fatal accidents in one set which applies to the catastrophic deaths in the middle age group or younger. A second set includes only the mortality categories for disease. That is just a very first cut, a preliminary attempt to stratify. I like Zeckhauser's use of "quality-of-life years." It has desirable priorities. Then it would be the next step if this were an acceptable yet crude method to set priorities then you'd like to refine it. I agree totally.

M. A. Schneiderman *(National Cancer Institute)*

Following up on Chauncey's remarks, I am concerned by this longevity measure because when I add up all those longevity increases obtained by eliminating all categories of risk, I find that the shortening of life for all coincidences of death comes to 15 years and that doesn't correspond to our expected life expectancy of 72 years. So there is something going on there. Eliminating all heart disease gives ten years longevity, cancer gives two years, and a couple of others give roughly two or three years. It adds up to only 16 or 17 years. What is the arithmetic problem that's going on there?

R. C. Schwing

Implicit in this empirical longevity increase model is the idea that the population already has some average life span. These longevity gains can be considered additions to the average life span. Summing the longevity gains will not give a value for an expected life span.

F. Parker *(Tenneco, Inc.)*

How do you add back into the models some of the other benefits that are difficult to address. For example, most of the transplant organisms come from young adults who are killed, not most but a very large segment of young adults are killed in automobile accidents. Now how do you add those benefits back into the quality of life.

R. C. Schwing

I won't. No way! I think these are crude tools. They're not going to be refined for that.

A. Kneese *(Resources for the Future)*

That was an intriguing analysis you did of the differential of cost in saving lives in different programs. Have you ever as an exercise tried to develop optimum programs from a cost minimizing standpoint where you would try to equate the marginal costs of saving lives across programs for a certain number of lives saved and if so, how did that look. If you wanted to save, let us say 200,000 lives a year, how would you optimally do that, simply given cost minimization?

R. C. Schwing

I haven't tried, but one would work through programs of the type shown here from the most efficient towards the less efficient until you reached 200,000 lives per year.

A. Kneese

I think it is an interesting point that apparently the marginal cost per life of different programs is vastly different over the programs.

R. C. Schwing

Similarly, in the case of air pollution abatement, the marginal costs goes up so rapidly that you back off quite a ways from that steep region of the cost curve if one uses efficiency criteria in allocating resources.

W. D. Rowe *(The American University)*

You had the concept that society can do something about risks and that the marginal costs to control certain risks might be quite different. You're just showing that the cost to control risks by society or by the individual varies all over the lot as a function of the ability to influence control.

R. C. Schwing

Yes, I think this relates to Allen's question. Each of these was a program studied with a certain focus and a certain scope and expected benefit value, if you will. Now in the air pollution case, with the dose response curve accepted as linear and an exponential cost curve as you approach total clean up, you have enough information for a marginal analysis. In the example I showed in the first couple figures, they represent discrete programs, not marginal. Ideally, you would do all programs in an incremental manner because obviously the first case of a detection scheme in a screening cancer program is a lot easier to detect than the remote cases. You can optimize by going properly after the vulnerable population in your screening process.

RISK-SPREADING THROUGH UNDERWRITING AND THE INSURANCE INSTITUTION

J. D. Hammond

*The Pennsylvania State University,
University Park, Pennsylvania*

ABSTRACT

The insurance underwriting decision process is often viewed as scientific, precise, and capable of accurate risk assessment. In fact, it is frequently influenced by competitive forces, the availability of reinsurance, and the judgment of underwriters. Inflation and changing social values also add to underwriting uncertainty by causing claims distributions to be unstable over time.

The financial stability of insurers to take risks is affected by the level of surplus and the stability of the underwriting portfolio. The latter is affected by the mix of insurance lines underwritten. Portfolio theory provides a helpful frame of reference for the evaluation of underwriting portfolios.

INTRODUCTION

Insurance is a key topic in any assessment of risk and the insurance institution is society's principal agency for the transfer of certain kinds of risks. Its presence is a major factor in corporate decisions relating to both safety and the transfer of risk. As the social order takes on more characteristics of what some identify as the "riskless society," [1] the insurance institution faces increased pressure to accept risks of increasing complexity and, in some lines and coverages, to make insurance available at prices which are "affordable" to buyers.

The underwriting pressure resulting from increasing complexities of risk and social demands, however, does not stand in isolation. The desire of diverse groups to transfer the financial consequences of loss exposures to another group seems a common part of current political and economic affairs. National immunization programs, environmental controls, safety standards, increased frequency and severity of liability claims, food and product bans, and even programs to protect national park visitors from bears [1] all, in one way or another, reflect strong

References pp. 175-176.

preferences for the reduction or transfer of risk.

In the analysis which follows, the insurance underwriting function will be reviewed in the context of the criteria by which risks are insurable or uninsurable. The second section contains an examination of the underwriting decision process as decision-making under uncertainty. The final section is devoted to analysis of the collective impact of underwriting decisions; the financial capacity of insurers to assume risk and provide insurance. Finally, suggestions for further research are noted.

THE INSURANCE TECHNIQUE

Sustained encounter with insurance in any form is typically through the insurance contract or policy. It is the document which specifies the contractual nature of the relationship between the insured and the insurer and which, from the standpoint of the insured, allows for the transfer of a risk to another party.

However, underlying all of the legal aspects of the insurance exchange is the technique itself. The technique appears to be one of society's oldest methods for reducing the impact of unforeseen losses on the individual. Legend describes the process of allocating merchandise over several boats instead of one or several caravans instead of one. Moreover, the Old Testament contains several observations about suretyship.*

Inherent in most description of the early techniques is the idea of risk-spreading. Some authorities identify it as essential to the presence of insurance. Kulp, perhaps one of the most influential insurance scholars, stated the position well:

> Insurance does not depend on the legal status of the insurer, the source or method of finances, or the application of particular statistical or actuarial techniques. If, in meeting hazard, average is substituted for actual loss, the result is insurance [3].

The averaging or loss-spreading technique of insurance is commonly associated with the familiar law of large numbers. Insurers, by dealing with several or large numbers of homogeneous observations are better able to forecast claims costs.** The improved forecasting accuracy associated with large numbers, however, is not always present. Even so, the actuarial process is frequently thought capable of developing accurate forecasts of claim costs. The tendency of layman and others to overestimate actuarial capabilities has been described as the "fallacy of actuarial perfection" [4].

---

*The early history of insurance is described well in [2].

**Technically, the standard deviation of a sample value, such as average claim frequency, will increase with sample size but not proportionately. The standard deviation is directly proportional to the square root of the number of sample observations. Claim forecasts, for example, are much more likely to be accurate in personal lines, such as auto collision insurance, than in unusual commercial lines, such as oil rig collapse.

INSURABILITY CRITERIA

Any risk can be insured if the rate is sufficiently high. The desire to transfer the risk under such conditions, however, is greatly reduced and may not be present at all.* Moreover, an underwriter would likely be quite skeptical about an insured who wanted to buy insurance at an extremely high rate. As a practical matter then, insurability or acceptability of a risk to an underwriter requires the presence of certain criteria to maintain the viability of the insurance transaction. Briefly stated, these are:

1. *A large number of homogeneous units exposed to loss.* The requirement clearly facilitates accurate claim prediction.
2. *A loss definite in time and place.* Claim prediction and settlement are greatly enhanced if the time and place of the event are easily identifiable. It would be difficult, for example, to write insurance coverage against the aging process or physical deterioration of a structure.
3. *Accidental loss.* The requirement parallels the randomness component of sample selection and also implies that the insured event be beyond the control of the insured. A highly probable loss is not an attractive candidate for transfer because the premiums required would likely be in excess of the claim value.
4. *Minimal catastrophe hazard.* Catastrophes can produce unstable insurance costs and, if the potential is extreme, financial ruin of the insurer.

The insurability criteria are obviously formulated in a way to minimize adverse fluctuations in claims experience. In practice, however, insurability is heavily influenced by competition and regulation so that the underwriting decision is affected by both objective and subjective factors. In fact then, the insurability criteria exist between two extremes; totally uninsurable and insurable without constraint under standard underwriting objectives.

Ultimately, insurability becomes a function of multiple factors, involving various actuarial and economic combinations. Of all economic factors, competition and the presence of an active insurance market likely exert the greatest impact upon the insurability decision. The actuarial factors of homogeneity, large numbers, definite and accidental losses, and minimal catastrophe hazard are seldom met perfectly; one or more of them are likely to be compromised in either the establishment of underwriting policy or in the underwriting decision.

UNDERWRITING: RISK-TAKING UNDER UNCERTAINTY

The decision on whether to insure and at what price affects the supply of insurance, individual or organizational financial security and planning, and insurer

---

The aviation hull rate for a transatlantic crossing was, in the early days of aviation, 105 percent of the value of the aircraft. See [3]. Underwriters have suggested to the author that a rate in excess of 10 percent for deep sea oil rigs would cause many insureds to retain the risk.

References pp. 175-176.

solidity. The decision process which deals with the selection and pricing of risks is commonly identified as underwriting. Underwriting decisions are essentially similar to other choices made under uncertainty. Some dimensions, however, are peculiar to insurance. Insurance pricing decisions, for example, are made before cost data are known.* Moreover, for most individuals, no satisfactory substitute for insurance exists and retention of the risk may severely alter resource allocations within society and perhaps be illegal as well. The ability of organizations to transfer at least the large loss potential also facilitates the flow of capital into productive activities. A stable supply of insurance, fairly priced, is an essential component of commerce.**

Recent research concerning the insurance purchase decision has cast considerable doubt upon the traditional analysis of that act. Kunreuther, in an extensive analysis of flood insurance purchase decisions of individuals, found that most of them apparently discounted the disaster potential to the extent that the relatively small insurance premiums were viewed as not worthwhile [6]. Indeed individuals apparently viewed the transaction as speculative; insurance would be purchased if it offered a sufficiently large chance of a return. Otherwise, insurance could be foregone and the premium amount spent on other needs. The negative decision, however, could apparently be influenced by loss experience of the individual or that of acquaintenances. In any event, large numbers of decisions are not explained by the traditional analysis of the expected utility model.

The Kunreuther findings on insurance buyers prompt several questions with respect to insurance sellers. Do underwriters also tend to discount the likelihood of large claims and if so, under what conditions? Does the misfortune of one insurer affect the underwriting decisions of another? Does the availability of insurance to the underwriter (reinsurance) alter the decision from what it would be without it? Research has not yet provided insights into all of these questions, but they can be better understood by a review of the underwriting process.

THE UNDERWRITING DECISION PROCESS

The underwriting decision process is fundamentally the same as any other. Information is collected, alternatives analyzed, action is taken, results are observed and corrective actions, if any, are undertaken. The first three steps form the primary element of the accept or reject choice. They are depicted in Fig. 1.***

At some point, the underwriting process must produce a decision to either

---

*Some rating plans provide for final premium determinations to reflect actual claims experience for the year in question. Retrospective rating in workers compensation is a prominent example. Even so, such plans may include a maximum premium specification to limit the impact of large or catastrophic losses.

**For a good discussion of the economic cost of uncertainty in an insurance context see [5].

***Fig. 1, as well as an extensive analysis of the products liability underwriting decision process is found in [7]. The Schweig analysis of underwriting decisions appears to be the most recent as well as being quite thorough.

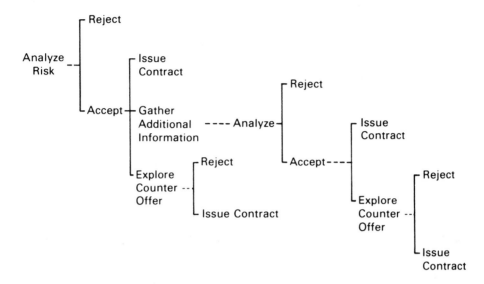

Fig. 1. Underwriting decision diagram.

insure or reject the risk. The actual decision, however, can range over several options as implied by Fig. 1. An insurer may choose to decline an individual risk, avoid an entire line of insurance, or withdraw from a particular territory.* Given a rate and system of risk classification, standards are developed to determine the risk classifications most consistent with the characteristiics of the loss exposure. Location, for example, is a key classification variable for many exposures, automobile insurance being one of the more prominent. Another underwriting option is to impose standards of conditional acceptance, such as requiring change in certain physical conditions affecting the risk. Further, contractual language might be made more restrictive by limiting the amount of insurance or excluding certain perils or hazards. Finally, price itself is a key control variable.

Errors or biases which produce wrong decisions result in individuals or organizations becoming uninsured or in some way altering the costs of insurance from those commensurate with the value of the risk. If a risk is misjudged as unacceptable it could, in a competitive market, likely be placed with another insurer, the underwriting error consequences simply being a loss of profit to the mistaken insurer and some inconvenience costs to the insured. Identification of the risk as unacceptable by one insurer may force the insurance buyer to retain the risk or enter a higher-priced market segment. Underwriting error can also undervalue risks, thereby producing at least a short-run advantage to the insured.

From the insurer viewpoint, incorrect underwriting choices can impair financial performance and therefore the development of financial capacity. The situation

---

For example, at least two major insurers have sought to withdraw from writing automobile insurance in New Jersey because of alleged rate inadequacy.

TABLE 1

Underwriting Decision Matrix

Underwriting Decision	Rate Level	
	Adequate	Inadequate
Accept Risk	Surplus Gain	Surplus Loss
Reject Risk	Surplus Opportunity Loss	Surplus Unchanged

can be characterized by the familiar decision matrix of Type I and Type II errors as shown in Table 1.

Favorable decisions, by contributing to surplus, help to enlarge the capacity to supply insurance. Others either reduce it or leave it unaffected.

UNDERWRITING UNDER UNCERTAINTY

Some outcomes of underwriting decisions may be highly probable but none are certain and some may be highly uncertain. Accurate pricing is clearly facilitated by conditions of low uncertainty. Where outcomes are highly uncertain, insurance may be very expensive or perhaps unavailable. The chance of underwriting error is magnified by uncertain conditions. The degree of uncertainty is affected by the quality of non-claims information, claims experience credibility, and the stability of claims distributions.

Non-Claims Information — Underwriters attempt to develop information relating observable characteristics of the risk to the probability of loss. For personal insurance lines, where individual habits or traits may affect the value of the risk, information is sought on such factors as alcohol or drug use, and physical impairment. The agent is expected to communicate important observations about applicants but may sometimes be unable to avoid the conflict potential involved.* Insurers may also employ specialists to develop such information.**

Where the risk is complex, informational reports are complex. Products liability underwriting, for example, requires credit and financial background, engineering and plant reports, product design, and past accident records of the firm [7]. For off-

---

Agents usually receive a commission on business which is accepted. In one incident related to the author, the agent had reported an applicant for automobile insurance as having "some speaking difficulties and as slightly hard of hearing." Further investigation revealed the applicant to be both deaf and mute.

**Some controversy currently exists between the insurer's need to develop underwriting information and individual rights of privacy. See [9].*

shore drilling rigs, identification and experience of the towing operator may significantly alter premium costs. Outside expertise such as engineering or architectural consultants is frequently employed.

Accurate and thorough descriptive information is accorded considerable weight in the formulation of the final underwriting decision.* If information is sparse or unreliable, it may lead to rejection of the risk.

Credibility, Large Numbers, and Homogeneity — Where large numbers of loss exposures have approximately the same probability of loss, insurance pricing is greatly simplified. Each member of the homogeneous class can fairly be charged the same rate. The class-rating associated with these conditions is employed in the personal insurance lines, automobile and homeowners coverage being dominant, or on some commercial lines where individual claim experience is too small to be accorded any credibility.

Insurance pricing and risk selection are reasonably straight-forward where experience is fully credible and class members have similar loss probabilities. The net or pure premium (net of administrative expenses) is simply the cost of all claims averaged over all exposure units so that:

$$\text{Average Pure Premium} = \frac{\text{Losses \& Loss Adjustment Expenses Incurred}}{\text{Number of Exposure Units}}$$

Some risks generate sufficient claims experience to warrant partial credibility. Individual claim costs and those of the class are then combined with appropriate credibility weights to derive a rate for the individual risk. The assignment of weights to individual claim experience can involve considerable underwriting judgment. The process is essentially Bayesian.

$$PP_E = ZPP_i + (1 - Z)PP^{**}$$

where:

PP_E = appropriate individual pure premium
PP_i = pure premium based on individual experience
Z = credibility weight where $0 \leqslant Z \leqslant 1$. As number of exposure units increase, $Z \rightarrow 1.0$

At some point, as the claims experience of the individual insured becomes fully credible, the final pure premium is based entirely on the claim experience of the single insured unit.*** Also at this point, the insured is essentially underwritten as a unique risk and the homogeneity required of class-rated risks is no longer pre-

---

*The subjectivity of the weight to be accorded such data has led to misunderstanding. Some lawmakers, for example, have argued that life insurers should not be able to use any risk evaluation measure for which there was not adequate and reliable claims data. The attempt to eliminate or at least reduce arbitrary or capricious judgment might also lead to insurance on uninsurable risks. Arbitrary and capricious decisions may exist, however. Again, see [7].

**The above formulation is one expressed in [10].

***The arrangement is usually not equivalent to "cost plus" pricing. Such formulations in workers compensation, for example, commonly specify a maximum premium, thereby requiring some risk charge by the insurer in exchange for stabilizing the loss costs of the insured.

References pp. 175-176.

sent. Indeed, with commercial fire insurance, the individual risk is evaluated on the basis of a schedule imposing credits and changes for various physical conditions and subjective underwriting assessments. Fire claims experience, because of its relatively low frequency, is evaluated more subjectively than the more frequent individual accident experience under worker compensation.

Stability of Claim Distributions — Where claims experience results from a large and homogeneous number of loss exposures, probability estimates can generally be made with improved confidence over those derived from insufficient data. Underwriting uncertainty, however, can be reduced further if claim distributions are characterized by small variances and remain stable over time. Departures from either of these aspects of stability amplify underwriting uncertainty.

Distributions of claim costs are combinations of the separate distributions of claim frequency and claim severity. A change in either alters the value of average claim cost and therefore the premiums to be charged. With most insurance coverages, the most likely outcome for a single insured is no loss at all. Distributions of claim frequency, therefore, can often be illustrated by distributions of the general nature shown in Fig. 2.

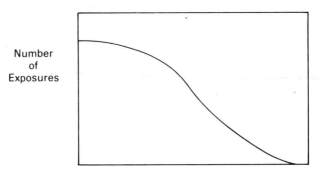

Number of Losses

Fig. 2. Hypothetical distribution of claim frequency.

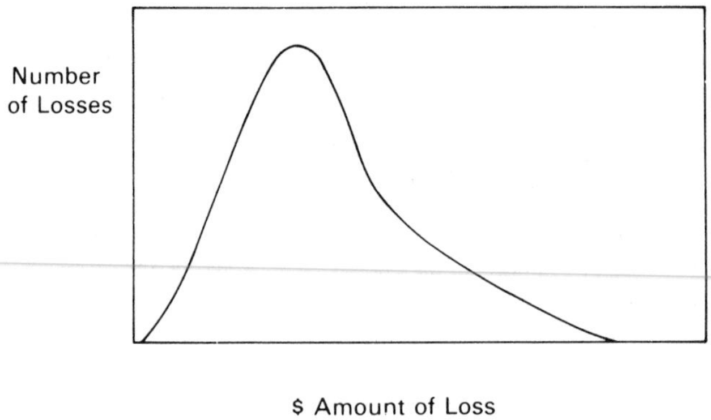

$ Amount of Loss

Fig. 3. Hypothetical distribution of claim severity.

Claim severity is limited by the amount of insurance while the lower limit can be any positive value, taking into account any deductible which might be present. Fig. 3 depicts a hypothetical distribution of claim severity. Generally, most losses tend to be partial, with values less than those of a total loss.

Finally, the means of the frequency and severity distributions can be multiplied to produce the average loss per exposure or pure premium. The pure premium distribution is illustrated in Fig. 4.

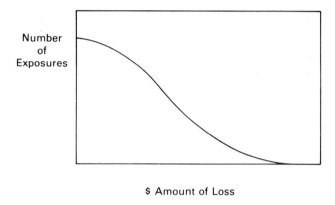

Fig. 4. Hypothetical pure premium distribution.

Again, most loss exposures will incur no loss at all during any reasonable interval of coverage while a small number will incur very large claims.*

Any circumstances which enlarge the variance of the pure premium distribution increases the uncertainty involved in underwriting the insurance risk. Combinations of large losses and small numbers of exposure units, for example, produce this effect. Distributional change over time can alter the expected value of the distribution or alter the variance but leave the expected value unchanged.

The various empirical distributions of claim frequency and severity confronting the underwriter are not conveniently available to researchers.** Nonetheless, certain kinds of data can be employed to construct distributions which complement those noted above. It is possible to construct, for example, a distribution of underwriting profit for any line of insurance and for various points in time, from industry-wide cross-sectional data.

---

*The distribution of claim frequency and severity are often difficult to describe with precision. They also differ among lines of insurance. For a thorough mathematical discussion of claim distributions, see [11].

**They frequently are difficult for insurers to construct since records are maintained to conform to statutory reporting requirements. These center upon aggregate results rather than the detail required to construct either claim frequency or severity distributions. Of course, where only small numbers of highly valued risks are insured, claim distribution data can be constructed with relative ease.

References pp. 175-176.

The most commonly used measure of underwriting gains and losses is a ratio relating losses and expenses to premiums. Specifically, the ratio is defined as:*

$$\frac{\text{Losses \& Loss Adjustment Expenses Incurred}}{\text{Net Premiums Earned}} + \frac{\text{Underwriting Expenses Incurred}}{\text{Net Premiums Written}}$$

Thus, if underwriting results develop a combined ratio of 98 percent, an underwriting gain of 2 percent has been incurred. The variability of underwriting results can serve as a convenient empirical approximation of the hypothetical distributions depicted earlier.**

Figs. 5 through 10 which follow are derived from underwriting data for nearly the complete universe of insurers operating in each line of insurance over the 1972-1975 interval. The number of insurers analyzed in each year, beginning with 1972, was 979, 998, 1003, and 1000.***

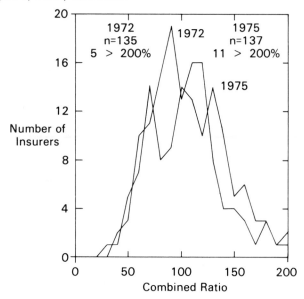

Fig. 5. 1972 and 1975 distributions of combined ratios for the smallest one-fifth of fire insurers.

---

*The combined ratio is a shortcut approach to by-passing the conventions of statutory accounting. Under statutory accounting, prepaid expenses cannot be accepted as assets and unearned premium reserves are calculated on the basis of gross premiums. Underwriting expenses, of course, are paid as incurred and constitute an immediate reduction in cash flow.

**Underwriting results for a given line of insurance will not be perfectly comparable across insurers. Line definitions are those specified in the insurance expense exhibit required under statutory accounting and frequently include diverse coverages. The Liability Other Than Auto line, for example, includes products liability, and all other liability risks except auto and medical malpractice. The latter was designated as a separate line in the expense exhibit in 1975. Also, insurers may have different underwriting objectives and face different regulatory constraints.

***The analysis of underwriting experience shown here is part of a larger study on insurer capital needs and solidity financed by the National Science Foundation and reported in detail in the technical report [12].

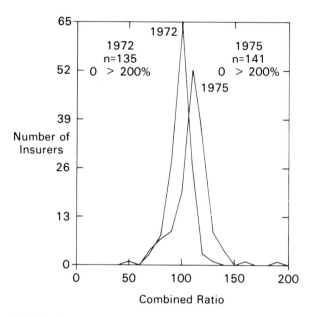

Fig. 6. 1972 and 1975 distributions of combined ratios for the largest one-fifth of fire insurance.

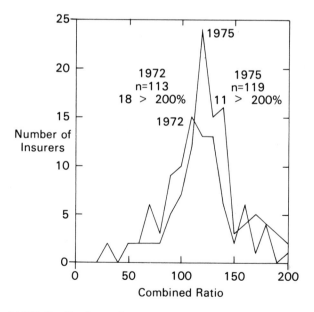

Fig. 7. 1972 and 1975 distributions of combined ratios for the smallest one-fifth of PPAL insurers.

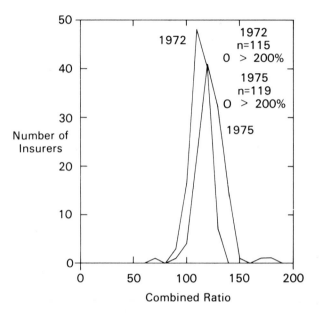

Fig. 8. 1972 and 1975 distributions of combined ratios for the largest one-fifth of PPAL insurers.

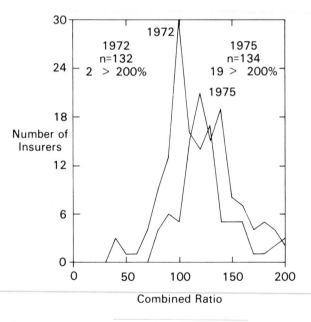

Fig. 9. 1972 and 1975 distributions of combined ratios for the smallest one-fifth of PPAPD insurers.

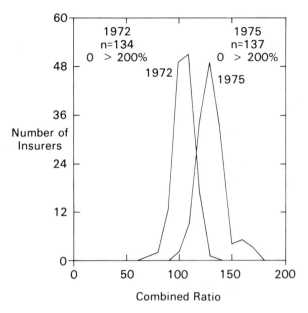

Fig. 10. 1972 and 1975 distributions of combined ratios for the largest one-fifth of PPAPD insurers.

The 1972 and 1975 distributions are depicted because of the greatly contrasting underwriting results between those two years. Underwriting gains in 1972 of $1.6 billion were record highs while the 1975 losses of $4.5 billion were record lows. Distributions are depicted for three lines of insurance for two different size classes of insurers; fire insurance, Private Passenger Automobile Physical Damage, Private Passenger Automobile Liability for the smallest and largest one-fifths of insurers.

Figs. 5 through 10 illustrate the important effects of size and stability. The fire insurance data of Fig. 5 clearly reveals great instability at small premium volumes; moreover, the distribution has clearly shifted across the 1972-1975 interval. Both aspects of instability magnify underwriting uncertainty. The two automobile coverages also exhibit distributional shifts over the three year interval, but each distribution is relatively compact.*

---

*The differences can be measured precisely by the standard deviation of the combined ratios for the data underlying Figs. 5 through 10. These are:

	1972	1975
Smallest One-Fifth Fire	0.37	0.95
Largest One-Fifth Fire	0.07	0.11
Smallest One-Fifth PPAL	1.21	0.50
Largest One-Fifth PPAL	0.08	0.10
Smallest One-Fifth PPAPD	0.28	0.43
Largest One-Fifth PPAPD	0.07	0.10

References pp. 175-176.

Underwriting uncertainty is clearly reduced by the claims distributions which are compact and stable over time. These attributes facilitate the formation of clear underwriting decision rules and reduce the chance of underwriting error.*

UNDERWRITING ERROR

Chance, temporal changes in claims distributions, and informational inaccuracies, can all result in underwriting errors with direct impact upon insurer capital and insurance capacity.

The various types of underwriting error are outlined in Table 2.**

TABLE 2

Classification of Underwriting Error

Underwriting Error			
Prognosis Error		Diagnosis Error	
Chance	Change	Type I Error	Type II Error

Prognosis — Even with perfect information from a large experience base, there is some chance that the actual outcome of underwriting decisions will be different from that expected, even where conditions affecting loss frequency and severity remain stable over time. Where such conditions do not remain stable, an additional source of underwriting error is introduced. Inflation and changing expectations are two prominent examples.*** While large numbers and compact claim distributions can reduce chance errors, they have no impact on the nonpoolable risks just noted.

Diagnosis — Risk evaluation involves pricing and the decision on whether or not it is acceptable. The Type I and Type II classification errors noted previously are directly affected by the quality of information received as well as by possible

---

Underwriting bias can exist quite independently of the nature of the claims distribution. For example, Schweig reported that the boss of one automobile liability underwriter once prohibited the underwriter from issuing insurance to anyone with one particular surname.

**The content and organization of Table 2 is adapted from [13]. Eichorn uses the phrase "underwriting risk" rather than underwriting error.*

***Insurance rates cannot adjust to immediately to changing costs so that for most lines of insurance, inflation increases claims costs prior to affecting rates. Changing expectations, as well as changing attitudes toward risks, have been reflected in increased liability judgments and added circumstances in which liability has been found by the courts as an appropriate remedy for personal and financial loss.*

biases of the underwriter or of the underwriting process itself.

Where underlying claims data are adequate, underwriting errors can still result from chance or from changes in the conditions which affect claim frequency or severity. Where data are inadequate, errors stemming from bias or wrong judgment become more prominent.* Arbitrary choices and bias, of course, can and do exist where data are plentiful. They appear, however, to be more associated with peculiarities of individual underwriters than from problems with the underlying data.

JUDGMENT UNDERWRITING

The underwriting process associated with adequate data, information, and stable loss distributions is frequently characterized by clear sets of decision rules designed to reduce reliance on individual underwriting judgment. The contrary conditions of sparce data and information and unknown or unstable loss distributions, however, place heavy demands upon individual judgment in risk pricing and evaluation. The frame of reference provided by decision research suggests a means of examining those underwriting decisions made under conditions of high uncertainty. Tversky and Kahneman have identified several heuristic principles which are commonly employed in the assignment of probabilities and values under uncertain conditions [14].** Many are directly relevant to underwriting.

Where uncertainty is high, decisions may reflect insensitivities to prior probabilities, undue weight to prominent events or small sample size information, or the use of information or data thought to be representative of an unknown or unavailable probability distribution. Similarly, information which is easily available or easily recalled may be accorded more weight than its real value would suggest. Individuals may also underestimate the probability of failure in complex systems and imagination may lead some individuals to assume the worst or mask difficulties with unbridled optimism.*** Research has not yet examined all details of the underwriting decision process, although general observation and some of the information developed by Schweig suggest that the decision patterns noted above can easily become part of underwriting judgment.

Various "crisis" environments have had observable impacts upon insurance availability and price. The medical malpractice and product liability difficulties, for example, have led to rapid changes in insurance price and availability. Large single claims or a wide-spread catastrophe may also alter insurance supply or price, even

---

*A broad discussion of possible biases in decisions is provided in [14]. Examples of bias in the underwriting process are noted in [7].

**Other helpful insight are noted in [15]. The March paper also lists an extensive bibliography. See also [16].

***Each of the decision heuristics noted, as well as several others, are explained in detail in [14].

References pp. 175-176.

if preceded by a relatively long claim-free period. In short, dramatic events may have the potential to alter perceptions of almost all other relevant information.*

Underwriters may also seek ways to employ representative information to evaluate complex loss exposures. Simon has observed that "the human information processor operates serially, being capable of dealing with only one or a few things at a time" [16]. There is no *a priori* reason to believe that underwriting judgment operates differently.

SOME EVIDENCE ON UNDERWRITING BEHAVIOR

The underwriting of personal automobile insurance has been found to be sufficiently routine as to be capable of computerization [17]. More recently, the underwriting decision process of product liability underwriters was observed to stress satisficing objectives, such as underutilization of insurer capacity. Under test circumstances, underwriters overwhelmingly chose to reject a product liability applicant with an approximately equal chance of acceptance or rejectance [7]. The results apparently suggested the least complicated way of deciding upon a difficult risk was to reject it for any of several reasons, a result consistent with satisficing objectives. Schweig also suggests that the minority of underwriters who reached an acceptance decision were more likely candidates for promotion and did not want to appear ineffectual by a negative decision [7]. If this proposition holds, the analysis of the underwriting decision process becomes more complex because of the presence of personal objectives of underwriters.**

Analysis of underwriting decisions to date have relied upon relatively formal research procedures, employed to examine a relatively limited dimension of the process. Less definitive but still valuable information can be secured through the experience of large insurance brokers.*** The observations on underwriting behavior which follow are based on conversations with senior officials in a prominant U.S. brokerage firm.**** While their insights are not conclusive, they do reflect extensive experience involving the placement of complex risks in international markets, over a considerable period of time, and for notable corporate clients.

This is not to suggest that an insurance "crisis" is unreal. If the underlying probability distributions of claims has actually shifted adversely, the crisis is real and market reactions will take place.

**The joint consideration of individual and corporate utility functions is involved. If risk-avoiding behavior is dominant, is it the result of corporate controls and supervision or is it the result of the underwriter's own risk-avoidance model? These questions are examined in a non-underwriting context in [18].*

***Technically, a broker is the agent of the insured and acts on the latter's behalf in attempting to secure insurance coverage. Insurance brokerage firms play a dominant role in placing insurance on complex risks and are thus very familiar with underwriting standards, behavior, and peculiarities.*

****The author wishes to thank these individuals and their firm for the generous interview and conversation time which they cheerfully provided.*

Underwriters frequently make decisions on risks with little or no claims experience. Even where more than one loss exposure is present (such as several drilling platforms), the claims experience may remain small. While such experience is not ignored, it inevitably leads underwriters to assess probabilities by what Tversky and Kahneman identify as the representativeness heuristic [14]. Probabilities are evaluated less by objective claims data and more on the basis of information thought to be representative of such data.

Non-claims information is prominent in the evaluation of complex risks, even where claims data are present. Depending upon the nature of the exposure, underwriters evaluate such factors as engineering and design features, architectural reputation, and the quality of management of the organization seeking insurance. Underwriters may meet directly with top management officials in an attempt to assess general attitudes and abilities. Similarly, underwriters can communicate information directly to management which might facilitate placement and acceptance of insurance coverage. The position or quality often sought is a high degree of mutual trust.* Underwriters may also rely heavily on the advice of expert consultants to evaluate those aspects of a risk in which their own experience and evaluative capabilities are deficient.

A form of representative information is particularly important for reinsurance underwriting.** Reinsurers must rely upon the information communicated by the primary insurer; direct inspection of the insured risk by the reinsurer would be unusual. Information developed by the primary insurer is accepted as wholly representative of the underlying risk. Because the relationship between the reinsurance and originating insurer is one of high trust, any breach of confidence may terminate or imperil future relationships.***

The cyclical nature of underwriting results provides a partial explanation of underwriting reactions to claims experience.**** Rate increases typically follow increases in incurred losses, usually producing improved underwriting profits. Improved profit levels ultimately generate increased price competition, thus setting the stage for another decline. Underwriting reactions to individual account experience or to the aggregate experience in a single line or class of coverage follows a similar pattern. A large loss to a single complex risk or single event losses to a number of risks is likely to promote a decision to increase rates to tighten underwriting standards.

---

*The insurance contract has always been held as one of utmost good faith where all relevant information is expected to be communicated. It is not a contract of caveat emptor nor of caveat venditor. The utmost good faith view of the exchange is probably most characteristic of the complex risk market.

**Reinsurance is the purchase of insurance by an insurance company. Reinsurers may in turn purchase insurance, a transaction termed retrocession.

***In one instance reported to the author, the originating insurer described a risk to the reinsurer as a hospital when in fact is was an institution housing a large number of pyromaniacs. The reinsurer was no longer interested in having the primary insurer as a client.

****A detailed description of the 1972-1975 decline phase of an underwriting cycle is given in [19].

References pp. 175-176.

Significant claim events have a direct impact upon the distribution of claim severity. This is especially characteristic of insurance coverages which generate reasonable bodies of claim experience. The rate increases on single or small groups of complex risks are likely to result from subjective underwriting assessments. Such events clearly produce information which is easy to recall, easily available, and frequently dramatic in its impact. It is an unusual underwriter indeed whose decisions are not impacted by dramatic and easily observable events.

Significant claim events in one line of insurance or for a single complex risk produce underwriting reactions for only that line of risk. A catastrophic products liability settlement, for example, would not prompt underwriters to raise liability rates for oil tankers.* It is possible and perhaps even likely, however, that a large loss settlement by a competitor for a given type of loss exposure, will prompt a pricing reassessment by other underwriters.

Underwriters apparently tend to imagine severe consequences in evaluating complex risks, particularly where the exposure is both new and large. The tendency may be a natural consequence of high uncertainty associated with inadequate or unreliable information. The reaction to pessimistic underwriting evaluations is either an outright refusal to insure or a relatively high insurance rate.**

The insurance market for complex risks and the large premiums which they develop is world-wide and appears quite competitive.*** The effect of competition on underwriting decisions is dramatic. Insurer growth objectives and competitive survival are generally not served by ultra-conservative underwriting of large account business. While traditional assessments of underwriting risk are probably not overwhelmed by competitive considerations, they are greatly diminished.

The availability of reinsurance and a competitive reinsurance market have the potential to alter the risk-taking behavior of underwriters. Small insurers are able to take on accounts not otherwise acceptable and large exposures can be accepted by a primary underwriter and then shared with reinsurers.

Aside from the actuarial benefits of loss-sharing provided by reinsurance, there is also some suggestion that it affects risk-taking attitudes of primary underwriters. Despite the utmost good faith basis of the reinsurance transaction, some primary underwriters may see potential gain in the opportunity to benefit at the expense of a reinsurer. If, for example, an undesirable risk or book of business can be shared or priced in a manner which benefits the primary insurer at the expense of the reinsurer, some underwriters have been known to succumb to the temptation. In those circumstances, the possibility of short-term gains apparently dominate the longer-

---

*This is not to deny the possible presence of the idea. A competitive marketplace, however, makes such a discriminatory pricing practice quite unlikely. Rate regulation in the U.S. also forbids rates which are unfairly discriminatory.

**Some insurance buyers seeking coverage on complex and hazardous risks have accepted even a 20 percent rate. As the claims-free interval widens and more exposures provide increased spread of risk, rates tend to decrease.

***Data are not conveniently available by which to assess world-wide competitive structures.

run consequences of violating the trust between the two insurers.*

Reinsurance is a dominant element in the underwriting decision process, frequently causing the underwriter to ignore prior probabilities.

Underwriters, where information is limited, appear to rely heavily upon the representativeness heuristic. Pricing and risk selection behavior generally run parallel to significant claim events and underwriting experience in general. Standards become "tighter" following large claims or generally bad experiences and become liberal in the absence of general or singularly large losses. Competition contributes heavily to the presence of liberal standards as does the availability of reinsurance.

Competition and reinsurance appear to be dominating factors in the underwriting of complicated loss exposures. These factors frequently create a condition in which underwriters are more prone to discount prior probability data and information representative of the probabilities.

UNDERWRITING CAPACITY

The financial ability of insurers to supply insurance is commonly referred to as capacity. The ability to supply insurance, given a price, is generally dependent upon the amount of surplus held by the insurer. If actual losses never exceeded those which were expected, the insurer underwriting function could be performed with no surplus at all; premium income would never be less than payments for claim and expenses. Such is clearly not the case, however, and prudence requires that premium volume always be dependent upon the presence of an adequate surplus. Other things equal, an insurer with a large surplus can supply more insurance than smaller counterparts.

Underwriting capacity is a function of three variables: the amount of capital, the variability of underwriting results, and the level of underwriting profit. Expansion of premium volume or an increase in the variability of underwriting results, without an increase in capital, increases the probability that capital will be eroded and elevates the probability of ruin. Alternately stated, an increase in the ratio of premiums to surplus will, other things equal, increase the probability of ruin. Also, an increased ratio is almost certain to produce conservative underwriting.

In recent years, research into capital adequacy of insurers has expanded considerably. Aside from the long-standing conceptual contributions of risk theory, advances were stimulated greatly by the linking of portfolio theory to the risk and return analysis of different lines of insurance.** Bachman [21] employed the

---

*The impact of reinsurance on underwriting decisions is illustrated by the following incident reported to the author. A large commercial property, located below a dam, was declined by a primary insurer, largely because reinsurance was not avaialable at that point. Subsequently, reinsurance capacity increased and became available to the insurer. The primary underwriter promptly accepted the risk.

**Ferrari [20] appears to have been the first to apply the Markowitz portfolio analysis in an insurance underwriting context.

References pp. 175-176.

optimization features of the portfolio model to estimate minimum capital requirements associated with optimal combinations of insurance lines, grouped in such a
way as to minimize the variance or risk of the entire underwriting portfolio.* Hammond and Shilling [22] also explored the application of portfolio theory towards
the estimation of capital required for conduct for the underwriting functions. The
National Association of Insurance Commissioners has also been active in developing and refining its "Early Warning System," a series of financial tests designed to
identify insurers with questionable solidity.

Both regulators and management have commonly relied upon rules-of-thumb to
assess safe upper limits to the ratio of premiums-to-surplus. Over the years a ratio
of 2.0 has been the most frequently applied guideline; net premium volume should
not exceed 200 percent of capital.** The ratio has usually been applied without
reference to insurer size or to the insurance line mix of the underwriting
portfolio.***

Premiums-to-Surplus: An Empirical Examination — The data underlying the
combined ratio distributions of Figs. 5 through 10 can be analyzed to estimate
maximum ratios of premium-to-surplus for each single line of insurance and for
any size classification of insurer. The most straight-forward estimation can be
made from a standard statistical formulation resting upon the assumption of normality in the underlying data; in this case the distribution of underwriting profit in
each line of insurance.****

Specifically:
$$S = P (Z \cdot \sigma_X + X - 1)^{*****}$$
where:

S = Capital
X = Expected Combined Ratio
P = Estimated Premium Volume
σ_X = Estimated Standard Deviation of Expected Combined Ratio
Z = Z value from normal distribution associated with a selected
probability of ruin value

For purposes of illustration, it is assumed that the highest probability of ruin
acceptable to regulators or management is 0.001 ($Z = 3.09$). For example, if the
data for a given line of insurance indicate an average combined ratio of 0.99 with a

---

*The bibliography of the monograph provides a thorough listing of publications dealing with insurer financial
analysis and also with its links to portfolio theory.*

**Net premiums reflect only the premiums actually retained by the insurer after reinsurance.*

***The "Early Warning System," developed by the National Association of Insurance Commissioners to help
detect insurers in poor financial condition, prescribes a ratio of 3.0 as an upper limit but still without direct
reference to size or business mix.*

****Underwriting profit is again defined as 1 - (loss ratio + expense ratio). Investment results are not part of
the analysis.*

*****The formulation was employed in this context a decade ago by Hofflander [23].*

standard deviation of 0.05, the maximum ratio for that line would be

S = $100 (3.09 · 0.05 - 0.01)

S = $14.45 for every $100 of premiums or an estimated maximum
      ratio of 6.92

In the above illustration, the combination of profitable and reasonably stable underwriting results produced a relatively high ratio of premiums to surplus.

Table 3 shows the maximum ratio of premiums-to-surplus that an insurer could have written, under the assumptions noted, in only one line of insurance and not have exceeded a ruin probability of 0.001. The values, therefore, do not reflect possible diversification benefits from writing more than a single line. Only in the values shown for all lines combined is there any indication of the direction and extent of possible diversification effects. Table 3, however, does show maximum premiums-to-surplus ratios for certain selected but standard groupings of insurance lines. It is not possible, however to attribute the relatively high values of these combinations directly to diversification.*

Apparent size effects are clearly observable from Table 3, with higher ratios and greater underwriting capacity associated with larger premium volumes. Small premium volumes with their apparently higher variability generally require more surplus per dollar of premium than larger volumes. In fact, the rule-of-thumb guideline of 2.0 and 3.0 even appears optimistic for small volumes.

Because 1972 was a year of record low combined ratios or high underwriting profits and 1975 was a year of record underwriting losses, the values of Table 3 provide interesting comparisons of underwriting profit differences and capacity. The general decline in the maximum ratios over the 1972-1975 interval reflect the changing claim distributions between 1972 and 1975.

Premiums to Surplus: A Portfolio Approach — In the previous section, cross-sectional data were employed to estimate maximum ratios of premiums to surplus for single lines of insurance and for selected groupings of lines. The resultant values, therefore, represent average relationships and do not necessarily portray the values which would obtain from the underwriting experience of a single insurer. A similar estimation approach could be employed, however, by using the temporal underwriting returns and variance for a single insurer.

It is intuitively clear that nearly any given insurer, with hindsight analysis, could identify alternate underwriting portfolios which, given their historical returns, would have produced improved underwriting returns. Insurer management may intuitively follow such analysis in formulating growth and underwriting policy objectives, but may not employ techniques beyond intuition and judgment in

---

Insurers with more than one line of insurance, to remain in the smallest size category, will have small premium volumes in each line thereby developing a possible increase in variability to go along with the possible decrease associated with diversification.

References pp. 175-176.

TABLE 3

Maximum Premium to Surplus Ratios for Monoline Underwriting and Selected Line
Groupings; for Smallest and Largest One-Fifths of Insurers; 1972-1975

	1972		1973		1974		1975	
	Smallest 1/5	Largest 1/5	Smallest 1/5	Largest 1/5	Smallest 1/5	Largest 1/5	Smallest 1/5	Largest 1/5
1. Fire	0.95	18.25	0.37	9.00	0.48	3.75	0.33	3.37
2. Allied Lines	1.46	8.67	0.33	5.83	0.35	2.04	0.48	2.59
3. Farm	0.61	4.69	0.19	3.04	0.26	1.56	0.34	1.83
4. Homeowners	0.74	6.89	0.44	7.47	0.26	2.47	0.49	3.09
5. Comm Mult Peril ...	1.48	10.33	0.70	7.48	0.29	3.18	0.28	3.26
6. Ocean Marine	0.61	4.60	0.37	3.85	0.73	2.40	0.17	2.30
7. Inland Marine	0.15	3.30	0.47	4.30	0.11	3.08	0.48	3.12
8. Earthquake	NA	NA	0.50	1.00	0.30	*	6.19	7.48
9. Group A & H	0.74	3.65	1.61	3.02	0.16	3.87	0.30	2.97
10. Other A & H	0.46	3.35	0.21	2.77	0.63	2.39	0.10	2.31
11. Workers' Comp. ...	0.46	4.58	0.32	4.32	0.12	3.06	0.05	2.74
12. Lb. Othr Thn Auto .	0.08	1.47	0.38	1.47	0.02	0.95	0.30	1.07
13. Pvt Pss Auto Lb ...	0.25	4.48	0.46	4.44	0.45	3.99	0.58	2.56
14. Comm Auto Liab ...	0.20	2.64	0.47	2.35	0.26	2.02	0.55	1.71
15. Pt Pss Aut Ph Dm ..	1.19	9.39	0.91	4.97	0.79	2.30	0.63	2.25
16. Comm Auto Phy Dm	0.49	7.25	0.68	5.19	0.15	3.58	0.56	2.55
17. Air	0.17	1.99	0.11	1.10	0.26	1.09	0.23	1.22
18. Fidelity	0.23	3.09	0.53	2.12	0.19	1.45	0.33	1.03
19. Surety	0.02	3.86	0.48	2.02	0.17	0.84	0.16	0.69
20. Glass	0.68	2.27	0.45	2.25	0.57	1.92	0.50	1.51
21. Burglary & Theft ...	0.51	6.63	0.28	5.77	0.15	2.89	0.46	3.08
22. Boiler & Machine ..	0.25	3.08	0.03	3.77	0.03	2.54	0.06	2.23
23. Credit	0.51	0.53	1.95	0.16	1.11	0.16	0.84	0.57
24. International	3.34	4.02	2.79	1.58	0.59	3.09	1.91	4.18
25. Reinsurance	0.20	3.33	0.18	3.03	0.07	1.41	0.07	2.06
26. Miscellaneous	0.38	5.15	0.69	6.40	0.31	1.82	0.92	1.51
27. All	1.65	9.94	0.15	7.06	0.99	3.97	0.71	3.62
28. Med. Malpractice ...	NA	NA	NA	NA	NA	NA	0.06	0.83
31. 1+2	1.06	33.16	0.42	13.33	0.59	4.23	0.38	3.88
32. 1+2+3+4	0.93	27.57	0.50	10.46	0.63	3.46	0.48	3.95
33. 13+15	0.59	6.62	0.67	4.95	0.64	4.42	0.71	2.62
34. 14+16	0.21	3.70	0.68	3.16	0.38	2.62	0.47	2.21
35. 13+14+15+16	0.41	6.69	0.81	5.30	0.94	4.36	0.62	2.88
36. 1+2+3+4+22	0.91	26.57	0.54	10.46	0.63	3.49	0.48	4.01
37. 1+2+3+4+12	0.36	5.14	0.07	3.47	0.41	1.87	0.60	2.80
38. 1+2+3+4+5	0.94	37.05	0.56	15.28	0.67	3.83	0.52	4.50
39. 11+12+13+14	0.12	4.10	0.64	3.73	0.47	2.79	0.80	2.66

*Because of a highly skewed distribution along with low combined ratios, the data on Earthquake
experience for the largest one-fifth of insurers in 1974 cannot accommodate the normality assumption
underlying the values of Table 30.

attempting to formalize such values.* Portfolio theory, however, provides such a mechanism.

Although it is beyond the scope of the present paper to provide a complete explanation of portfolio theory, it essentially provides a mechanism for computing an optimal risk and return position from a series of investment activities with individual risk and return values.** Unless the returns from all investment activities are perfectly correlated, the optimal risk and return position will offer an improvement over the actual result by: (1) identifying a higher return for the same risk or (2) identifying a lower risk for the same return. These positions are said to be efficient and are characterized in Fig. 11.

Curve AB forms a "frontier of alternate efficient positions and the 'X's" illustrate inefficient positions. Inefficient positions can be improved because (1) for the same risk, a higher return is possible and (2) for the same return, a lower risk is possible. Each point on the frontier represents an optimal combination of activities for given levels of risk.

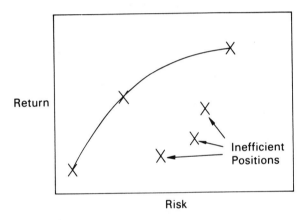

Fig. 11. Hypothetical efficient frontier and risk-return positions.

Underwriting results from insurance lines can be assumed to vary in random fashion producing returns and variations analogous to those of equity securities and other forms of investment. As with investment portfolios, variation in underwriting returns can be reduced by successful diversification of insurance lines.

---

*Formal attempts to improve precision in forecasting premium growth, alternate results from different underwriting mixes, and to estimate carefully underwriting risk appear to be increasingly employed by larger numbers of insurers. The use may reflect at least in part, the increased research and financial model development of the past decade as well as the appearance of individuals in top management groups whose educational backgrounds reflect these developments.

**The development of portfolio theory is credited to H. M. Markowitz and many of its principal extensions to W. F. Sharpe [24]. The Sharpe book is especially helpful to explaining the theory's applications to financial analysis.

References pp. 175-176.

While it is easy to imagine portfolio combinations and diversification benefits involving two activities, combinations involving large numbers are more complex. Computer capabilities, however, are sufficient to enable large numbers of activities to be arranged so that for any given level of risk, return is maximized or, for any given level of return, risk is minimized. Together, such combinations form an efficient frontier as illustrated in Fig. 11.

Consider now, the application of these ideas to the actual underwriting experience of two insurers.

Table 4 shows the actual underwriting portfolio of a large insurance company over a ten year interval. Underwriting return is measured by 100 percent less the combined loss and expense ratio and the standard deviation is the standard deviation of underwriting returns over the ten year time interval. Table 5 shows the same type of information but for an insurer with an unstable underwriting profile.

TABLE 4

Actual Underwriting Portfolio, Ten Year Interval
Insurer "A"

Line	Underwriting Return	Standard Deviation	Percentage of Total Premiums
Fire	15.52	4.43	.066
Allied Lines	24.51	5.66	.024
Homeowners	2.46	5.71	.171
Com. Mult. Peril	28.10	9.32	.041
Workers Comp.	21.95	4.70	.056
Misc. Liability	7.75	3.92	.053
Auto Liability	-5.68	3.43	.349
Auto Physical Dmg.	8.45	5.05	.172
Burglary & Theft	17.76	10.28	.008
TOTAL	5.70	2.42	100.0

TABLE 5

Actual Underwriting Portfolio, Ten Year Interval
Insurer "B"

Line	Underwriting Return	Standard Deviation	Percentage of Total Premiums
Workers Compensation	-4.3125	20.01	.497
Auto Liability	-11.6125	17.42	.205
Auto Physical Damage	-12.99	9.98	.158
Miscellaneous	-17.83	10.37	.139
TOTAL	-9.175	10.896	100.0

Risk and Return: Two Illustrations — Insurer "A" averaged an underwriting return over the interval of 5.7 percent and an underwriting risk, or standard deviation of return, of 2.42 percent. That enviable position contrasts greatly with that of Insurer "B" which averaged a negative return of -9.175 percent with a standard deviation of 10.9 percent. The actual ratio of premiums-to-surplus for Insurer "A" was 2.08 and for Insurer "B", 4.06. Using the same assumptions and formulations as before, it can be shown that the underwriting returns for "A" were sufficiently high and stable that a premiums-to-surplus ratio of 58.14 could have been maintained at a probability of ruin level of .001.* For Insurer "B", the corresponding maximum ratio is only 2.33, a value well below the actual ratio.

Although even the casual observer can identify "B" as the riskier insurer, it is difficult to assess how the risks of the two insurers impact surplus and whether the actual results are close to being "efficient," that is, would reapportionment of the book of business have offered improvement in the risk-return position. Figs. 12 and 13 depict the relevant risk-return points and frontiers of Insurers "A" and "B".

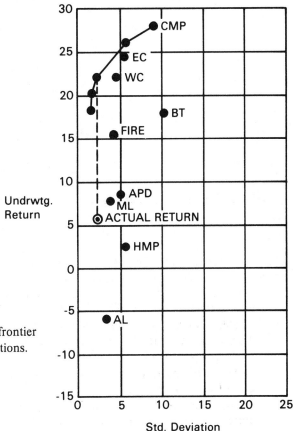

Fig. 12. Efficient underwriting frontier insurer "A" and risk-return positions.

It is important to note that the value of 58.14 is an estimated maximum value and in no way implies that the insurer should have performed in that manner.

References pp. 175-176.

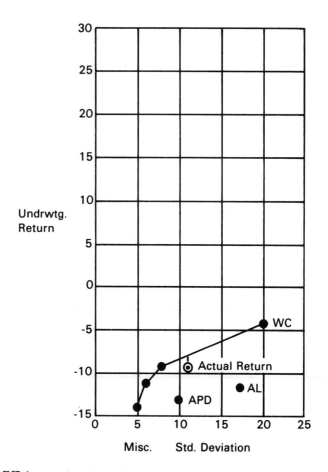

Fig. 13. Efficient underwriting frontier insurer "B" and risk-return positions.

The differences between "A" and "B" raise several questions associated with underwriting capacity and insurer solicity:

1. How do the actual risk and return points of the companies compare to the most efficient or optimal risk and return possibilities; how good or bad is the "book of business" represented by the underwriting portfolio?

2. Assuming that the risk of ruin for an insurance company should be at a relatively low level, how much surplus is required to support the actual underwriting portfolio and how does that compare to the actual surplus position of the companies.

3. What is the minimum surplus associated with the lowest risk position of the underwriting frontier of each company?

The application of portfolio theory to these questions provides additional information to assess the underwriting capacity and solidity of the insurers.

A comparison of Figs. 12 and 13 shows the inferior position of Insurer "B".

Even an optimal reassignment of its past portfolio of risks could not develop a positive return.

The absolute minimum capital level is that associated with the lowest risk value of the frontier; 1.67 for Insurer "A" and 5.11 for Insurer "B". Again, it can be shown that the lowest risk underwriting portfolio (risk = 1.67 and return = 18.43) could be written with no surplus at all. Returns are sufficiently high to absorb the relatively small fluctuations. The optimal reordering of Insurer "B's" portfolio, however, produces a risk value of 10.90 and a negative return of -13.87 percent. Even the minimum risk level would accommodate a maximum premium-to-surplus ratio of 3.37, a value still below that actually written.

The preceding formulations, given the underwriting risk and return values noted, suggest that greater underwriting capacity could have been achieved by a restructuring of the underwriting portfolios. If a lower risk position can be obtained, a greater ratio of premiums to surplus can be written, given the same probability of ruin.

The application of portfolio theory to insurance is not as direct as with the security market, the principal application of the theory to date. Securities are characterized by a market price which is generally unaffected by the number of shares owned. If the holdings of Securities "A" and "B" are altered from 10 and 90 percent of the investment portfolio to 40 and 60 percent, the prices of "A" and "B" are essentially independent of the propostions held. The independence of returns is not likely to be true with insurance underwriting [22].

The concepts of pooling and large numbers suggest that risk or variation is larger with small premium volumes than with large. Thus, if proportions of insurance lines in the total underwriting portfolio are altered, thereby developing different premium volumes, underwriting risk values are likely to be altered so that the historical returns do not necessarily remain as observed. Nonetheless, the model's application does point to those directions of change likely to result in improved efficiency and therefore enlarged underwriting capacity. Moreover, it does focus upon a portfolio while traditional insurance analysis tends to center on single insurance lines and rules-of-thumb. Risk, return, and their interrelationships are not easily accommodated in the latter framework. A portfolio approach, has clear potential for use in the development of underwriting policy, primarily as an information tool for management decision-making.*

OBSERVATIONS AND SUGGESTIONS FOR RESEARCH

The use of the insurance technique helps societies to function more efficiently. Loss costs of individuals and firms are stabilized over time. Individual savings and corporate capital are freed for alternate uses. When insurance is not available or its

---

A few insurers now employ this and related analysis in assessing premium growth targets, and premium volume within lines.

References pp. 175-176.

supply restricted, attempts are made to identify alternate sources of security. Increases in insurance costs promote increased demands for new systems of claim-cost allocation.*

The capacity restrictions directly impact risk-retention policies of firms and some individuals. Risk-retention exacerbates cash-flow and asset management problems, particularly for firms with large complex risk exposures. At the personal level, individuals may choose to do without insurance, sometimes in violation of the law. Corporations may operate insurance with larger than normal deductibles or with inadequate coverage. They may also form their own "captive" insurance companies with the hope of gaining financial or marketplace advantages.**

The availability of insurance capacity is affected by underwriting decisions but basically it is linked to stability and profitability of underwriting results. Capacity can likely improve, however, by efficient management and balancing of underwriting portfolios. Portfolio theory offers considerable insight and application towards this end. The continued use of rules-of-thumb by regulatory authorities to control premiums-to-surplus relationships ignores size and portfolio effects and probably contributes to inefficient use of insurer capital.

Insurance capacity restrictions prompt changing relations between private and public sectors. Capacity restrictions on the private insurance market have led to guarantees, subsidies, and government insurance. Demands for the "riskless society" [1] have placed new pressures on the private insurance system and expanded the number of statutes oriented towards protection and safety and their administration by public bodies.

Despite numerous and useful efforts to apply portfolio theory to insurance underwriting, considerable work remains. It is important to identify the function relating premium volume and variation in underwriting results more clearly.*** Improvements in the efficiency of underwriting portfolios cannot achieve the potential offered through portfolio theory until this relationship is better understood.

Management nearly always can be improved by better understanding of decision processes. The Kunreuther study [6] strongly suggested that traditional models of decision-making under uncertainty may not be appropriate for many insurance purchase decisions. At least equal doubt surrounds their applicability to the insurance-selling decisions of underwriters, especially where reliable decision information is scant or absent.

---

*As noted previously, this situation now prevails in personal automobile insurance. Urban dwellers pay higher rates than their suburban and rural neighbors. They argue that the location factor of the rating classification system is unfair and, if not eliminated, should at least be restricted in its impact. Pricing equity concerns are much more dominant in class-related markets than with individual risk pricing.

**Bermuda is the principal domicile of corporate captives with estimates suggesting over 700 with home offices there. Recent rulings by the Internal Revenue Service have made it more difficult for parent firms to deduct premiums to a captive insurer. Marketplace advantages, however, may still be present. Captives, for example, have direct access to the world-wide reinsurance market.

***Some additional size effect evidence is offered in [25].

To the extent that traditional utility analysis and other choice models offer inadequate or inconsistent explanations of insurance or risk-taking behavior, analysis occurs without a reliable frame of reference. The academic gain from further examination of insurance decisions appears large and so does the potential improvement in the underwriting decision process likely to be associated with that line of inquiry. The analysis of choices leading to underinsurance, for example, may lend additional support for the Kunreuther evidence. The insurance and risk-retention decisions of corporate risk managers may offer interesting parallels.

REFERENCES

1. J. L. Athearn, "The Riskless Society," The Journal of Risk and Insurance, 565-573, December 1978.
2. I. Pfeffer and D. R. Klock, Perspectives on Insurance, Prentice-Hall, Englewood Cliff, New Jersey, 3-13, 1974.
3. C. A. Kulp, Casualty Insurance, Ronald Press, New York, 10, 1957.
4. I. Pfeffer, Insurance and Economic Theory, Richard D. Irwin, Homewood, Illinois, 5, 1955.
5. C. A. Williams, Jr. and R. M. Heins, Risk Management and Insurance, Third Edition, McGraw-Hill Book Co, New York, 12-16, 1976.
6. H. Kunreuther, Disaster Insurance Protection Public Policy Lessons, John Wiley & Sons, New York, 1978.
7. B. A. Schweig, An Analysis of the Effectiveness of Products Liability Underwriters, unpublished doctoral dissertation, University of Pennsylvania, 1977. (Forthcoming as a monograph of the S. S. Huebner Foundation.)
8. C. A. Kulp and J. W. Hall, Casualty Insurance, Fourth Edition, Ronald Press, New York, 43-45, 1968.
9. H. Skipper, Jr., "The Privacy Implications of Insurers' Information Practices," The Journal of Risk and Insurance, 9-32, 1969.
10. D. B. Houston, "Risk, Insurance, and Sampling," The Journal of Risk and Insurance, 535-538, December 1964.
11. H. Buhlmann, Mathematical Methods in Risk Theory, Springer-Verlag, New York, 3-34, 1970.
12. J. D. Hammond, principal investigator, A. F. Shapiro and N. Shilling, The Regulation of Insurer Solidity Through Capital and Surplus Requirements, APR 75-16550 A01.
13. W. Eichorn, "The Various Forms of Underwriting Risk: Definitions, Comparisons, and Examples," Proceedings of the XV International Insurance Seminar, 41-52, June 1979.
14. A. Tversky and D. Kahneman, "Judgment Under Uncertainty: Heuristics and Biases," Science, 185: 1124-1131, 1974.
15. J. G. March, "Bonded Rationality, Ambiguity, and the Engineering of Choice," The Bell Journal of Economics, 9: No. 2., 587-608, 1978.
16. H. A. Simon, "On How to Decide What to Do," The Bell Journal of Economics, 9: No. 2, 494-507, 1978.
17. C. R. Smith, A Study to Analyze and Model the Decision-Making Process of Automobile Insurance Underwriters. Unpublished doctoral dissertation, the Pennsylvania State University, 1972.
18. R. O. Swalm, "Utility Theory — Insights Into Risk-Taking," Harvard Business Review, 123-136, November-December 1977.

19. J. D. Hammond, A. F. Shapiro, and N. Shilling, *"Analysis of an Underwriting Decline,"* CPCU Journal, 66-74, June 1979.
20. J. R. Ferrari, *"A Theoretical Portfolio Selection Approach for Insuring Property and Liability Lines,"* Proceedings of the Casualty Actuarial Society, 33-54, 1967.
21. J. E. Bachman, Capitalization Requirements for Multiple Line Property-Liability Companies, Huebner Foundation Monograph No. 6., Richard D. Irwin, Homewood, Illinois, 1978.
22. J. D. Hammond and N. Shilling, *"Some Relationships of Portfolio Theory to the Regulation of Insurer Solidity,"* Journal of Risk and Insurance, September 1978; also reprinted by the American Bar Foundation as Research Contributions of the American Bar Foundation, No. 2, 1978.
23. A. E. Hofflander, *"Minimum Capital and Surplus Requirements for Multiple Line Insurance Companies: A New Approach,"* Insurance, Government and Social Policy: Studies in Insurance Regulation, S. L. Kimball and H. S. Denenberg, eds., S. S. Hueber Foundation, Richard D. Irwin, Homewood, Illinois, 80-88, 1969.
24. H. M. Markowitz, Portfolio Selection: Efficiency Diversification of Investment, John Wiley & Sons, New York, 1959; W. F. Sharpe, Portfolio Theory and Capital Markets, McGraw-Hill, New York, 1970.
25. J. D. Hammond and A. F. Shapiro, *"Capital Requirements for Entry Into Property and Liability Underwriting: An Empirical Examination,"* paper presented at the Boston Federal Reserve Bank Conference on Regulation of Financial Institutions, October 4 & 5, 1979.

DISCUSSION

W. Lowrance *(Stanford University)*

I've always wondered whether there were incentives for insurers to encourage or assist their insureds in reducing risks, that is, somebody who owns a big building or something of that sort. Is it simply that in the matter of internal competition for the contract, firms try to offer some risk reduction advice and so on?

J. D. Hammond

Yes, I think I can comment on that. It depends a lot on the size of the premium. If the premium is large enough, loss prevention activities are much more likely to be undertaken, not only by insurance companies but by large insurange brokerage firms as well. There is a gentlemen here who represents a brokerage firm and he may comment upon what they do. For you and I with our homeowners insurance we may be fortunate to receive a calendar in the mail one a year. And then there are some segments of the insurance industry which actually make a living, so to speak, selling loss prevention. Such companies are highly selective as to who they insure. Because they try to select only the best risks, their loss ratio is low. They can therefore spend a lot more money on loss prevention. For folks such as you and I, however, there is simply not enough premium money to do it. Perhaps the gentlemen from Johnson and Higgins can also comment.

E. V. Anderson *(Johnson and Higgins)*

Well, there is the factor of experience rating and Joe Ferriera from MIT can also explain the economics of that technique.

J. D. Hammond

J. D. Hammond

That's right. In many risks, workmen's compensation is a prominent example, experience rating is common and so there is a direct incentive from the firm to keep losses down. However, that is an expenditure by the firm and not by the insurer.

W. C. Chapman *(GM Industry-Govt. Relations Staff)*

There are some of us in this room who are confused as to why Allstate, for instance, tried to lower premiums for heavier bumpers as opposed to lighter ones. Most of the data we have says lighter ones are better. Is there an explanation as to why insurance companies would do that on the basis of data which shows the experience to run the other way?

J. D. Hammond

There may be from someone, but not from me.

W. A. Wallace *(Rensselaer Polytechnic Inst.)*

Have you given any thought about what's coming up in the way of public insurance funds, such as the super fund on oil and the international funds on what we might call catastrophic losses? Do those kinds of activities have relationships between the variance and the mean that would make the kind of analysis we're doing today not possible?

J. D. Hammond

That's right. It won't always be possible. The kind of analysis which I described breaks down because of inadequate data. I have not tried to extend my work into that and so I really have nothing to say that would be anything beyond totally speculative. There are some things which are related to that that I thought you were going to ask which are also speculative but I think they are interesting. I didn't feel free to speculate about them in the paper. What I thought that you perhaps were going to get into is the presence of guaranty funds, which every state now has and the implication of that for solvency analysis of companies. The concern, although it's still too early for empirical people like me to analyze, is that the presence of guaranty funds may cause supervision to become more lax because perhaps regulation pays less attention to solvency. It may be that the chance of insurer insolvency increases because the cost of failure is simply passed on to everybody else in the industry. Failure is not a small concern. Everybody in the room is aware of GEICO, a large, large company that almost went under a few years ago. In the decade ending with 1975 there were about 170 failures. That does not even include those insurers which may have come close to failure and that were rehabilitated which were acquired by or merged into other firms. And so it's a signficant concern.

A. Kneese (*Resources for the Future*)

I just wanted to say that as I was listening to these stimulating talks today a paradee of MacArthur's famous parting remarks occurred to me and it says, "Old scholars who've run out of fresh ideas in their field never die, they become conference chairmen."

SESSION III
"ACCEPTABILITY" IN A DEMOCRACY — WHO SHALL DECIDE?

Session Chairman
William D. Rowe

The American University
Washington, D. C.

INTRODUCTORY REMARKS
SESSION III

William D. Rowe

The American University
Washington, D.C.

The session we are to address this morning is called "Acceptability" in a Democracy — Who Shall Decide? Well, first it seems to me that there is no such thing as a decision maker for our decision process in the United States. There are a multitude of decision makers and in many cases I dare you to find one who will take the responsibility for a decision. In fact, there are people who affect decisions all the way up and down the line, from the engineer who decides what to include and what not include in environmental impact analysis all the way down to the public at the end of the process; there are many players throughout the process. The problem is that anyone of them can say no and it takes a preponderance of yesses to really overcome a single no. So, in my opinion, we really can't say that we have such a thing as a single decision maker; but, perhaps, a decision making process or, worse, a decision inhibiting process, I'm not sure which.

There were some other remarks I was going to make, but since we're behind I'm going to keep them to a minimum. I did want to point out in response to yesterday's discussion on decision analysis that one man's utility function may be another man's anathema. The best way to illustrate this is a little story: In recent years an explorer in the wilds of Borneo discovered a white, female, albino gorilla and brought it back into captivity. It became the hit of the zoo and people were lining up to see it. The zoo never had so much business before. After several years the zoo found that the gorilla was getting a little older, and they thought that they would like to perpetuate the attraction, so they went on a major worldwide hunt for a male albino gorilla. After many years they were unable to discover one. However, the scientists, a geneologist, suggested that being an ape, white, and albino the ape may be very close to man. Maybe it would be possible to mate it with man at the lower end of the human scale. They put an ad in the paper indicating that the mating with this white albino gorilla was sought, and they posted a fee of $1,000. For weeks there were no takers and finally a young man with an ethnic background whose origin we'll leave unnamed came in. He said, "I'll be glad to mate with the gorilla, but I have three requests first that you must honor." "All right. What's the first one?" He responded, "First, no kissing on the lips." "O.K., what's the next one?" "Secondly, any children will have to be raised catholic and lastly, I can't do it for four weeks." "Why not?" He replied, "It will take me that long to raise the $1,000."

FACTS AND FEARS:
UNDERSTANDING PERCEIVED RISK

Paul Slovic, Baruch Fischhoff and Sarah Lichtenstein

Decision Research, A Branch of Perceptronics,
Eugene, Oregon

ABSTRACT

Subjective judgments, whether by experts or lay people, are a major component in any risk assessment. If such judgments are faulty, efforts at public and environmental protection are likely to be misdirected. The present paper begins with an analysis of biases exhibited by lay people and experts when they make judgments about risk. Next, the similarities and differences between lay and expert evaluations are examined in the context of a specific set of activities and technologies. Finally, some special issues are discussed, including the difficulty of reconciling divergent opinions about risk, the possible irrelevance of voluntariness as a determinant of acceptable risk, the importance of catastrophic potential in determing perceptions and triggering social conflict, and the need to facilitate public participation in the management of hazards.

INTRODUCTION

People respond to the hazards they perceive. If their perceptions are faulty, efforts at public and environmental protection are likely to be misdirected. For some hazards, extensive statistical data are readily available; for example, the frequency and severity of motor vehicle accidents are well documented. The hazardous effects of other familiar activities, such as the consumption of alcohol and tobacco, are less readily discernible; their assessment requires complex epidemiological and experimental studies. However, even when statistical data are plentiful, the "hard" facts can only go so far towards developing policy. At some point human judgment is needed to interpret the findings and determine their relevance.

Still other hazards, such as those associated with recombinant DNA research or nuclear power, are so new that risk assessment must be based on complex theoretical analyses such as fault trees (see Fig. 1), rather than on direct experience. Despite their sophistication, these analyses, too, include a large component of judgment. Someone, relying on educated intuition, must determine the structure of the problem, the consequences to be considered, and the importance of the various branches of the fault tree.

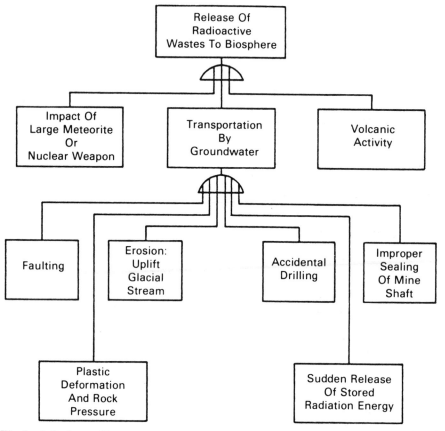

Fig. 1. A fault tree illustrates ways in which radioactive wastes might be released from a nuclear waste repository in bedded salt. Source: P. E. McGrath, *"Radioactive Waste Management,"* Report EURFNR 1204, Karlsruhe, Germany, 1974.

Once the analyses have been performed, they must be communicated to the various people who actually manage hazards, including industrialists, environmentalists, regulators, legislators, and voters. If these people do not see, understand, or believe these risk statistics, then distrust, conflict and ineffective hazard management are likely.

In this paper, we shall explore some of the psychological elements of the risk assessment process that are critical to the management of hazards. Our basic premises are that both the public and the experts are necessary participants in that process, that assessment is inevitably subjective, and that understanding judgmental limitations is crucial to effective decision making.

JUDGMENTAL BIASES IN RISK PERCEPTION

When laypeople are asked to evaluate risks, they seldom have statistical evidence on hand. In most cases, they must rely on inferences based on what they remember hearing or observing about the risk in question. Recent psychological research has identified a number of very general inferential rules that people seem to use in such situations [1]. These judgmental rules, known technically as *heuristics*, are employed to reduce difficult mental tasks to simpler ones. Although they are valid in some circumstances, in others, they lead to large and persistent biases with serious implications for risk assessment.

Availability — One heuristic that has special relevance for risk perception is called *availability* [2]. People using this heuristic judge an event as likely or frequent if instances of it are easy to imagine or recall. Frequently occurring events are generally easier to imagine and recall than are rare events. Thus availability is often an appropriate cue. However, availability is also affected by numerous factors unrelated to frequency of occurrence. For example, a recent disaster or a vivid film such as "Jaws" could seriously distort risk judgments.

Availability-induced errors are illustrated by several recent studies in which we asked college students and members of the League of Women Voters to judge the frequency of various causes of death such as smallpox, tornadoes and heart disease [3]. In one study, these people were told the annual death toll for motor vehicle accidents in the United States (50,000) and asked to estimate the frequency of 40 other causes of death. In another study, participants were asked to judge which of two causes of death was more frequent. Both studies showed the judgments to be moderately accurate in a global sense; i.e., people usually knew which were the most and least frequent lethal events. Within this global picture, however, there was evidence that people made serious misjudgments, many of which seemed to reflect availability bias.

To illustrate, Fig. 2 compares the judged number of deaths per year with the actual number as reported in public health statistics. If the frequency judgments were accurate, they would equal the statistical rates, and all data points would fall on the 45° line. While more likely hazards generally evoked higher estimates, the points are best described as being scattered about a curved line that lies sometimes above and sometimes below the line of accurate judgment. In general, rare causes of death were overestimated and common causes of death were underestimated. As a result, while the actual death toll varied over a range of one million, average frequency judgments varied over a range of only a thousand.

In addition to this general bias, many important specific biases were evident. For example, accidents were judged to cause as many deaths as diseases, whereas diseases actually take about 15 times as many lives. Homicides were incorrectly judged more frequent than diabetes and stomach cancer. Homicides were also judged to be about as frequent as stroke, although the latter actually claims about 11 times as many lives. Frequencies of death from botulism, tornadoes and pregnancy (including childbirth and abortion) were also greatly overestimated.

Table 1 lists the lethal events whose frequencies were most poorly judged in our studies. In keeping with availability considerations, overestimated items were dra-

References pp. 212-214.

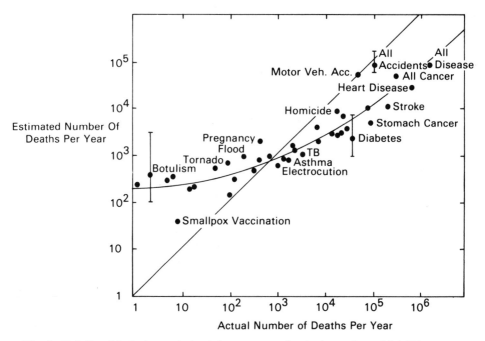

Fig. 2. Relationship between judged frequency and actual number of fatalities per year for 41 causes of death. Source: Lichtenstein et al. [3].

TABLE 1

Bias in Judged Frequency of Death

Most Overestimated	Most Underestimated
All accidents	Smallpox vaccination
Motor vehicle accidents	Diabetes
Pregnancy, childbirth, and abortion	Stomach cancer
Tornadoes	Lightning
Flood	Stroke
Botulism	Tuberculosis
All cancer	Asthma
Fire and flames	Emphysema
Venomous bite or sting	
Homicide	

matic and sensational, whereas underestimated items tended to be unspectacular events, which claim one victim at a time and are common in nonfatal form.

The availability heuristic points up the vital role of experience as a determinant of perceived risk. If one's experiences are biased, one's perceptions are likely to be inaccurate. Unfortunately, much of the information to which people are exposed provides a distorted picture of the world of hazards. As a follow-up to the studies reported above, we have examined the reporting of causes of death in two newspapers on opposite coasts of the United States [4]. We found that the two newspapers were similar in their reporting habits and both were biased in their coverage of life-threatening events, that is, the number of reports was not closely related to statistical frequencies of occurrence. All forms of disease appeared to be relatively neglected whereas violent, often catastrophic events such as tornadoes, fires, drownings, homicides, and accidents were reported disproportionately often. For example, even though all diseases take about 100 times as many lives as do homicides, there were about three times as many articles about homicides as about all diseases. Perhaps most interesting was the strong similarity between the amount of newspaper coverage given to a particular cause of death and the judgmental biases shown in Fig. 2 and Table 1.

In the public arena, the availability heuristic has a particularly important implication: the biasing effects of memorability and imaginability may pose a barrier to open discussions of risk. Consider an engineer demonstrating the safety of subterranean nuclear waste disposal by pointing out the improbability of each branch of the fault tree in Fig. 1. Rather than reassuring the audience, the presentation could encourage such thoughts as: "I didn't realize there were so many things that could go wrong." Simply discussing a low-probability hazard may increase its judged probability regardless of what the evidence indicates.

In other situations, lack of availability may lull people into complacency. In a recent study [5], we presented people with various versions of a fault tree showing the "risks" to starting a car. Participants were asked to judge the completeness of the representation (reproduced in Fig. 3). Their estimate of the proportion of no-starts falling in the category labeled "all other problems" was about the same when looking at the full tree of Fig. 3 or at versions in which half of the branches were deleted. Such pruning should have dramatically increased the judged likelihood of "all other problems." However, it did not. In keeping with the availability heuristic, what was out of sight was effectively out of mind.

Overconfidence — A particularly pernicious aspect of heuristics is that people are typically very confident in judgments based upon them. For example, in a follow-up to the study of causes of death, participants were asked to indicate the odds that they were correct in choosing the more frequent of two lethal events [6]. Odds of 100 : 1 or greater were given frequently (25% of the time). However, about one out of every eight answers associated with such extreme confidence was wrong (fewer than 1 in 100 would have been wrong if the odds had been appropriate). About 30% of the judges gave odds greater than 50 : 1 to the incorrect assertion that homicides are more frequent than suicides. The psychological basis for this unwarranted certainty seems to be people's insensitivity to the tenuousness of the assumptions upon which their judgments are based (in this case, they may have assumed that frequent media reports of homicide were valid reflections of relative frequency). Such overconfidence can keep us from realizing how little we know and how much additional information is needed about the various problems and risks we face.

References pp. 212-214.

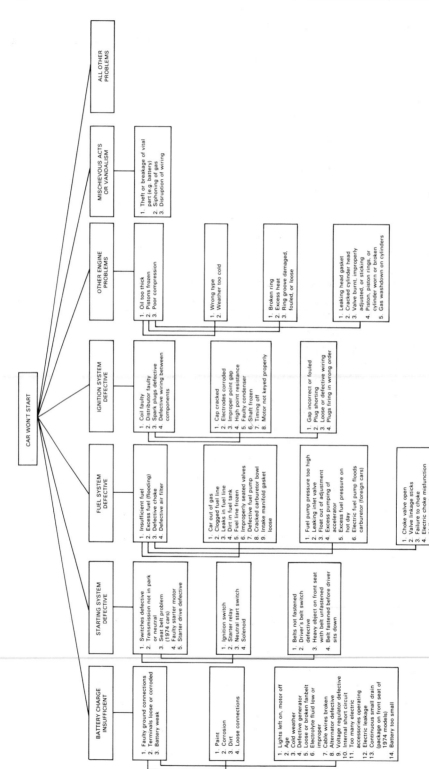

Fig. 3. A fault tree indicating the ways in which a car might fail to start. Source: Fischhoff, Slovic & Lichtenstein [5].

Overconfidence manifests itself in other ways as well. A typical task in estimating uncertain quantities like failure rates is to set upper and lower bounds such that there is a 98% chance that the true value lies between them. Experiments with diverse groups of people making many different kinds of judgments have shown that, rather than 2% of true values falling outside the 98% confidence bounds, 20-50% do so [7]. Thus, people think that they can estimate such values with much greater precision than is actually the case.

Unfortunately, experts seem as prone to overconfidence as laypeople. When the "fault tree" study described above was repeated with a group of professional automobile mechanics, they, too, were insensitive to how much had been deleted from the tree [5]. Hynes and VanMarcke [8] asked seven "internationally known" geotechnical engineers to predict the height of an embankment that would cause a clay foundation to fail and to specify confidence bounds around this estimate that were wide enough to have a 50% chance of enclosing the true failure height. None of the bounds specified by these experts actually did enclose the true failure height.

The 1976 collapse of the Teton Dam provides a tragic case in point. The Committee on Government Operations has attributed this disaster to the unwarranted confidence of engineers who were absolutely certain they had solved the many serious problems that arose during construction. Indeed, in routine practice, failure probabilities are not even calculated for new dams even though about 1 in 300 fails when the reservoir is first filled.

Further anecdotal evidence of overconfidence may be found in many other technical risk assessments [9]. Some common ways in which experts may overlook or misjudge pathways to disaster include the following:

- Failure to consider the ways in which human errors can affect technological systems. Example: Due to inadequate training and control room design, operators at Three Mile Island repeatedly misdiagnosed the condition of the reactor and took inappropriate corrective actions. A minor incident thus became a major accident.
- Overconfidence in current scientific knowledge. Example: The failure to recognize the harmful effects of X rays until societal use has become widespread and largely uncontrolled.
- Insensitivity to how technological systems function as a whole. Example: The rupture of a liquid natural gas storage tank in Cleveland in 1944 resulted in 128 deaths, largely because no one had appreciated the need for a dike to contain spillage. The DC10 failed in several early flights because its designers had not realized that decompression of the cargo compartment would destroy vital parts of the plane's control system which ran through it.
- Slowness in detecting chronic, cumulative environmental effects. Example: Although accidents to coal miners have long been recognized as one cost of operating fossil-fueled plants, the effects of acid rains on ecosystems were slow to be discovered.
- Failure to anticipate human response to safety measures. Example: The partial protection afforded by dams and levees gives people a false sense of security and promotes development of the flood plain. When a rare flood does exceed the capacity of the dam, the damage may be considerably greater

References pp. 212-214.

than it would have been had the flood plain been unprotected.

- Failure to anticipate "common-mode failures." Example: Because electrical cables controlling the multiple safety systems of the reactor at Browns Ferry, Alabama, were not spatially separated, all five emergency core cooling systems were incapacitated by a single fire.

Desire for Certainty — Every technology is a gamble of sorts and, like other gambles, its attractiveness depends on the probability and size of its possible gains and losses. Both scientific experiments and casual observation show that people have difficulty thinking about and resolving the risk/benefit conflicts even in simple gambles. One way to reduce the anxiety generated by confronting uncertainty is to deny that uncertainty. Such denial represents another source of overconfidence in addition to those described earlier. Denial is illustrated by people faced with natural hazards who often view their world as either perfectly safe or predictable enough to preclude worry. Thus, some flood victims interviewed by Kates [10] flatly denied that floods could ever recur in their area. Some thought (incorrectly) that new dams and reservoirs in the area would contain all potential floods, while others attributed previous floods to freak combinations of circumstances unlikely to recur. Denial, of course, has its limits. Many people feel that they cannot ignore the risks of nuclear power. For these people, the search for certainty is best satisfied by outlawing the risk.

Scientists and policy makers are often resented for the anxiety they provoke when they point out the gambles inherent in societal decisions. Borch [11] noted that corporate managers get annoyed with consultants who give them the probabilities of possible events instead of telling them exactly what will happen. Just before hearing a blue-ribbon panel of scientists report being 95% certain that cyclamates do not cause cancer, former Food and Drug Administration Commissioner Alexander Schmidt said, "I'm looking for a clean bill of health, not a wishy-washy, iffy answer on cyclamates" [12]. Senator Muskie has called for "one-armed" scientists who do not respond "on the one hand, the evidence is so, but on the other hand . . . " when asked about the health effects of pollutants [13].

It Won't Happen to Me — Another important and potentially tragic form of overconfidence is the fact that people tend to consider themselves personally immune to many hazards whose societal risks they would readily acknowledge. In a report titled, "Are We All Among the Better Drivers?", Svenson [14] showed that most people tend to rate themselves as among the most skillful and safe drivers in the population. This effect does not seem to be limited just to driving. Rethans [15] found that people rated their own personal risk from each of 29 hazardous consumer products (e.g., knives, hammers) as lower than the risk to others in the society. In an extension of Svenson's study, ninety-seven percent of Rethans' respondents judged themselves average or above average in their ability to avoid both bicycle and power mower accidents. Weinstein [16] found that people were unrealistically optimistic when evaluating the chances that a wide variety of good and bad life events (e.g., living past 80, having a heart attack) would happen to them.

Although the determinants of such personal optimism are not well understood, we believe that several contributing factors can be identified. First, the hazardous activities for which personal risks are underestimated tend to be seen as under the individual's control. Second, they tend to be familiar hazards whose risks are low enough that the individual's personal experience is overwhelmingly benign. Automobile driving is a prime example of such a hazard. Despite driving too fast, tailgating, etc., poor drivers make trip after trip without mishap. This personal experience demonstrates to these drivers their exceptional skill and safety. Moveover, their indirect experience via the media shows them that accidents do happen — to others. Given such misleading experiences, people may feel quite justified in refusing to take protective actions such as wearing seat belts [17].

Reconciling Divergent Opinions about Risk — Both observation of risk debates and empirical data to be presented below suggest that experts and laypeople have quite different perceptions about the riskiness of various technologies. Given that they are prisoners of rather different experiences, such divergence is to be expected. One would like to believe that, as evidence accumulates and the public and the experts come to share a common experience, their perceptions would converge towards one "appropriate" view. Unfortunately, this is not likely to be the case. A great deal of research indicates that once formed, people's beliefs change very slowly, and are extraordinarily persistent in the face of contrary evidence [18]. Initial impressions tend to structure the way that subsequent evidence is interpreted. New evidence appears reliable and informative if it is consistent with one's initial belief, whereas contrary evidence is dismissed as unreliable, erroneous, or unrepresentative. Thus, depending on one's predispositions, intense effort to reduce a hazard may be interpreted as meaning either that the risks are great or that the technologists are responsive to public concerns. Similarly, whereas opponents of nuclear power viewed the accident at Three Mile Island as proof that reactors are unsafe, proponents claimed that it demonstrated the effectiveness of the multiple safety and containment systems. Nelkin's case history of a nuclear-siting controversy [19] provides a good example of the inability of technical arguments to change opinions. In that debate, each side interpreted technical ambiguities in ways that reinforced its own position.

DETERMINANTS OF PERCEIVED RISK

In order to aid the hazard management process, a theory of perceived risk must explain people's extreme aversion to some hazards, their indifference to others, and the discrepancies between these reactions and experts' recommendations. Why, for example, do some communities react vigorously against locating a liquid natural gas terminal in their vicinity despite the assurances of experts that it is safe? Why, on the other hand, do many communities situated on earthquake faults or below great dams show little concern for experts' warnings? Such behavior is doubtless related to the perceived probability of possible consequences from these hazards. The role of judgmental processes in probability assessment was discussed in the preceding section. The studies reported below broaden the discussion. They ask, when people judge the risk inherent in a technology, are they referring only to the (possibly misjudged) number of people it could kill or also to other, more qualitative features of the risk it entails?

Quantifying Perceived Risk — In one study, we asked four different groups of people to rate 30 activities (e.g., smoking, fire fighting), substances (e.g., food coloring), and technologies (e.g., railroads, aviation) according to the present risk of death from each [20, 21]. Three groups were from Eugene, Oregon; they included 30 college students, 40 members of the League of Women Voters (LOWV), and 25 business and professional members of the "Active Club." The fourth group was composed of 15 persons selected nationwide for their professional involvement in risk assessment. This "expert" group included a geographer, an environmental policy analyst, an economist, a lawyer, a biologist, a biochemist, and a government regulator of hazardous materials.

All these people were asked, for each of the 30 items, "to consider the risk of dying (across all U.S. society as a whole) as a consequence of this activity or technology." In order to make the evaluation task easier, each activity appeared on a 3" x 5" card. Respondents were told first to study the items individually, thinking of all the possible ways someone might die from each (e.g., fatalities from non-nuclear electricity were to include deaths resulting from the mining of coal and other energy production activities as well as electrocution; motor vehicle fatalities were to include collisions with bicycles and pedestrians). Next, they were to order the items from least to most risky and, finally, to assign numerical risk values by giving a rating of 10 to the least risky item and making the other ratings accordingly. They were also given additional suggestions, clarifications and encouragement to do as accurate a job as possible.

Table 2 shows how the various groups ranked these 30 activities and technologies according to riskiness. There were many similarities between the three groups of laypeople. For example, each group believed that motorcycles, motor vehicles and handguns were highly risky, while vaccinations, home appliances, power mowers, and football posed relatively little risk. However, there were strong differences as well. Active Club members viewed pesticides and spray cans as relatively much safer than did the other groups. Nuclear power was rated as highest in risk by the LOWV and student groups, but only eighth by the Active Club. The students viewed contraceptives as riskier and mountain climbing as safer than did the other lay groups. Experts' judgments of risk differed markedly from the judgments of laypeople. The experts viewed electric power, surgery, swimming and X rays as more risky than did the other groups and they judged nuclear power, police work and mountain climbing to be much less risky.

What Determines Risk Perception? — What did people mean, in this study, when they said that a particular technology was quite risky? A series of additional studies was conducted to answer this question.

Perceived Risk Compared to Frequency of Death — When people judge risk, are they simply estimating frequency of death? To answer this question, we collected the best available technical estimates of the annual number of deaths for the activities included in our study. For some, such as commercial aviation and

TABLE 2

Ordering of Perceived Risk
for 30 Activities and Technologies[a]

	Group 1 LOWV	Group 2 College Students	Group 3 Active Club Members	Group 4 Experts
Nuclear power	1	1	8	20
Motor vehicles	2	5	3	1
Handguns	3	2	1	4
Smoking	4	3	4	2
Motorcycles	5	6	2	6
Alcoholic beverages	6	7	5	3
General (private) aviation	7	15	11	12
Police work	8	8	7	17
Pesticides	9	4	15	8
Surgery	10	11	9	5
Fire fighting	11	10	6	18
Large construction	12	14	13	13
Hunting	13	18	10	23
Spray cans	14	13	23	26
Mountain climbing	15	22	12	29
Bicycles	16	24	14	15
Commercial aviation	17	16	18	16
Electric power	18	19	19	9
Swimming	19	30	17	10
Contraceptives	20	9	22	11
Skiing	21	25	16	30
X rays	22	17	24	7
High school & college football	23	26	21	27
Railroads	24	23	20	19
Food preservatives	25	12	28	14
Food coloring	26	20	30	21
Power mowers	27	28	25	28
Prescription antibiotics	28	21	26	24
Home appliances	29	27	27	22
Vaccinations	30	29	29	25

[a]*The ordering is based on the geometric mean risk ratings within each group. Rank 1 represents the most risky activity or technology.*

handguns, there is good statistical evidence based on counts of known victims. For others, such as nuclear or fossil-fuel power plants, available estimates are based on uncertain inferences about incompletely understood processes, such as the effect of low doses of radiation on latent cancers. For still others, such as food coloring, we could find no estimates of annual fatalities.

For the 25 cases for which we found technical fatality estimates, we compared these estimates with perceived risk. The experts' judgments of risk were so closely

related to these statistical or calculated frequencies that it seems reasonable to conclude that they both knew what the technical estimates were and viewed the risk of
an activity or technology as synonymous with them. The risk judgments of laypeople, however, were only moderately related to the annual death rates, raising the
possibility that, for them, risk may not be synonymous with fatalities. In particular,
the perceived risk of nuclear power was remarkably high compared to its estimated
number of fatalities.

Lay Fatality Estimates — Before concluding that perceived risk does not mean
annual fatalities, we investigated the possibility that laypeople based their risk
judgments on subjective fatality estimates which were inaccurate. To test this
hypothesis, we asked additional groups of students and LOWV members "to estimate how many people are likely to die in the U.S. in the next year (if the next year
is an average year) as a consequence of these 30 activities and technologies."
    These subjective fatality estimates are shown in columns 2 and 3 of Table 3. If
laypeople really equate risk with annual fatalities, their own estimates of annual
fatalities, no matter how inaccurate, should be very similar to their judgments of
risk. There was, however, only a low to moderate agreement between these two
sets of judgments (r = .60 for LOWV and .26 for students). Of particular importance was nuclear power, which had the *lowest* fatality estimate and the *highest* perceived risk for both LOWV members and students. Overall, laypeople's risk perceptions were no more closely related to their own fatality estimates than they were
to the technical estimates. Thus we can reject the idea that laypeople wanted to
equate risk with annual fatalities, but were inaccurate in doing so. Apparently,
laypeople incorporate other considerations besides annual fatalities into their concept of risk.

Disaster Potential — One striking result is the fact that the LOWV members and
students assigned nuclear power the highest risk values and the lowest annual
fatality estimates. One possible explanation is that they expected nuclear power to
have a low death rate in an average year, but considered it to be a high risk technology because of its potential for disaster.
    In order to understand the role played by expectations of disaster in determining
lay people's risk judgments, we asked these same respondents to indicate for each
activity and technology "how many times more deaths would occur if next year
were particularly disastrous rather than average." The geometric means of these
multipliers are shown in columns 4 and 5 of Table 3. For most activities, people
saw little potential for disaster. The striking exception is nuclear power, with a
mean disaster multiplier in the neighborhood of 100.
    For any individual, an estimate of the expected number of fatalities in a dis-
astrous year could be obtained by applying the disaster multiplier to the estimated
fatalities for an average year. When this was done for nuclear power, almost 40% of
the respondents expected more than 10,000 fatalities if next year were a disastrous
year. More than 25% expected 100,000 or more fatalities. An additional study [22],
in which people were asked to describe their mental images of the consequences of
a nuclear accident, showed an expectation that a serious accident would likely

TABLE 3

Fatality Estimates and Disaster Multipliers
for 30 Activities and Technologies

Activity or Technology	Techincal Fatality Estimates	Geometric Mean Fatality Estimates Average Year		Geometric Mean Multiplier Disastrous Year	
		LOWV	Students	LOWV	Students
1. Smoking	150,000	6,900	2,400	1.9	2.0
2. Alcoholic beverages	100,000	12,000	2,600	1.9	1.4
3. Motor vehicles	50,000	28,000	10,500	1.6	1.8
4. Handguns	17,000	3,000	1,900	2.6	2.0
5. Electric power	14,000	660	500	1.9	2.4
6. Motorcycles	3,000	1,600	1,600	1.8	1.6
7. Swimming	3,000	930	370	1.6	1.7
8. Surgery	2,800	2,500	900	1.5	1.6
9. X rays	2,300	90	40	2.7	1.6
10. Railroads	1,950	190	210	3.2	1.6
11. General (private) aviation	1,300	550	650	2.8	2.0
12. Large construction	1,000	400	370	2.1	1.4
13. Bicycles	1,000	910	420	1.8	1.4
14. Hunting	800	380	410	1.8	1.7
15. Home appliances	200	200	240	1.6	1.3
16. Fire fighting	195	220	390	2.3	2.2
17. Police work	160	460	390	2.1	1.9
18. Contraceptives	150	180	120	2.1	1.4
19. Commercial aviation	130	280	650	3.0	1.8
20. Nuclear power	100[a]	20	27	107.1	87.6
21. Mountain climbing	30	50	70	1.9	1.4
22. Power mowers	24	40	33	1.6	1.3
23. High school & college football	23	39	40	1.9	1.4
24. Skiing	18	55	72	1.9	1.6
25. Vaccinations	10	65	52	2.1	1.6
26. Food coloring	—[b]	38	33	3.5	1.4
27. Food preservatives	—[b]	61	63	3.9	1.7
28. Pesticides	—[b]	140	84	9.3	2.4
29. Prescription antibiotics	—[b]	160	290	2.3	1.6
30. Spray cans	—[b]	56	38	3.7	2.4

[a] Technical estimates for nuclear power were found to range between 16 and 600 annual fatalities. The geometric mean of these estimates was used here.

[b] Estimates were unavailable.

result in hundreds of thousands, even millions, of immediate deaths. These extreme estimates can be contrasted with the Reactor Safety Study's conclusion that the maximum credible nuclear accident, coincident with the most unfavorable

References pp. 212-214.

combination of weather and population density, would cause only 3,300 prompt fatalities [23]. Furthermore, that study estimated the odds against an accident of this magnitude occurring next year to be about 3,000,000 : 1.

Disaster potential seems to explain much of the discrepancy between the perceived risk and the annual fatality estimates for nuclear power. Yet, because disaster plays only a small role in most of the other activities and technologies, it provides only a partial explanation of the perceived risk data.

Qualitative Characteristics — Are there other determinants of risk perceptions besides frequency estimates? We asked experts, students, LOWV members and Active Club members to rate the 30 technologies and activities on nine qualitative characteristics that have been hypothesized to be important [24]. These rating scales are described in Table 4.

The "risk profiles" made from mean ratings on these characteristics showed nuclear power to have the dubious distinction of scoring at or near the extreme on all of the characteristics associated with high risk. Its risks were seen as involuntary, delayed, unknown, uncontrollable, unfamiliar, potentially catastrophic, dreaded, and severe (certainly fatal). Fig. 4 contrasts its unique risk profile with non-nuclear electric power and another radiation technology, X rays, both of whose risks were judged to be much lower. Both electric power and X rays were judged more voluntary, less catastrophic, less dreaded, and more familiar than nuclear power.

Across all 30 items, ratings of dread and of the severity of consequences were closely related to lay judgments of risk. In fact, the risk judgments of the LOWV and student groups could be predicted almost perfectly from ratings of dread and severity, the subjective fatality estimates, and the disaster multipliers in Table 3. Experts' judgments of risk were not related to any of the nine qualitative risk characteristics.

Judged Seriousness of Death — In a further attempt to improve our understanding of perceived risk, we examined the hypothesis that some hazards are feared more than others because the deaths they produce are much worse than deaths from other activities. We thought, for example, that deaths from risks imposed involuntarily, from risks not under one's control, or from hazards that are particularly dreaded might be given greater weight in determining people's perceptions of risk.

However, when we asked students and LOWV members to judge the relative seriousness to society of a death from each of the 30 activities and technologies, the differences were slight. The most serious forms of death (from nuclear power and handguns) were judged only about 2 to 4 times worse than the least serious forms of death (from alcoholic beverages and smoking). Furthermore, across all 30 activities, judged seriousness of death was not closely related to perceived risk of death.

An Extended Study of Risk Perception — Our recent work extends these studies of risk perception to a broader set of hazards (90 instead of 30) and risk characteristics (18 instead of 9). Although data have thus far been collected only

TABLE 4

Risk Characteristics Rated by LOWV Members and Students

Voluntariness of risk
Do people face this risk voluntarily? If some of the risks are voluntarily undertaken and some are not, mark an appropriate spot towards the center of the scale.

| risk assumed voluntarily | 1 | 2 | 3 | 4 | 5 | 6 | 7 | risk assumed involuntarily |

Immediacy of effect
To what extent is the risk of death immediate — or is death likely to occur at some later time?

| effect immediate | 1 | 2 | 3 | 4 | 5 | 6 | 7 | effect delayed |

Knowledge about risk
To what extent are the risks known precisely by the persons who are exposed to those risks?

| risk level known precisely | 1 | 2 | 3 | 4 | 5 | 6 | 7 | risk level not known |

To what extent are the risks known to science?

| risk level known precisely | 1 | 2 | 3 | 4 | 5 | 6 | 7 | risk level not known |

Control over risk
If you are exposed to the risk, to what extent can you, by personal skill or diligence, avoid death?

| personal risk can't be controlled | 1 | 2 | 3 | 4 | 5 | 6 | 7 | personal risk can be controlled |

Newness
Is this risk new and novel or old and familair?

| new | 1 | 2 | 3 | 4 | 5 | 6 | 7 | old |

Chronic-catastrophic
Is this a risk that kills people one at a time (chronic risk) or a risk that kills large numbers of people at once (catastrophic risk)?

| chronic | 1 | 2 | 3 | 4 | 5 | 6 | 7 | catastrophic |

Common-dread
Is this a risk that people have learned to live with and can think about reasonably calmly, or is it one that people have great dread for — on the level of a gut reaction?

| common | 1 | 2 | 3 | 4 | 5 | 6 | 7 | dread |

Severity of consequences
When the risk from the activity is realized in the form of a mishap or illness, how likely is it that the consequence will be fatal?

| certain not to be fatal | 1 | 2 | 3 | 4 | 5 | 6 | 7 | certain to be fatal |

from college students, the results appear to provide further insights into the nature of risk perception. In addition, they suggest that some accepted views about the importance of the voluntary-involuntary distinction and the impact of catastrophic losses may need revision.

Design of the Study — The extended study is outlined in Table 5. The 90 hazards were selected to cover a very broad range of activities, substances, and tech-

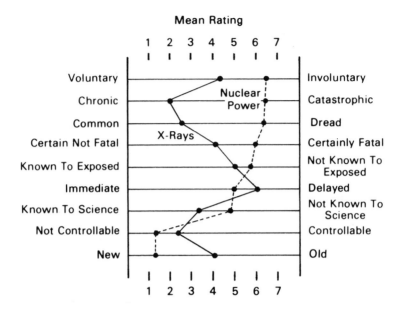

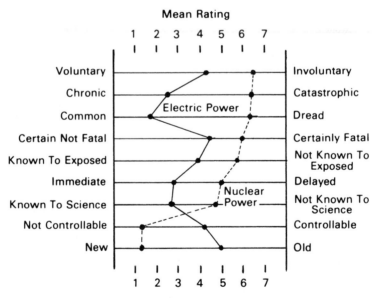

Fig. 4. Rated characteristics of risk for nuclear power and related technologies. Source: Fischhoff et al. [17].

nologies. To keep the rating task to a manageable size, some people judged only risks, others judged only benefits and others rated the hazards on five of the risk characteristics. Risks and benefits were rated on a 0-100 scale (from "not risky" to "extremely risky" and from "no benefit" to "very great benefit").

TABLE 5

Overview of the Extended Study

I. The Hazards

1 Home Gas Furnaces	31 Food Coloring	61 Darvon
2 Home Appliances	32 Saccharin	62 Morphine
3 Home Power Tools	33 Sodium Nitrite	63 Oral Contraceptives
4 Microwave Ovens	34 Food Preservatives	64 Valium
5 Power Lawn Mowers	35 Food Irradiation	65 Antibiotics
6 Handguns	36 Earth Orbit Satellite	66 Prescription Drugs
7 Terrorism	37 Space Exploration	67 Boxing
8 Crime	38 Lasers	68 Downhill Skiing
9 Nerve Gas	39 Asbestos	69 Fireworks
10 Nuclear Weapons	40 Police Work	70 Football
11 National Defense	41 Firefighting	71 Hunting
12 Warfare	42 Christmas Tree Light	72 Jogging
13 Bicycles	43 Cosmetics	73 Mountain Climbing
14 Motorcycles	44 Flourescent Lights	74 Mushroom Hunting
15 Motor Vehicles	45 Hair Dyes	75 Recreational Boating
16 Railroads	46 Chem. Disinfectants	76 Roller Coasters
17 General Aviation	47 DNA Research	77 Scuba Diving
18 SST	48 Liquid Natural Gas	78 Skateboards
19 Jumbo Jets	49 Smoking	79 Sunbathing
20 Commercial Aviation	50 Tractors	80 Surfing
21 Anesthetics	51 Chem. Fertilizers	81 Swimming Pools
22 Vaccinations	52 Herbicides	82 Fossil Elec. Power
23 Pregnancy, Childbirth	53 DDT	83 Hydroelectric Power
24 Open-Heart Surgery	54 Pesticides	84 Solar Electric Power
25 Surgery	55 Aspirin	85 Non-Nuc. Elec. Power
26 Radiation Therapy	56 Marijuana	86 Nuclear Power
27 Diagnostic X Rays	57 Heroin	87 Dynamite
28 Alcoholic Beverages	58 Laetril	88 Skyscrapers
29 Caffeine	59 Amphetamines	89 Bridges
30 Water Fluoridation	60 Barbiturates	90 Dams

II. Judgments Made for Each Hazard
   A. Perceived Risk of Death (0 - 100 scale)
   B. Perceived Benefit (0 - 100 scale)
   C. Risk Adjustment (necessary to make perceived risk equal acceptable risk)
   D. Ratings on 18 Risk Characteristics
      1. — 8. The characteristics in Table 4 excluding ''control over risk''
      9. Can mishaps be prevented?
      10. If a mishap occurs, can the damage be controlled?
      11. How many people are exposed to this hazard?
      12. Does the hazard threaten future generations?
      13. Are you personally at risk from this hazard?
      14. Are the benefits equitably distributed among those at risk?
      15. Does the hazard threaten global catastrophe?
      16. Are the damage-producing processes observable as they occur?
      17. Are the risks increasing or decreasing
      18. Can the risks be reduced easily?

References pp. 212-214.

After rating the hazards with regard to risk, respondents were asked to rate the degree to which the present risk level would need to be adjusted to make the risk level acceptable to society. The instructions for this adjustment task read as follows:

The acceptable level of risk is not the ideal risk. Ideally, the risks should be zero. The acceptable level is a level that is "good enough," where "good enough" means you think that the advantages of increased safety are not worth the costs of reducing risk by restricting or otherwise altering the activity. For example, we can make drugs "safer" by restricting their potency; cars can be made safer, at a cost, by improving their construction or requiring regular safety inspection. We may, or may not, believe such restrictions are necessary.

If an activity's present level of risk is acceptable, no special action need be taken to increase its safety. If its riskiness is unacceptably high, *serious action*, such as legislation to restrict its practice, should be taken. On the other hand, there may be some activities or technologies that you believe are currently safer than the acceptable level of risk. For these activities, the risk of death could be higher than it is now before society would have to take serious action.

On their answer sheets, participants were provided with three columns labeled: (a) "Could be riskier: it would be acceptable if it were＿times riskier;" (b) "It is presently acceptable;" and (c) "Too risky; to be acceptable: it would have to be＿ times safer."

The 18 risk characteristics included eight from the earlier study. The ninth characteristic from that study, controllability, was split into two separate characteristics representing control over the occurrence of a mishap (preventability) and control over the consequences given that something did go wrong. The remaining characteristics were selected to represent additional concerns thought to be important by risk assessment researchers. As in the earlier study, all characteristics were rated on a bipolar 1-7 scale representing the extent to which the characteristics described the hazard. For example:

15. To what extent does pursuit of this activity, substance or technology have the potential to cause catastrophic death and destruction across the whole world?

Very Low   Very High
Catastrophic   1 2 3 4 5 6 7   Catastrophic
Potential   Potential

Results —

Risk Characteristics — The mean ratings for the eighteen risk characteristics revealed a number of interesting findings. For example, the risks from most of these hazards were judged to be at least moderately well known to science (63 had mean ratings below 3 where 1 was labeled "known precisely"). Most risks were thought to be better known to science than to those who were exposed. The only risks for which those exposed were thought to be more knowledgeable than scientists were those from police work, marijuana, contraceptives (judged relatively unknown to both science and those exposed), boxing, skiing, hunting, and several other sporting activities.

Only 25 of the hazards were judged to be decreasing in riskiness; two of them (surgery and pregnancy-childbirth) were thought to be decreasing greatly. Risks from sixty-two hazards were judged to be increasing, thirteen of these markedly so. The risks from crime, warfare, nuclear weapons, terrorism, national defense, herbicides and nuclear power were judged to be increasing most. None of the hazards was judged to be easily reducible. The lowest of the 90 means on this characteristic was 3.2 (where 1 was labeled "easily reduced"); it was obtained for home appliances and roller coasters.

The ratings of the various risk characteristics tended to be rather highly intercorrelated, as shown in Table 6. For example, risks with catastrophic potential were also judged as quite dread ($r = .83$). Application of a statistical technique known as factor analysis showed that the pattern of intercorrelations could be represented by three underlying dimensions or factors. The nature of these factors can be seen in Table 6 in which the characteristics are ordered on the basis of the factor analysis. The first 12 characteristics represent the first factor; they correlate highly with one another and less highly with the remaining six characteristics. In other words, these data suggest that risks whose severity is believed not to be controllable tend also to be seen as dread, catastrophic, hard to prevent, fatal, inequitable, threatening to future generations, not easily reduced, increasing, involuntary, and threatening to the rater personally. The nature of these characteristics suggests that this factor be called "Dread." The second factor primarily reflects five characteristics that correlate relatively highly with one another and less highly with other characteristics. They are: observability, knowledge, immediacy of consequences, and familiarity (see Table 6). We have labeled this factor "Familiarity." The third factor is dominated by a single characteristic, the number of people exposed. This characteristic can be seen in Table 6 to be relatively independent of the other characteristics.

Just as each of the 90 hazards has a mean score on each of the 18 risk characteristics, each hazard also has a score on each factor. These scores give the location of each hazard within the factor space. Fig. 5 plots the hazards on Factors 1 and 2. Items at the high end of Factor 1 are all highly dreaded. Items at the negative end of Factor 1 are seen as posing risks to individuals and being injurous rather than fatal. The placement of items on the vertical dimension, Factor 3, intuitively fits the theme of familiarity and observability associated with the dimension label. Hazards lying at the extremes on Factor 2 (number exposed) are shown in Table 7.

This three-dimensional factor structure is of interest because it differs considerably from the two-dimensional structure obtained from ratings of 30 hazards on 9 characteristics [20]. That structure, in which Factor 1 was labeled "severe" (i.e., certain to be fatal) and Factor 2 was labeled "high technology," has been found to be remarkably consistent across four different groups of lay and expert respondents [21]. The present results indicate that the particular set of hazards and the particular set of risk characteristics under study can have an important effect on the nature of the observed "dimensions of risk."

One point of commonality between the present analysis and the previous one is that nuclear power is an isolate in both. Although activities such as crime, nerve gas, warfare and terrorism are seen as similarly dreaded (Factor 1), none of these is

TABLE 6

Intercorrelations Among 18 Risk Characteristics in the Extended Study

	1	2	3	4	5	6	7	8	9	10	11	12	13	14	15	16	17	18
1. Severity not controllable																		
2. Dread	.82																	
3. Globally catastrophic	.78	.83																
4. Little preventive control	.86	.72	.71															
5. Certain to be fatal	.80	.82	.73	.77														
6. Risks & benefits inequitable	.75	.76	.84	.63	.65													
7. Catastrophic	.77	.66	.77	.74	.67	.77												
8. Threatens future generations	.62	.76	.86	.52	.64	.81	.63											
9. Not easily reduced	.59	.67	.64	.67	.66	.59	.56	.59										
10. Risks increasing	.51	.63	.76	.48	.58	.76	.57	.75	.60									
11. Involuntary	.68	.58	.69	.52	.42	.77	.74	.62	.33	.44								
12. Affects me personally	.57	.64	.77	.43	.50	.71	.64	.78	.35	.67	.61							
13. Not observable	-.04	-.14	.04	-.19	-.28	.08	.00	.24	-.15	-.10	.33	.10						
14. Unknown to those exposed	.14	.05	.22	-.05	-.12	.28	.24	.35	-.20	.05	.63	.31	.79					
15. Effects immediate	.21	.15	.00	.34	.36	-.08	.11	-.29	.22	.01	-.25	-.14	-.87	-.77				
16. New (unfamiliar)	.32	.29	.32	.22	.20	.25	.18	.44	.13	.06	.36	.20	.63	.56	-.48			
17. Unknown to science	-.10	-.10	-.04	-.07	-.15	-.02	-.04	.02	-.23	.02	.13	.00	.43	.50	-.44	.32		
18. Many people exposed	-.04	.04	.23	-.14	-.11	.32	.23	.47	.03	.26	.34	.56	.37	.46	-.52	.07	-.01	

CONCLUSION: *Characteristics 1-12 and 13-17 form clusters. Items within each cluster are highly correlated with one another. Correlations are low between items from different clusters. Thus, although these characteristics are distinct, there is much commonality among them.*

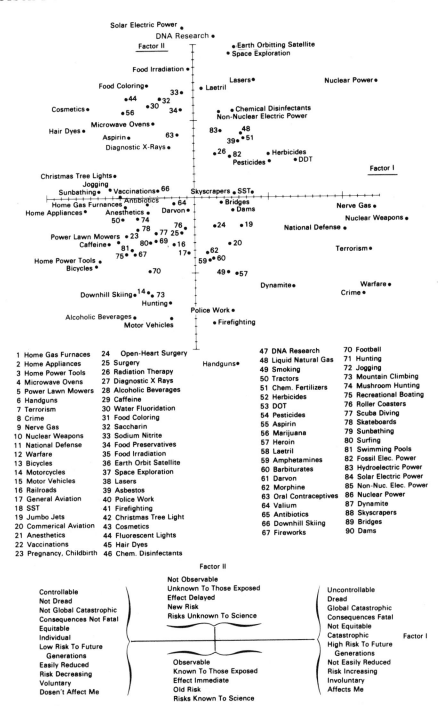

Fig. 5. Factors 1 and 2 of the three-dimensional structure derived from interrelationships among 18 risk characteristics in the extended study. Factor 3 (not shown) reflects the number of people exposed to the hazard.

TABLE 7

Extreme Scores on Factor 3
(Degree of Exposure to the Hazard)

High Exposure	Factor Score	Low Exposure	Factor Score
Alcoholic Beverages	2.9	Nerve Gas	−1.2
Caffeine	2.0	SST	−1.2
Smoking	1.8	Surfing	−1.4
Food Preservatives	1.8	Laetril	−1.5
DDT	1.6	Boxing	−1.5
Herbicides	1.5	Roller Coasters	−1.6
Motor Vehicles	1.5	Scuba Diving	−1.8
Food Irradiation	1.4	Open-heart Surgery	−1.8
Pesticides	1.3	Lasers	−1.8
Sunbathing	1.2	Space Exploration	−2.0
		Solar Electricity	−2.0

judged to be as new or as unknown (Factor 2) as nuclear power.

Risks and Benefits — Table 8 shows the mean judgments of perceived risk, perceived benefit, and need for risk adjustment. Rather than discuss the details of the ratings for specific hazards, we shall focus on their intercorrelations and their relation to the risk characteristics and factor scores.

Our earlier study showed that perceived risk could be predicted from knowledge of an item's judged dread and severity. Table 9 shows this pattern was repeated with perceived risk also being closely related to threat to future generations, potential for global catastrophe, personal threat, and inequitability. Among the factors, Factor 1 was the best predictor of perceived risk. Perceived risk was inversely related to perceived benefit, a finding that has been obtained consistently in previous studies [e.g., 21]; our respondents do not believe that societal mechanisms have worked to limit risks from less beneficial activities, contrary to claims by Starr and others [25].

The greater the perceived risk, the larger the adjustment judged necessary to bring the risk to an acceptable level (r = .91). Because of this close relationship, some risk characteristics that predicted perceived risk also predicted the adjustment ratings. Note also that the more beneficial items were thought to need less risk adjustment (r = -.54).

Perceived risk is obviously the primary determiner of the risk adjustment rating. Would the adjustment rating continue to correlate with the other risk characteristics if the influence of perceived risk were partialed out? The right-hand column of Table 9 indicates that the answer is yes. Although correlations between the risk characteristics and the adjustment ratings are much reduced once perceived risk is held constant, dread, global catastrophe, equity, Factor 1 and perceived benefit still show moderate relationships. In other words, adjustments are mostly

TABLE 8

Mean Risk and Benefit Judgments
for 90 Activities, Substances and Technologies

		Perceived Risk	Perceived Benefit	Adjusted Risk
1.	Nuclear Weapons	78	27	49.1
2.	Warfare	78	31	26.2
3.	DDT	76	26	8.8
4.	Handguns	76	27	15.2
5.	Crime	73	9	19.2
6.	Nuclear Power	72	36	22.2
7.	Pesticides	71	38	4.2
8.	Herbicides	69	33	6.3
9.	Smoking	68	24	6.6
10.	Terrorism	66	6	26.3
11.	Heroin	63	17	7.6
12.	National Defense	61	58	4.7
13.	Nerve Gas	60	7	17.4
14.	Barbiturates	57	27	4.8
15.	Alcoholic Beverages	57	49	2.9
16.	Chemical Fertilizers	55	48	2.8
17.	Motor Vehicles	55	76	3.1
18.	Amphetamines	55	27	4.9
19.	Open-Heart Surgery	53	50	2.1
20.	Morphine	53	31	4.4
21.	Radiation Therapy	53	36	3.6
22.	Darvon	52	38	3.9
23.	Oral Contraceptives	51	67	3.8
24.	Asbestos	51	51	4.9
25.	Liquid Natural Gas	50	56	1.5
26.	Chemical Disinfectants	49	47	2.2
27.	Valium	48	37	2.6
28.	Surgery	48	64	2.4
29.	Dynamite	47	30	2.0
30.	Diagnostic X Rays	44	54	2.2
31.	Fire Fighting	44	83	2.7
32.	Motorcycles	43	43	2.5
33.	Police Work	43	75	2.5
34.	Lasers	42	34	1.5
35.	Food Preservatives	42	36	2.7
36.	DNA Research	41	41	2.2
37.	Prescription Drugs	41	73	1.5
38.	Fossil Electric Power	40	65	1.7
39.	Food Irradiation	39	42	2.1
40.	Sodium Nitrite	38	31	3.0
41.	Microwave Ovens	36	34	1.7
42.	Power Lawn Mowers	35	39	1.5
43.	Laetril	35	30	2.6
44.	Saccharin	35	25	1.9
45.	Home Power Tools	33	52	1.1

References pp. 212-214.

TABLE 8 (cont.)

		Perceived Risk	Perceived Benefit	Adjusted Risk
46.	Hunting	33	47	1.8
47.	SST	33	30	1.5
48.	Jumbo Jets	32	48	1.4
49.	Dams	31	64	1.5
50.	Aspirin	31	63	1.1
51.	Commercial Aviation	31	54	1.4
52.	Fireworks	31	42	1.4
53.	General Aviation	30	39	1.6
54.	Caffeine	30	42	1.0
55.	Hydroelectric Power	30	66	1.1
56.	Football	30	54	1.7
57.	Antibiotics	30	68	1.5
58.	Preganancy, Childbirth	30	60	1.1
59.	Anesthetics	29	55	1.5
60.	Home Gas Furnaces	29	49	1.0
61.	Railroads	29	48	1.1
62.	Food Coloring	29	19	1.6
63.	Tractors	29	61	.9
64.	Mountain Climbing	28	47	1.3
65.	Bridges	27	69	1.4
66.	Christmas Tree Lights	27	44	1.0
67.	Skateboards	27	37	1.7
68.	Scuba Diving	26	41	1.3
69.	Skyscrapers	26	39	1.5
70.	Non-nuclear Electric Power	26	75	1.1
71.	Swimming Pools	26	57	1.2
72.	Home Appliances	26	72	.8
73.	Downhill Skiing	26	57	1.1
74.	Space Exploration	25	51	.9
75.	Bicycles	24	68	.9
76.	Vaccinations	24	77	1.0
77.	Water Fluoridation	24	44	1.2
78.	Hair Dyes	23	28	1.2
79.	Mushroom Hunting	23	32	1.1
80.	Boxing	23	34	1.2
81.	Recreational Boating	22	45	1.2
82.	Earth Orbit Satellite	22	54	.7
83.	Fluorescent Lights	21	46	1.0
84.	Surfing	21	41	1.0
85.	Marijuana	21	53	1.1
86.	Roller Coasters	20	33	1.3
87.	Cosmetics	20	49	.9
88.	Sunbathing	20	49	.9
89.	Jogging	14	65	.6
90.	Solar Electric Power	12	56	.6

determined by perceived risk but they are somewhat sensitive to benefit and certain risk characteristics.

Table 9 shows that certain characteristics can do a good job, by themselves, of

TABLE 9

Correlations Between Perceived Risk, Risk Adjustment
and Risk Characteristics

Perceived Risk		Adjusted Risk		Correlations with Adj. Risk Holding the Effects of Perceived Risk Constant	
Dread	.83	Dread	.87	Dread	.47
Future generations	.80	Global catast.	.82	Global catast.	.41
Global catast.	.78	Future generations	.77	Inequitable	.38
Fatal	.74	Increasing	.76	Increasing	.35
Increasing	.73	Fatal	.74	Not easily reduced	.34
Affects me	.70	Inequitable	.73	Not preventable	.30
Inequitable	.68	Not easily reduced	.69	Uncontrollable	.25
Not easily reduced	.63	Affects me	.65	Catastrophic	.24
Uncontrollable	.63	Uncontrollable	.65	Fatal	.23
Not preventable	.51	Not preventable	.57	Immediate	.18
Catastrophic	.50	Catastrophic	.54	Involuntary	.18
Involuntary	.39	Involuntary	.42	Future generations	.15
Many exposed	.25	Immediate	.17	Unknown to science	.07
New	.17	New	.16	Affects me	.05
Immediate	.10	Many exposed	.14	New	.02
Unknown to exposed	−.06	Unknown to exposed	−.09	Unknown to exposed	−.09
Not observable	−.19	Unknown to Science	−.22	Not observable	−.16
Unknown to science	−.27	Not observable	−.23	Many exposed	−.23
Factor 1	.74	Factor 1	.79	Factor 1	.43
Factor 2	−.22	Factor 2	−.22	Factor 2	−.04
Factor 3	.41	Factor 3	.29	Factor 3	−.21
Perceived Benefit	−.42	Perceived Benefit	−.54	Perceived Benefit	−.44
		Perceived Risk	.91		

predicting perceived and adjusted risk. We have also used multiple regression analysis to develop simple equations involving combinations of characteristics. These produced multiple correlations in the range of .89 to .95. In other words, perceived and adjusted risk are quite predictable from knowledge of other risk characteristics.

Special Issues

The Voluntariness Hypothesis — By examining statistical and economic indicators of benefit and risk for eight hazards, Starr [25] proposed several hypotheses about the nature of acceptable risk:
- The acceptable level of risk is roughly proportional to the third power (cube) of the benefits.

References pp. 212-214.

- The public seems willing to accept risks from voluntary activities, such as ski-ing, that are roughly a thousand times greater than it would tolerate from involuntarily imposed hazards providing the same level of benefit.

Although Starr acknowledged the preliminary nature of his data and hypotheses, the voluntary/involuntary distinction has been widely cited as relevant for stan-dard setting [26, 27, 28]. Attempts to derive quantitative criteria for acceptable levels of risk often recommend stricter standards on hazards imposed involuntari-ly.

The judgments of current and acceptable risks in our own studies provide a test of Starr's hypotheses with very different methodology and data. Our first study [20], in which members of the League of Women Voters rated the risks and benefits of 30 hazards, produced results supportive of Starr's (see Fig. 6). Our res-pondents appeared to believe that greater risk should be tolerated for more beneficial activities and that a double standard is appropriate for voluntary and involuntary activities. However, these people also seemed to desire similar double standards based on characteristics such as controllability, knowledge, familiarity, and immediacy. We concluded that, in addition to benefits and voluntariness, a number of other psychological and physical characteristics of risk might need to be incorporated into risk standards.

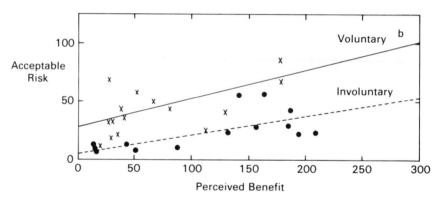

Fig. 6. Comparison between judgments of acceptable risk levels and judgments of benefits for voluntary and involuntary activities. Source: Fischhoff et al. [17].

The results of our extended (90 hazard) study have clarified these tentative con-clusions. Consider first the role of voluntariness: the correlation matrix in Table 6 shows that involuntariness is closely related to many other risk characteristics and particularly to lack of control ($r = .68$), global catastrophe ($r = .69$), inequity ($r = .76$), and catastrophe ($r = .74$). For example, six of the ten most involuntary hazards (nuclear weapons, nerve gas, terrorism, warfare, nuclear power, and DDT) are also among the 10 most catastrophic hazards. These relationships sug-gest that much, if not all, of the observed aversion to involuntary risks may be due to other characteristics that are closely associated with voluntariness.

Support for this interpretation comes from the following data analyses con-ducted on the extended study. An estimate of the acceptable level of risk was calculated for each of the 90 hazards by dividing the hazard's mean risk level by its mean adjustment factor. This index of acceptable risk was found to correlate positively with perceived benefit ($r = .58$) as Starr hypothesized and as our previous studies had also demonstrated. Furthermore, deviations from the best-fit line relating benefit and acceptable risk were significantly correlated with voluntariness, in the direction predicted by Starr's hypothesis. However, these deviations were much more strongly related to characteristics such as catastrophic potential, dread, and equity than to voluntariness. In fact, when the effects of any of these other characteristics were removed statistically, voluntariness no longer was related to acceptable risk.

These doubts about the importance of voluntariness have been reinforced by other considerations, such as the following thought experiment. Suppose that you own and ride in two automobiles. One is needed for your work and when you ride in it, you are chauffeured (thus your exposure to risk is involuntary and relatively uncontrollable). The second auto is driven by you for pleasure, thus the risks you take while driving it are voluntary and somewhat controllable. Consider the standards of safety you would wish for the two vehicles. Would you demand that standards be much safer for the first car than for the second? If you are like us, then you would desire equivalent safety standards for both vehicles.

Lave [29] expressed similar doubts about the intrinsic significance of involuntary risks. He noted that involuntary hazards typically affect larger numbers of people so that stricter safety standards for such hazards merely reflect the greater amount of money that groups would be willing to pay for safety, relative to what an individual would be willing to pay.

We conclude that society's apparent aversion to involuntary risks may be mostly an illusion, caused by the fact that involuntary risks are often noxious in more important ways, such as being inequitable or potentially catastrophic.

The Importance of Catastrophic Potential — Whereas too much significance may have been attributed to voluntariness, our research suggests that more attention needs to be focused on the role played by catastrophic potential in determining societal response to hazards. As noted above, catastrophic potential appears to be a major determiner of judgments of perceived and acceptable risk and may account for the presumed importance of voluntariness. In similar fashion, it seems likely that catastrophic potential may also account for some of the other double standards observed by Fischhoff et al. [20]. That is, the finding that, at any constant level of benefit, acceptable risk was lower for unknown, unfamiliar, and uncontrollable activities may have been due to the expectation that such activities could lead to a great many deaths. This hypothesis would seem to merit careful examination.

Nuclear Power — The importance of catastrophic potential may be seen most clearly in the case of nuclear power. Our early studies with students and the League of Women Voters showed that these groups believed nuclear power to pose greater risks of death than any of the 29 other hazards under consideration. Further research linked this perception to the perceived potential for disaster (which was extremely high).

Like nuclear power, diagnostic X rays produce invisible and irreversible contamination which leads to cancer and genetic damage. However, X rays are not similarly feared. One reason for this difference may be found in the risk profile in Fig. 4 and the disaster multipliers of Table 3 which show that X rays are not seen as having much potential for catastrophe.

We are led to conclude that beliefs about the catastrophic nature of nuclear power are a major determinant of public opposition to that technology. This is not a comforting conclusion because the rarity of catastrophic events makes it extremely difficult to resolve disagreements by recourse to empirical evidence. Demonstrating the improbability of catastrophic accidents requires massive amounts of data. In the absence of definitive evidence, weaker data tend to be interpreted in a manner consistent with the individual's prior beliefs. As a result, the "perception gap" between pro-nuclear experts and the anti-nuclear public is likely to persist, leaving frustration, distrust, conflict, and costly hazard management as its legacy. We suspect that the potential for similar disputes exists with other low-probability, high-consequence hazards such as LNG, pesticides, industrial chemicals, and recombinant DNA research.

Weighing Catastrophes — Any attempt to control accidents must be guided by assessments of their probability and severity. As we have seen, probability assessment poses serious difficulties, particularly for rare events. Unfortunately, weighing of the severity or social cost of an accident is also problematic.

Society appears to react more strongly to infrequent large losses of life than to frequent small losses. This has led analysts to propose a weighting factor that accomodates the greater impact of N lives lost at one time relative to the impact of one life lost in each of N separate incidents. Risk/benefit analysts would then apply this weighting factor when evaluating the expected social costs of a proposed activity.

The precise nature of the fatality weighting factor has been the subject of some speculation. Wilson [30] suggested that N lives lost simultaneously were N^2 times more important than the loss of one life. Ferreira and Slesin [31] hypothesized, on the basis of observed frequency and severity data, that the function might be a cubic one.

We believe that a single weighting function cannot adequately explain, predict, or guide social response to catastrophe. For one, we have found that people hold, simultaneously, several conflicting attitudes about the weighting function [32]. They believe that the function relating social impact to N lives lost should be (a) concave, because they recognize that the same additional number of lives lost seems more important in a small accident than in a large accident; (b) linear, because each unidentified life is equally important; and (c) convex, because large losses of life have important higher order consequences and may even threaten the resilience of a community or society. Clearly any attempt to model the impact of a multiple fatality event will need to consider how situational factors will interact with these multiple values (for a start on this problem see Keeney [33]).

Signal Value — Another complication is that the occurrence of a rare, catastrophic event contains information regarding the probability of its reoccurrence. As a result, the impact of an accident may be determined as much by its signal value as by its toll of death and destruction.

The importance of accidents as signals is demonstrated by a study in which we asked 21 women (median age = 37) to rate the seriousness of 10 hypothetical accidents. Several aspects of seriousness were rated including:

a. The total amount of suffering and grief caused by the loss of life in each mishap,
b. The number of people who need to be made aware of the mishap via the media,
c. The amount of effort (and money) that should be put into investigating the cause of the mishap and preventing its recurrence, and
d. The degree to which hearing about the mishap would cause one to be worried and upset during the next few days.

The accident scenarios were constructed so as to vary on total fatalities and informativeness (see Table 10). The five uninformative accidents represented single incidents, generated by reasonably well-known and understood processes, and limited to a particular time and locale. The high-information mishaps were designed to signal a change in riskiness, potential for the proliferation of similar mishaps, or some breakdown in the system controlling the hazard. Thus a bus skidding on ice represented a low-information mishap because its occurrence did not signal a change in motor-vehicle risks, whereas an accident caused by a poorly designed steering system in a new model automobile would be informative about all such vehicles. To check our intuitions our respondents also judged informativeness, defined as the degree to which the mishap told them (and society) something that may not have been known about the hazardousness of the specific activity.

All ratings were on a seven-point scale. The means ratings are shown in Table 10. Note that the five mishaps designed to be high in signal value were all judged more informative than the most informative mishap in the low-information category. As expected, the judged amount of suffering and grief was closely related to the number of people killed. However, all other aspects of seriousness were more closely related to the information content of the accident. Accidents signaling either a possible breakdown in safety control systems or the possibility that the mishap might proliferate were judged more worrisome and in need of greater awareness and greater public effort to prevent reoccurrences. The number of people killed appeared to be relatively unimportant in determining these aspects of seriousness.

For this study we conclude that risk analyses, designed to anticipate public reaction or to aid in decision making, need to consider what accidents indicate about the nature and controllability of risk. An accident that takes many lives may have little or no impact on perceived risk if it occurs as part of a familiar, well-understood and self-limiting process. In contrast, a small accident may greatly enhance perceived risk and trigger strong corrective action because it increases the judged probability of future accidents. The great impact of the accident at Three Mile Island (which caused no immediate fatalities) would seem to reflect such considerations.

References pp. 212-214.

TABLE 10

Effect of Informativeness on the Impact of Catastrophic Mishaps

	Inform-ativness	Suffering and Grief	Need for Awareness	Effort to Prevent Recurrence	Worry
Less Informative Mishaps					
Bus skids on ice and runs off road (27 killed)	1.8	4.4	2.5	3.1	1.8
Dam collapse (40 killed)	4.7	4.9	4.7	5.9	3.8
Two jumbo jets collide on runway (600 killed)..........	4.8	6.1	5.8	6.5	4.5
Hundred year flood (2,700 killed)	2.8	6.1	5.3	3.5	2.7
Meteorite hits stadium (4,000 killed)	2.2	6.2	5.7	2.1	2.5
More Informative Mishaps					
Nuclear reactor accident: Partial core meltdown releases radiation inside plant but not to outside (1 killed)	6.5	4.5	6.5	7.0	6.1
Botulism in well-known brand of food (2 killed)	5.7	3.7	5.2	6.1	4.6
New model auto steering fails (3 killed)	5.2	3.8	5.2	6.3	4.6
Recombinant DNA workers contract mysterious illness (10 killed)	6.1	4.6	5.9	6.3	5.1
Jet engine falls off on takeoff (300 killed)	5.7	6.0	6.1	6.9	5.5

CONCLUSIONS

Although research into the nature of perceived risk is still incomplete, we offer the following tentative conclusions:

1. Perceived risk is quantifiable and predictable.
2. Groups of laypeople sometimes differ systematically in their perceptions. Experts and lay persons also differ, particularly with regard to the probability and consequences of catastrophic accidents.
3. Cognitive limitations, biased media coverage, misleading experience, and the anxieties generated by the gambles life poses cause uncertainty to be denied, risks to be misjudged, and judgments to be believed with unwarranted confidence.
4. Experts' risk assessments are also susceptible to bias, particularly underestimation due to omitting important pathways to disaster.
5. Differences in the judged seriousness of various modes of death are slight and do not seem responsible for discrepancies between public and expert assessments of risk.
6. The degree of adjustment judged necessary to make risk levels "acceptable" is strongly determined by the perceived level of current risk; the greater the perceived risk, the greater the desired reduction. Perceived benefit plays a secondary role; all else being equal, somewhat less reduction in risk is deemed necessary to make highly beneficial activities "acceptable."
7. Many of the eighteen characteristics of risk hypothesized to be important to the public do correlate highly with perceived risk and desire for risk reduction. Certain clusters of characteristics are highly interrelated across hazards, indicating that they can be combined into higher-order characteristics or "factors." Three factors, labeled Dread, Familiarity, and Exposure, seem able to account for most of the interrelations among the eighteen characteristics.
8. The perceived potential for catastrophic loss of life emerges as one of the most important risk characteristics, responsible for:
   a. the erroneous belief that people want involuntary hazards to be less risky than voluntary hazards of equal benefit,
   b. the irresolvable disputes between experts and the public which lead to frustration, distrust, conflict, and ineffective hazard management, and
   c. the fear of nuclear power.
9. Disagreements about risk should not be expected to evaporate in the presence of evidence. Definitive evidence, particularly about rare hazards, is difficult to obtain. Weaker information is likely to be interpreted so as to reinforce existing beliefs.
10. The social impact of losing N lives in a single mishap cannot be adequately represented by a simple weighting function. People believe simultaneously in several different functional relationships. In addition, accidents serve as signals regarding the probability of further mishaps. The alarm created by an accident signal is a strong determiner of its social impact and is not necessarily related to the number of people killed.

References pp. 212-214.

POSTSCRIPT: WHO SHALL DECIDE?

The research described in this paper demonstrates that judgment of risk is fallible. It also shows that the degree of fallibility is often surprisingly great and that faulty estimates may be put forth with much confidence. Since even well-informed laypeople have difficulty judging risks accurately, it is tempting to conclude that the public should be removed from the risk assessment process. Such action would seem to be misguided on several counts. First, we have no assurance that experts' judgments are immune to biases once they are forced to go beyond hard data. Although judgmental biases have most often been demonstrated with laypeople, there is evidence that the cognitive functioning of experts is basically like that of everyone else.

Second, in many if not most cases, effective hazard management requires the cooperation of a large body of laypeople. These people must agree to do without some things and accept substitutes for others; they must vote sensibly on ballot measures and for legislators who will serve them as surrogate hazard managers; they must obey safety rules and use the legal system responsibly. Even if the experts were much better judges of risk than laypeople, giving experts an exclusive franchise for hazard management would mean substituting short-term efficiency for the long-term effort needed to create an informed citizenry.

For those of us who are not experts, these findings pose an important series of challenges: to be better informed, to rely less on unexamined or unsupported judgments, to be aware of the qualitative aspects that strongly condition our risk judgments, and to be more open to new evidence.

For experts, our findings pose what may be a more difficult challenge: to recognize and admit one's own cognitive limitations, to temper risk assessments with the important qualitative aspects of risk that influence the responses of laypeople, and somehow to create ways in which these considerations can find expression in hazard management without, in the process, creating more heat than light.

ACKNOWLEDGMENTS

The authors wish to express their appreciation to Christoph Hohenemser, Roger Kasperson, and Robert Kates for their helpful comments and suggestions on portions of this manuscript. This work was supported by the National Science Foundation under Grant PRA79-11934 to Clark University and under subcontract to Perceptronics, Inc. Any opinions, findings, and conclusions or recommendations expressed in this publication are those of the authors and do not necessarily reflect the views of the National Science Foundation.

REFERENCES

1. A. Tversky and D. Kahneman, "Judgment Under Uncertainty: Heuristics and Biases," Science, 185: 1124-1131, 1974.
2. A. Tversky and D. Kahneman, "Availability: A Heuristic for Judging Frequency and Pro-

bability," *Cognitive Psychology*, 4: 207-232, 1973.

3. S. Lichtenstein, P. Slovic, B. Fischhoff, M. Layman and B. Combs, "*Judged Frequency of Lethal Events*," *Journal of Experimental Psychology: Human Learning and Memory*, 4: 551-578, 1978.

4. B. Combs and P. Slovic, "*Causes of Death: Biased Newspaper Coverage and Biased Judgments*," *Journalism Quarterly*, in press.

5. B. Fischhoff, P. Slovic and S. Lichtenstein, "*Fault Trees: Sensitivity of Estimated Failure Probabilities to Problem Representation*," *Journal of Experimental Psychology: Human Perception and Performance*, 4: 342-355, 1978.

6. B. Fischhoff, P. Slovic and S. Lichtenstein, "*Knowing With Certainty: The Appropriateness of Extreme Confidence*," *Journal of Experimental Psychology: Human Perception and Performance*, 3: 552-564, 1977.

7. S. Lichtenstein, B. Fischhoff and L. D. Phillips, "*Calibration of Probabilities: The State of the Art*," *Decision Making and Change in Human Affairs*, H. Jungermann and G. de Zeeuw, eds., D. Reidel, Dordrecht, the Netherlands, 1977.

8. M. Hynes and E. VanMarcke, "*Reliability of Embankment Performance Prediction*," *Proceedings of the ASCE Engineering Mechanics Division Specialty Conference*, University of Waterloo Press, Waterloo, Ontario, Canada, 1976.

9. B. Fischhoff, "*Cost-Benefit Analysis and the Art of Motorcycle Maintenance*," *Policy Sciences*, 8: 177-202, 1977.

10. R. W. Kates, "*Hazard and Choice Perception in Flood Plain Management*," *Research Paper 78*, Department of Geography, University of Chicago, Chicago, 1962.

11. K. Borch, *The Economics of Uncertainty*, Princeton University Press, Princeton, N. J., 1968.

12. *Eugene Register-Guard*, "*Doubts Linger on Cyclamate Risks*," January 14, 1976.

13. E. E. David, "*One-Armed Scientists?*" *Science*, 189: 891, 1975.

14. O. Svenson, "*Are We All Among the Better Drivers?*" Unpublished report, Department of Psychology, University of Stockholm, 1979.

15. A. Rethans, "*An Investigation of Consumer Perceptions of Product Hazards*," Ph.D. dissertation, University of Oregon, 1979.

16. N. D. Weinstein, "*It Won't Happen to Me: Cognitive and Motivational Sources of Unrealistic Optimism*," Unpublished paper, Department of Psychology, Rutgers University, 1979.

17. P. Slovic, B. Fischhoff and S. Lichtenstein, "*Accident Probabilities and Seat Belt Usage: A Psychological Perspective*," *Accident Analysis and Prevention*, 10: 281-285, 1978.

18. L. Ross, "*The Intuitive Psychologist and His Shortcomings*," *Advances in Social Psychology*, L. Berkowitz, ed., Academic Press, New York, 1977.

19. D. Nelkin, "*The Role of Experts on a Nuclear Siting Controversy*," *Bulletin of the Atomic Scientists*, 30: 29-36, 1974.

20. B. Fischhoff, P. Slovic, S. Lichtenstein, S. Read and B. Combs, "*How Safe is Safe Enough? A Psychometric Study of Attitudes Towards Technological Risks and Benefits*," *Policy Sciences*, 8: 127-152, 1978.

21. P. Slovic, B. Fischhoff and S. Lichtenstein, "*Expressed Preferences*," *Decision Research Report 80-1*, Eugene, Oregon, 1980.

22. P. Slovic, S. Lichtenstein and B. Fischhoff, "*Images of Disaster: Perception and Acceptance of Risks from Nuclear Power*," *Energy Risk Management*, G. Goodman and W. D. Rowe, eds., Academic Press, London, in press.

23. U. S. Nuclear Regulatory Commission, *Reactor Safety Study: An Assessment of Accident Risks in U.S. Commercial Nuclear Power Plants*, *WASH 1400 (NUREG-75/014)*, Washington, D.C., October 1975.

24. W. Lowrance, *Of Acceptable Risk: Science and the Determination of Safety*, William Kaufmann Co., Los Altos, California, 1976.

25. C. Starr, "*Social Benefit vs. Technological Risk*," *Science*, 165: 1232-1238, 1969.

26. Council for Science and Society, *The Acceptability of Risks*, London, 1977.

27. G. H. Kinchin, "*Assessment of Hazards in Engineering Work*," *Proceedings of the Institute*

of Civil Engineers, Part I, 64: 431-438, 1978.
28. W. D. Rowe, The Anatomy of Risk, John Wiley, New York, 1977.
29. L. Lave, "Risk, Safety, and the Role of Government," Perspectives on Benefit-Risk Deci-
    sion Making, The National Academy of Engineering, Washington, D.C., 96-108, 1972.
30. R. Wilson, "The Costs of Safety," New Scientist, 68: 274-275, 1975.
31. J. Ferreira and L. Slesin, Observations on the Social Impact of Large Accidents, Technical
    Report No. 122, Operations Research Center, Massachusetts Institute of Technology, Octo-
    ber 1976.
32. B. Fischhoff, P. Slovic and S. Lichtenstein, "Knowing What You Want: Measuring Labile
    Values," Cognitive Processes in Choice and Decision Behavior, T. Wallsten, ed., Hillsdale,
    N. J., Erlbaum, in press.
33. R. L. Keeney, "Evaluation Involving Potential Fatalities," Unpublished report, Woodward
    Clyde Consultants, San Francisco, California 1977.

DISCUSSION

P. Slovic

C. Starr (Electric Power Research Institute)

I am a great admirer of the work that's just been described and I want to encourage it. I just want to straighten out one point. The original paper I wrote a long time ago really addressed a different topic than the one you're describing and I don't think they ought to be confused. I tried to find out what the actual accident rates and the risk rates were, and what people were actually doing, not what they were saying. So my paper did *not* deal with perceptions, and I don't think that the kind of data which I accumulated and used can be compared with yours. What people say and tell you and what they perceive may be quite different than what they develop empirically over a period of 30 or 40 years, which is what I tried to plot, in terms of real living conditions. I think the two things are different and I don't think they really ought to be compared on the same basis.

P. Slovic

I would agree that there's a difference between the analysis of behavior that you did and the analysis of perceptions and expressed preferences that we've done. We think that both types of analyses have a role in understanding, and perhaps influencing, risk policy and that they are complimentary rather than competing methods. They each have their advantages and their disadvantages.

Nevertheless, with regard to the effect of voluntariness, the important point is that involuntary hazards tend also to be inequitable and catastrophic. Therefore, careful analysis is required to determine whether or not society's observed (or expressed) concern about involuntary risks is merely an artifact of the high correlation between voluntariness and other aversive characteristics. We have attempted such an analysis based on expressed preferences. We look forward to someone doing the comparable analysis on a large set of hazards with revealed preferences data such as you used. If such a study is done, we would bet that the observed importance of the voluntary-involuntary dimension would be rather small.

C. R. Plott *(California Institute of Technology)*

Some of your findings apparently are that people are resistant to change and it takes a great deal of evidence to change their minds. On the other hand, studies by Kahneman and Tversky suggest that people put too much weight on some types of evidence, namely that from a sample of data, and they neglect prior information. Well, on the face of it, these would seem to be entirely inconsistent conclusions, so I was wondering if you might be able to clarify this inconsistency.

P. Slovic

When we talk about resistance to change as it applies to perceived risk, it usually involves people who have very strong prior opinions. Such people tend to interpret new evidence in a manner consistent with their prior opinions and, therefore, they tend not to change their views. So, for example, the accident at Three Mile Island convinced people who were already pro-nuclear that the safety systems really did work and it convinced anti-nuclear people that their worst fears were valid. I think this is really a very different kind of situation than what the Kahneman and Tversky studies dealt with. In their studies, people were asked to make judgments about things for which they had no strong prior opinions one way or the other. In those tasks, people were very sensitive to evidence that they felt was relevant to the required judgment.

J. Ferreira *(Massachusetts Institute of Technology)*

Some of your numbers seem to focus on measures of perceived risk with respect to total numbers of deaths, say nationally, and others measure riskiness in terms of some normalization by exposure, chances of dying if you ski for one day, say. It seems as though the choice of the standardized cases, risk per time that you try it, would matter. I wonder if you would tell us how you view this distinction?

P. Slovic

I think that all of our perceptual studies have focused on just one of those characteristics, the risk for U.S. society as a whole and not standardized by unit of exposure. Starr's work did standardize by exposure. We're not sure what is the best procedure. They're different measures; they might have different results, and there needs to be some research to determine the effects of the different measures and to indicate whether one measure is better than the other.

L. B. Lave *(The Brookings Institution)*

Your newspaper study could either be interpreted to suggest that the media are the culprit, fostering all of these terrible beliefs or else that the media are simply reflecting what people think is important anyway. Even without newspapers, people may have these perceptions. Which of those two views do you think is true?

P. Slovic

I think they are both true. We've talked with some people in the media and they feel that they have a responsibility to direct peoples' perceptions to what is important. Also, it may be that public interest and response to what the media presents influences the media and so forth. I think there is really a two-way interaction going on here.

ETHICS, ECONOMICS AND THE VALUE OF SAFETY*

William D. Schulze

University of Wyoming, Laramie, Wyoming

ABSTRACT

Economists view private risk as a commodity — riskier jobs receive higher wages to compensate workers for voluntarily accepting job related risks of death. In the economist's view, it is then a simple task to determine a money-risk trade-off derived from the job market and apply this trade-off in making decisions on social or public risks (as opposed to private risks). However, whereas private risks are typically compensated (payment is made *a priori* to accept risk), social or public risks are typically uncompensated. This distinction, compensated versus uncompensated risks, has long been observed, but in terms of a parallel terminology, voluntary versus involuntary risks. Society seems much more averse to the latter as opposed to the former. Economics as a behavioral science does not provide any clues as to this distinction. However, ethics, or rather the study of ethical systems, does provide a rationale for distinguishing between compensated and uncompensated risks. In this paper, four ethical systems — Utilitarian, Nietzschean, Rawlsian, and Libertarian — are examined as they apply to the question of the right or wrong of imposing or accepting risks. The conclusion of this analysis is that all of the four ethical systems examined accept compensated risks but reject uncompensated risks under at least some circumstances. The technical implication for economic analysis is that if any of these four ethical systems are used to construct a social welfare function, that function may imply lexicographic social preferences between public and private safety.

This paper is drawn in great part from Chapter Two of "A Study of Ethical Issues in Benefit Cost Analysis," by Shaul Ben-David, Allen Kneese and William D. Schulze. This study was funded in part by the National Science Foundation Program in Ethics and Values in Science and Technology. Additional support in risk assessment for the author has been obtained from Los Alamos Scientific Laboratory and from Resources for the Future. Thanks go to Douglas McClean, David Brookshire, Todd Sandler, Mark Thayer and Bruce Forster for helpful comments. All opinions, conclusions and errors are, of course, the sole responsibility of the author.

References pp. 229-230.

INTRODUCTION

A number of ethical questions persistently trouble economic policy analysis — which usually takes the form of a benefit-cost study. One of the most vexing is the problem of public safety. Many "costs" take the form of risks to human life. Benefits result from reducing such risks or increasing public or private safety. But, how are such safety benefits to be estimated in dollar terms without raising ethically objectionable notions such as a dollar value for human life? Economists have approached this issue recently by scrupulously avoiding any notion of valuing a *particular* human life. In fact, if a willingness to pay measure is used, the value of a particular life can be taken as, from economic theory, infinity since to induce a particular rational individual to accept a risk of *certain* death would require infinite compensation (no value is large enough). However, individuals do, in fact, accept compensation for small risks of (uncertain) death. Examples abound. Driving to work entails a small risk presumably compensated as part of wages or salaries. Jobs which are riskier than others can usually be shown to be paid more, all other things equal. Thus, a TV antenna installer voluntarily risks falling from rooftops, but presumably demands *a priori* compensation for the additional risk he undertakes. Studies of wages and risk have shown that differing groups and individuals require between $340 (Thaller and Rosen, 1976) to as much as $1,000 (Smith, 1974) more in annual income to *voluntarily* accept increased annual job related risks of death of about one in one thousand (.001). Economists then argue that if a government project or program reduces risk to 1000 people by .001 that we would expect to save one life (.001 X 1000) and that the increased safety is worth between $340,000 and $1 million as a benefit ($340 X 1000 people or $1,000 X 1000 people). Note, this is an *a priori* measure of the value of safety to 1000 people, of whom one is expected to die, not an *ex post* measure of how much should or would be paid to compensate for a particular death.

However, grave difficulties remain. First, in observing actual expenditures for safety, enormous differences are apparent. Many traffic safety programs — more barricades, traffic lights, etc. — could save an expected life at a cost of less than $100,000. Given the numbers above, it would seem that the public would clamor for better roads or at least buckle their own seat belts more often than about 25% of the time. On the other hand, public airline safety may cost more than one million dollars per expected life saved, yet public pressure still exists for safer airline operations. It appears then, that the public rather readily accepts individuals killing themselves in their own automobiles, but views public airline safety in a very different way. Economists tend to make no distinction, but it may well be that a significant ethical difference exists between public and private safety, i.e., knowingly imposing a risk on oneself is "right," while imposing a risk on someone else is "wrong."

In focusing on an analysis of safety in this paper, we first develop and define the notion of an ethical system in the next section. The following section then treats safety in an ethical context and concludes with an example discussion of how benefit-cost analysis can be modified to incorporate weighting structures consistent with alternative ethical systems. These weightings allow alternative specifications for a particular benefit-cost analysis of a safety program to test the sensitivity of outcomes to the ethical assumptions embodied in the analysis.

ETHICAL SYSTEMS

Ethical systems attempt to provide a mechanism for answering the question: "Is a contemplated action right or wrong?" An ethical system can take the form of a list of rules. Examples of this class of ethical systems include such specific lists as the Ten Commandments and Kant's Categorical Imperative, which states: "Act only on the maxim whereby thou canst at the same time will that it should become a universal law." Note that the Ten Commandments provides a list of specific behavioral rules while Kant provides a mechanism for generating such a list. The difficulties with lists, however, are first that some of the rules may well come into conflict (be inconsistent) under some circumstances, as a result requiring a hierarchical ordering of rules to resolve conflicts. Second, such lists, if explicit, may fail to cover certain eventualities.

An alternative specification of an ethical system can take the form of a criterion for evaluation. Thus, for example, "do unto others as you would have them do unto you," can be applied to nearly all ethical decisions. Similarly, the statement, "turn the other cheek" and "individuals should have freedom of choice where no one else is bothered" imply that ethical behavior involves not harming others under any circumstances and yield a general criterion or decision rule.

The latter approach, ethical systems based on ethical criteria, can generally be incorporated into economic analysis by reweighting benefits and costs according to the particular criterion. The former (i.e., lists) is potentially much more difficult to treat, in that, a list of rules would become a set of mathematically specified constraints to a decision process such as benefit-cost analysis. Thus, in this initial exploration of ethics, economics and risk, we will focus on ethical systems which take the form of general criteria.

A second difficulty in merging ethics and economics is the possibility that an ethical criterion and a basic assumption of economic analysis are incompatible. An example is the democratic ethic — what is "right" is what the majority approves. Arrow has shown in his impossibility theorem that majority voting may imply intransitive social preferences. Transitivity is a fundamental assumption of economic theory necessary even for the basic concept of economic efficiency. Economic analysis thus requires that "if situation A is socially preferred to situation B and B to C., then A is preferred to C." However, even if individuals are transitive in their preferences, Arrow has shown that majority votes can result in social preferences of the "situation A is preferred to B, B is preferred to C, but C is preferred to A" (intransitive) type. Thus, we also require that the ethical criteria employed be transitive.

This requirement, that an ethical system can provide a transitive criterion for social choice leaves at least four ethical systems (and probably more) for analysis. These four are described in a little detail below.

References pp. 229-230.

Utilitarian (Benthamite) — A Utilitarian Ethical system requires "the greatest good for the greatest number" as expressed by Jeremy Bentham (1789), John Mill (1863), and others. The social objective is to maximize the sum of the cardinal (measurable) utilities of all individuals in a society. Thus, for an individual to take an ethically "correct" action, all consequences of that action must be considered. Thus, the Utilitarian Ethic has a pragmatic consequentialist character which in a matter of fact way is quite appealing: if the utility gain of an action exceeds the utility loss across society, the action is "right." If the utility gain is less than the utility loss across society, the action is "wrong."

In addition to the obvious difficulty in making the requisite calculations necessary for moral choices, a fundamental problem afflicts Utilitarians — measuring utility. The problem of distributing income will serve to demonstrate the problem of measurable or cardinal utility. First, we will make the assumption, consistent, for example, with the view of Pigou (1920) that all individuals have the same utility function. Thus, turning to Fig. 1, Mr. A and Mr. B have the same relationship between utility (U_A and U_B, respectively) and income (Y_A and Y_B, respectively).

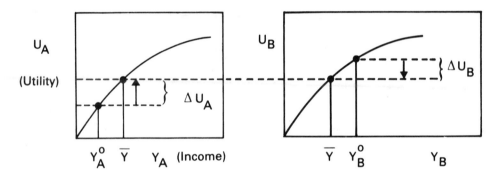

Fig. 1. Increasing A's income and decreasing B's income increases total utility. Identical utility functions implies identical incomes.

If the initial position of society is such that Mr. B is wealthier than Mr. A, $Y_B^o > Y_A^o$, then B has a higher utility level than A. Given the traditional Utilitarian assumption of diminishing marginal utility — that the utility curves in Fig. 1 "flatten out", it is easy to show that we can improve society's total utility by giving A and B the same income, $\overline{Y}$. This is true because, by raising A's income from Y_A^o to $\overline{Y}$, we get a gain in utility of ΔU_A to A compared to the loss in utility ΔU_B to B resulting from lowering B's income from Y_B^o to $\overline{Y}$. Note that $Y_B^o - \overline{Y} = \overline{Y} - Y_A^o$, so we take the money away from B to give to A and get a gain in total utility, $U_A + U_B$, since $|\Delta U_A| > |\Delta U_B|$ or A's gain exceeds B's loss.

The same solution can be obtained by solving the following problem:

$$\text{maximize } U_A(Y_A) + U_B(Y_B) \tag{1}$$

$$\text{subject to } Y_A + Y_B = Y_A^o + Y_B^o \tag{2}$$

which implies at the optimum that $\Delta U_A/\Delta Y_A = \Delta U_A^o/\Delta Y_B^o$ or that the rate of increase of utility with income (marginal utility) must be equal for the two individuals. Since the two individuals in the example have the same utility function, marginal utilities are equated where incomes are the same, $Y_A = Y_B = \overline{Y}$.

If, on the other hand, we assume different individuals have different utility functions, e.g., Edgeworth in *Mathematical Psychics* argues that the rich have more sensitivity and can better enjoy money income than the poor, we end up with a situation like that shown in Fig. 2. Y_A^* and Y_B^* are optimal incomes for A and B because the marginal utilities of income are equated. Mr. A gets more income than Mr. B because he obtains more utility from income than B does. In Edgeworth's view, Mr. A by his sensitivity should have more money to be used in appreciating fine wine than B, who is satisfied with common ale.

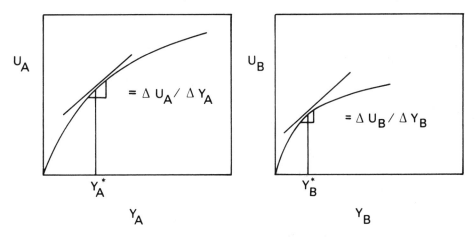

Fig. 2. If A and B have different utility functions an unequal distribution of income is "best".

Obviously, then, depending on beliefs about measurable utility functions, any distribution of income can be justified, ranging from a relatively egalitarian viewpoint (Pigou) to a relatively elitist viewpoint (Edgeworth).

There do exist ethical systems which are totally egalitarian and totally elitist. These diametrically opposed systems, developed extensively by John Rawls and Friedrich Nietzsche, respectively, are described next.

Totally Egalitarian (Rawlsian-Kantian) — Rawls proposes in his book, *A Theory of Justice* (1971), that the well being of a society can be measured by the well being of the worst off person in that society. This simple notion would lead, if adopted, to a totally egalitarian distribution of income.

The Rawlsian criterion can be expressed mathematically as follows: For two individuals A and B, where utility is denoted U, if $U_A < U_B$ we maximize U_A subject to $U_A \lesseqgtr U_B$; if $U_B < U_A$ then we maximize U_B subject to $U_B \lesseqgtr U_A$. If we reach a state where $U_A = U_B$, then we maximize U_A subject to $U_A = U_B$. The implication of this for redistribution of income is that we begin by adding income to the worst off individual (taking income away from wealthier individuals) until he catches up with the next worst off individual. We then add income to both individuals until their utility levels (well being) have caught up to the third worst off, etc. Eventually, this process must lead to a state where $U_A = U_B = U_C = U_D \ldots$ for all individuals in a society, where all utilities are identical. This criterion can be written more compactly for a two person society as max min $\{U_A, U_A\}$, so we are always trying to maximize the utility of the individual with the minimum utility. Implicit also in Rawls' arguments (e.g., the veil of ignorance) is the assumption that individual's utility functions with respect to income are about the same. Thus, a Rawlsian ethic would work towards a relatively equal distribution of income based on need. Rawls claims that he is Kantian and that his criterion is consistent with Kant's Categorical Imperative as well. Thus, we will assume that as a weighting scheme for benefit-cost analysis, the totally Egalitarian view embodies Kant as well.

Totally Elitist (Nietzschean) — A Nietzschean criterion can be derived as the precise opposite of the Rawlsian criterion discussed above. The well being of society is measured by the well being of the best off individual. Every act is "right" if it improves the welfare of the best off and "wrong" if it decreases the welfare of the best off.

Lest the reader dismiss the Nietzschean criterion as irrelevant for a Western democratic society, a number of elitist arguments should be mentioned. The gasoline shortage of the summer of 1979 moved Senator Hayakawa of California to comment, "The important thing is that a lot of the poor don't need gas because they're not working." Economic productivity can, in this sense, rationalize a defined "elite." Thus, concepts of merit can be elitist in nature, e.g., those who produce the most "should" have the largest merit increases in salary (even though they may already have the highest salaries).

Income distributional questions are a bit more complex in that the solution is not simply to give all of society's wealth to the best off. This occurs because, if between two individuals A and B, we are attempting to

$$\text{max max } \{U_A, U_B\} \tag{3}$$

or to maximize the utility of the individual who can attain the greatest utility, we must first find the solution for max U_A, and then separately for max U_B. and then pick whichever solution gives the greatest individual utility. Obviously, it will usually be better to keep B alive to serve A, i.e., to contribute to his well being than to give B nothing if A is to be best off. Thus, subsistence is typically required for B. Similarly, if we have two succeeding generations, it may well be "best" for the first generation to save as much as possible to make the next generation better off. Thus, an elitist viewpoint may support altruistic behavior.

Libertarian (Paretian, Christian) — The last of our ethical systems is an amalgam of a number of ethical principals embodied both in the Libertarian view (see Nozick, 1974) that any individual action is moral if no one else is harmed, and in a Christian ethic, "turn the other cheek." Thus, we are not concerned with changing the initial position of individuals in society to some ideal state, but rather in benefitting all, or at least preventing harm to others, even if they are better off. This ethic has been embodied often by economists in the form of requiring "Pareto Superiority," that all persons be made better off or at least as well off as before. Any act is then immoral or wrong if anyone else is harmed. Any act which improves an individual's or several individuals' well being and harms no one is moral or "right." The acceptance of the initial social position, even if elitist or highly unequal in income distribution as part of a Libertarian ethic does not, however, imply consistency with an elitist ethic. Nietzsche, for example rejects this view as a "slave mentality" in attacking Christianity.

A Libertarian Ethic does not define a best distribution of income. Rather, the criterion requires that any change from the existing social order harm no one. If, for example, Mr. A and Mr. B initially have incomes Y_A^o and Y_B^o, we then require for any new distribution of wealth (Y_A, Y_B) — which would be the case, for example, if more wealth becomes available — that

$$U_A(Y_A) \gtreqless U(Y_A^o) \tag{4}$$
and
$$U_B(Y_B) \gtreqless U(Y_B^o) \tag{5}$$

or each individual must be at least as well off as he initially was. Any redistribution, e.g., from wealthy to poor or visa versa, is specifically proscribed by this criterion. Thus, this criterion while seemingly weak, i.e., it does not call for redistribution, can block many possible actions if they do, as a side effect, redistribute income to make *anyone* worse off, however slight the effect may be. Often then, to satisfy a Libertarian criterion requires that gainers from a particular social decision must actually compensate losers (for a discussion of compensation, see E. J. Mishan, *Introduction to Cost-Benefit Analysis*, 1971).

The four ethical systems presented above are, as noted, by no means exhaustive. While some, such as a democratic ethic, have been excluded on technical grounds, others must await future treatment in our analysis. However, a Darwinian ethic — survival of the fittest — may imply that the existing distribution of income is ideal — derived from competition between individuals in society and be consistent with benefit-cost analysis as it is currently performed — in unweighted dollars (see Ayn Rand, 1964 for discussion of a "Darwinian" view). Of course, the rules under which competition occurs may come under question. For example, is lying permissable in a business contract? In any case, the four ethics chosen here plus unweighted addition of benefits and costs do have the advantage of simplicity, but may in turn represent, at least in their mathematically specified forms — necessary to utilize them in benefit-cost analysis — considerable oversimplications. All ethical systems as logical constructs may, however, suffer from this charge.

ECONOMICS, ETHICS AND SAFETY

The economics of safety has developed rapidly over the last several years. Unfortunately, earlier misguided attempts at measuring the value of safety programs have given economists a "black eye" for supposedly advocating that individual human lives could be valued as the lost economic productivity associated with a shortened life span. This view, pursued in great detail by Dorothy Rice (1966) and used by Lave and Seskin (1970) and others, implied that the value of the life of, for example, a 50 year old carpenter would be the remaining earnings to retirement age. The value of the life of a retired female (somebody's grandmother) was by the same argument taken to be zero. Similarly, small children, since many years would pass before they could begin to earn productive income, were valued at next to zero, given the notion of discounting future earnings at a market rate of interest. Elaborate calculations were made for different individuals on the basis of age, occupation, sex, etc., to determine the value of remaining earnings as a measure of the "value of life." On economic theoretical grounds all of these calculations have been shown to be nonsense. However, permanent harm remains in that many decisionmakers now shy away from any attempt to value the benefits of safety programs in dollar terms.

The economists' notion that individuals do voluntarily trade off safety for monetary compensation in no way attempts to value life (Mishan, 1971). Rather, the question is asked, how much do individuals require as *a priori* compensation to voluntarily accept a small additional risk of death? Note then, that in studying the problem of risk, economists utilize data on behavior where monetary compensation is actually paid for accepting risk. We will find in analyzing ethics and risk, that the economist is utilizing a rather special situation to derive estimates of the value of safety. But, for the moment, let us follow through on the notion of a trade-off between safety and monetary compensation.

Imagine a game of Russian Roulette in which an individual is offered sums of money to participate voluntarily. If, for example, the risk of the gun firing when the trigger was pulled were only one in ten thousand, and the compensation for accepting the risk was $1000, current economic studies would suggest that most people would accept the risk (this is a much better proposition — risk versus compensation — than driving to work for a typical day's pay!). However, economic theory suggests that as the probability of death increases, monetary compensation would have to increase dramatically. Fig. 3 shows the expected kind of relationship between compensation and risk. Clearly as the probability of death approaches unity, compensation approaches infinity — odds are that the participant won't survive the game to enjoy the proceeds, so no amount of money is sufficient. Note, however, that for small increases in annual risk of death such as those associated with risky jobs (typically less than .001 per year) annual job compensation is increased by $340 - $1000, as shown in Fig. 3. Thus, economists now focus on the far left hand side of Fig. 3, only dealing with the dollar values necessary to compensate individuals for small voluntary risks. Total benefits for a safety program may, of course, sum up to a large dollar figure if many people obtain small reductions in risk through a public policy action.

However, this method of valuing risk seems at least in part out of accord with observed human behavior. Just as the old value of life measure used by Rice and Lave and Seskin leads to counter intuitive results (e.g., grandmothers do not take

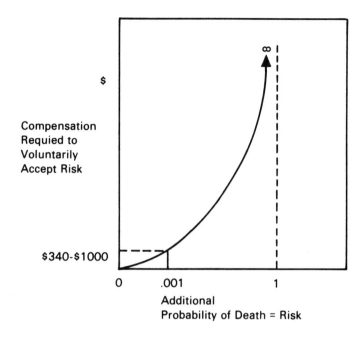

Fig. 3. Compensation required to accept additional risk of death.

up hang gliding because their lives are "worthless" nor does society place a near zero value on the lives of children) the new measure of the value of safety seems to ignore the special importance many individuals place on involuntary or uncompensated risks. The risks associated with nuclear reactor accidents or nuclear waste storage, risks associated with public transportation including airlines, risks of flood, fire, earthquakes, or other disasters, all seem to be treated differently both in a social and individual perspective than do voluntary-compensated risks (for discussion on this point, see Starr, 1968). Economic theory, and consequently, empirical estimates of the value of safety have notably failed to account for these differences. Rather, economists have argued that placing a different value on safety in different situations is economically inefficient.

The logic behind this argument is as follows: given a fixed safety budget, if we have a program that can save one life for $50,000 and another program which can save ten lives for $50,000, we should pick the second program. Further, given a larger total budget, we would wish to pick the combination of programs which we would expect to save the greatest number of lives. We should pursue each of the programs to the extent that each additional life saved by a program costs the same at that margin.

The obvious counter to this argument is that individuals may well value safety differently in differing settings. Are uncompensated risks less ethically acceptable than compensated risks? We can use the four ethical systems presented above to analyze uncompensated risk as follows: assume we have two identical individuals,

A and B with the same utility functions. If A imposes a small risk on B and receives a benefit equal to B's incremental or marginal value of safety we have a situation which satisfies traditional benefit-cost analysis, i.e., benefits to A are equal to costs on B at the margin in dollar terms, so the situation is accepted. However, from an ethical perspective, A is imposing an uncompensated risk on B.

Technically, we define A's expected utility as:

$$E_A = (1-\pi_A^o) \, U_A(Y_A^o + G\Delta\pi_B) \tag{6}$$

where π_A^o is A's risk of death, U_A is A's utility which is a function of his initial income, Y_A^o, plus the monetary gain G, for imposing risk on B, times the increase in risk imposed on B, $\Delta\pi_B$. An example of this situation would be Mr. A building a dam upstream of Mr. B for his own gain. Mr. A receives a net benefit of G for each unit of risk (of failure of the dam which could drown Mr. B) which he imposes on Mr. B. This net gain would result from the value of irrigation water net of the cost of constructing the dam. In this context, benefit-cost analysis as traditionally practiced would argue that marginal net benefits to A must exceed or equal marginal costs (of risk) to B, as we have defined the problem, so the dam is acceptable.

Mr. B's expected utility where he is not compensated for risk imposed by A, is then defined as:

$$E_B = (1-\pi_B^o - \Delta\pi_B) \, U_B \, (Y_B^o) \tag{7}$$

where π_B^o is Mr. B's original risk of death, $\Delta\pi_B$ is the additional risk imposed by A and U_B is Mr. B's utility, a function of his income, Y_B^o.

If we were to compensate Mr. B for voluntarily accepting risk from Mr. A at a rate of C dollars per unit risk, Mr. B would maximize his compensated expected utility,

$$(1-\pi_B^o - \Delta\pi_B) \, U_B \, (Y_B^o + C \cdot \Delta\pi_B), \tag{8}$$

with respect to $\Delta\pi_B$, which implies

$$C = \left. \frac{U_B}{U_B^{'} \, (1-\pi_B^o)} \right|_{\Delta\pi_B \,=\, 0} \tag{9}$$

or that C equals B's marginal value of safety, the definition of the right hand side of the expression above. Thus, traditional benefit-cost analysis requires that G, Mr. A's marginal gain, exceed or equal C, Mr. B's marginal loss, both valued in dollar terms for the first unit of risk to be imposed. However, Mr. A does not have to actually compensate Mr. B at a rate equal to or exceeding C, as defined above.

How, alternatively would the four ethical systems described above view the situation where B's risk is uncompensated? Where social welfare, W, for the first three criteria is defined as follows:

Utilitarian $\qquad W = E_A + E_B \tag{10}$

Totally Egalitarian $\quad W = \max \min \{E_A, E_B\} \tag{11}$

$$\text{Totally Elitist} \quad W = \max \max \{E_A, E_B\} \tag{12}$$

The condition for acceptability is that $[dW/d\pi_B]_{\Delta\pi_B} = 0 \geq 0$. If $[dW/d\pi_B]_{\Delta\pi_B=0} < 0$, then the risk, $\Delta\pi_B$ is rejected. For the Libertarian Ethic, we require:

$$\text{Libertarian} \quad E_A \geq E_A^\circ \equiv (1-\pi_A^\circ) \, U_A \, (Y_A^\circ) \tag{13}$$

$$E_B \geq E_B^\circ \equiv (1-\pi_B^\circ) \, U_B \, (Y_B^\circ) \tag{14}$$

or that both parties be better off than before the initial state preceding consideration of the risk $\Delta\pi_B$ on Mr. B.

Table 1 presents a summary of an analysis for uncompensated risk where we assume that A's incremental gain, G, is equal to B's marginal cost of risk, C in dollar terms; that the *a priori* risks π_A° and π_B° are identical and that utility functions U_A and U_B are identical as well. (A detailed theoretical treatment of this model is available by request from the author.) The example is, of course, structured so that traditional benefit-cost analysis just accepts the imposition of uncompensated risk. However, the Utilitarian ethic rejects the situation if Mr. B is worse off than A, i.e., B has a lower income, $Y_A^\circ > Y_B^\circ$. Similarly, the Totally Egalitarian or Rawlsian Ethic rejects the imposition of risk by Mr. A on Mr. B if B is worse off initially. The Totally Elitist Ethic implies the converse: that if Mr. A is better off, he has the ethical right to impose a risk on Mr. B. Finally, the Libertarian Ethic rejects the notion of uncompensated risk no matter what the initial distribution of wealth, since, by definition, an uncompensated risk makes someone worse off.

Table 1 is surprising in several respects. First, it is often supposed that benefit-cost analysis can be justified, or at least is supported by the Utilitarian Ethic. Clearly, at least for uncompensated risks, this is not necessarily the case.

TABLE 1

Uncompensated Risk
"A" Imposes Risk on "B"
"A's" Gain Equal to "B's" Loss

Ethic	$Y_A^\circ > Y_B^\circ$	$Y_A^\circ < Y_B^\circ$
Utilitarian	Reject	Accept
Totally Egalitarian	Reject	Accept
Totally Elitist	Accept	Reject
Paretian	Reject	Reject
Traditional B/C	Accept	Accept

Second, all of the four ethical systems examined reject uncompensated risks at least some of the time. This may in part explain the failure of the traditional

economic view of a uniform-smooth risk trade-off, as expressed in Fig. 3, to predict the observed societal aversion to uncompensated public or social risks. All of
the ethics consider some uncompensated risks "wrong," implying no trade-off
exists at all! Individuals, may, of course, adopt "ethical" preferences in accordance with the ethical systems we have described. In cases where individual
preferences reject uncompensated risks as "wrong," the preferences of individuals
become lexicographic. This term implies, in the case of safety, that among a set of
alternatives, the alternative with the most public safety (least uncompensated risk)
is always preferred to all other alternatives. Fig. 4 shows a mapping of the
preferences of an individual with lexicographic preferences between public and private (uncompensated and compensated) safety. If the individual initially has a
combination of public and private safety denoted z^0, all points to the right of z^0,
marked as the shaded area, are preferred because they have more public safety,
i.e., less uncompensated risk. Thus, any point like z^2 which has less private safety
but more public safety than z^0, is preferred to z^0. Note that z^1, which has more private safety but less public safety is clearly inferior to z^0. The only points which do
not increase public safety but which are still preferable to z^0 are those marked
above z^0 by the solid line,

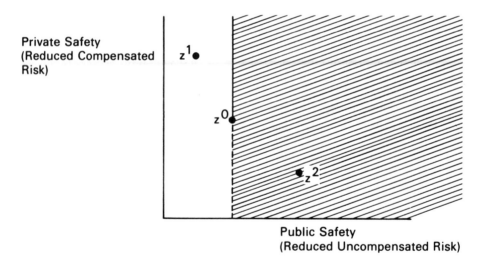

Fig. 4. Lexicographic preferences for safety.

i.e., those which maintain the same level of public safety but which increase private safety. Thus, with lexicographic preferences, private safety is still desired, but
only secondarily to public safety. Clearly, where individuals or society invokes an
ethical objection to uncompensated risks, social or individual preferences are lexicographic. Economists have always assumed preferences of the sort shown in Fig.
5, where points such as z^1 (or those in the shaded area which are all preferred to z^0)
are bounded below by a smooth curve (or straight line) as shown, implying a continuous trade-off exists. If Table 1 were to be reworked on the assumption of *a*

priori compensation for risk imposed on Mr. B by Mr. A we would find that all of the ethical systems considered accept compensated risk. Thus, the economists' notion of a money-risk trade-off is perfectly acceptable if *a priori* compensation *actually* occurs.

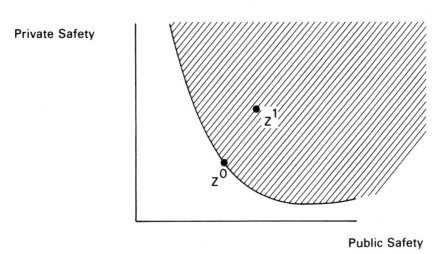

Fig. 5. A trade-off for safety preferences.

Thirdly, the Libertarian Ethic, with its close relationship to the Christian Ethic is of special importance for Western Democracies. This ethical system, although seemingly weak in not requiring redistribution of income, rejects all uncompensated risks, implying lexicographic preferences as described above.

It is perhaps this last point when applied in a broad context which goes furthest in explaining the opposition that benefit-cost analysis has received in analyzing certain issues. For example, in many cases, lawyers and environmentalists have argued that polluting the environment is "wrong" and that economic policies such as selling permits for the right to pollute or taxing pollution are unethical. Similarly, advocates of public safety argue that no amount of effort on the part of airlines, operators of nuclear reactors or large hydroelectric dams is sufficient when uncompensated risks are involved for the general public. These attitudes which are reflected in enormously higher expenditures per life saved in some safety programs as opposed to others seem consistent both with a Libertarian Ethic and with Lexicographic preferences.

REFERENCES

1. K. Arrow, *Social Choice and Individual Values*, John Wiley and Sons, New York, 1963.
2. J. Bentham, *An Introduction to the Principles of Morals and Legislation*, 1789.
3. S. Bok, *Lying*, Random House, New York, 1978.

4. F. Edgeworth, *Mathematical Psychics: An Essay on the Application of Mathematics to the Moral Science*, A. M. Kelley, New York, 1967.
5. I. Kant, *Fundamental Principles of the Metaphysic of Morals*, 1785.
6. L. Lave and E. P. Seskin, "Air Pollution and Human Health," *Science*, 109: 723-732, August 1970.
7. J. Mill, *Utilitarianism*, 1863.
8. E. Mishan, *Introduction to Cost Benefit Analysis*, Praegar, New York, 1971.
9. F. Nietzsche, *Beyond Good and Evil*, 1886.
10. R. Nozick, *Anarchy, State, and Utopia*, Basic Books, New York, 1974.
11. T. Page, *Conservation and Economic Efficiency*, Johns Hopkins, Baltimore, 1977.
12. A. C. Pigou, *The Economics of Welfare*, 1920.
13. A. Rand, *The Virtue of Selfishness*, American Library, New York, 1964.
14. J. Rawls, *A Theory of Justice*, Belknap Press, Cambridge, 1971.
15. D. Rice, *Estimating the Cost of Illness*, U.S. Department of Health, Education and Welfare, Public Health Service, Home Economics Series No. 6, May, 1966.
16. W. D. Rowe, *An Anatomy of Risk*, John Wiley and Sons, 1977.
17. R. S. Smith, "The Feasibility of an 'Inquiry Tax' Approach to Occupational Safety," *Law and Contemporary Problems*, Summer-Autumn, 1974.
18. C. Starr, "Social Benefit vs. Technological Risk," *Science*, 165: 1969.
19. R. Thaler and S. Rosen, "The Value of Saving a Life: Evidence from the Labor Market," in *Household Production and Consumption*, N. E. Terleckyj, ed., Columbia University Press, New York, 1976.

DISCUSSION

R. A. Howard *(Stanford University)*

First of all, I'm glad to see that I agree with you in terms of your bottom line, that just about every ethical approach rules out prospective analysis. I would point out, however, that it seems to me that the set of ethical approaches that you have depicted here are what we might call "end state" ethics. That is they look at society as a snapshot, and say what the people like and then attempt to say which way is up. Well, I think I'd be more impressed by process ethics as well. That is, we have to look not only where we are now, but how we got here and the ethical choices that were made in that process. For example, if somebody went to Las Vegas and put all his money on the line and lost, you might have a different attitude toward his being poor than of a person who had made a lot of, what we would say quite reasonable, but productive, choices in society and ended up in that state. I guess the most dramatic example of this is the old story about the person who killed his parents and then threw himself at the mercy of the court because he's an orphan.

W. D. Schulze

Two brief responses. First, let me identify compensated risk as opposed to uncompensated risk. Now imagine there was a circumstance where we're talking about a TV antenna installer who's getting a wage to go up there and undertake that risk. Either we have indifference or acceptance all the way across. So the problem with traditional benefit cost analysis is that it is only appropriate for com-

W. D. Schulze

pensated risks. If we are actually compensating, then none of the ethics really reject the imposition of risk.

My second comment is yes, I agree with you in terms of process ethics as opposed to this cartoon, but this represents a first, a very preliminary step to look at benefit cost analysis and ethics. So it was just the first step, a lot more can be done.

R. A. Howard

I would just suggest that people who want to pursue this might want to read Nozick's book. Nozick is a contemporary and colleague of Rawls and has a quite different view of the ethical system.

H. Joksch *(Center for the Environment & Man)*

I wonder whether you used, as a definition of egalitarian, the maximum or the minimum. Why not the difference? You might get a quite different result.

W. D. Schulze

Certainly. If I didn't, then I should have said, "I tried four and there are an endless number." Some of them get bounced by some of the fundamental axioms of choice which are necessary to have a consistent criterion like transitivity. So, for example, a democratic ethic won't work in a benefit-cost framework because it may imply intransitive social preferences. If you don't know what I'm saying, don't worry about it. If you do you'll know what I mean.

PROBLEMS AND PROCEDURES IN THE REGULATION OF TECHNOLOGICAL RISK*

Dorothy Nelkin and Michael Pollak

Cornell University, Ithaca, New York

ABSTRACT

Until recently, evaluating the risks of technology has been considered a technical problem, not a political issue; a problem relegated to expertise, not to public debate. But disputes have politicized the issue of risk and led to the development of procedures to enhance public acceptability of controversial projects. Our paper reviews various types of hearings, public inquiries, advisory councils, study groups and information forums that have been formed in the United States and Western Europe to resolve disputes over science and technology. We analyze the assumptions about the sources of conflict and the appropriate modes of decision making that underly these procedures. And we suggest some reasons for their rather limited success in reducing political conflict and achieving public consensus.

INTRODUCTION

The public bureaucracies responsible for regulating complex technologies are increasingly challenged by scientists calling public attention to the risks of technology-based projects, and by citizen groups demanding greater accountability and participation in technical policy decisions. Indeed, decisions about science and technology, long considered the domain of expertise, are more and more the targets of political controversy. And the persistence of conflict — over nuclear power plant siting, airport expansion, food additives, toxic substances, chemical carcinogens; the list is long and growing — indicates that existing procedures for

A version of this paper appeared as D. Nelkin and M. Pollak, "Public Participation in Technological Decisions: Reality or Grand Illusion," Technology Review, September 1979, 55-64. The authors would like to acknowledge the German Marshall Fund and the National Science Foundation EVIST Program for their research support.

References pp. 247-248.

mediating among competing interests and resolving political conflicts are far from adequate.

Until recently, the risks of technology have mainly been perceived as a technical problem, not a political issue; a problem to be relegated to expertise, not to public debate. But controversies have politicized the issue of risk, called attention to the interests and the question of power that are involved. Several features of these disputes have contributed to this politicization and to the difficulties of bureaucratic decision-making.

First is the nature of the risks that are characteristic of many new technologies. These risks are often "invisible" and vaguely understood. How does one know if a lethal gene has been produced by recombinant DNA research? Or if a nuclear waste storage facility is adequately protected against radiation leakage? Often it is the fear of an unlikely but possibly devastating catastrophe (e.g. a nuclear accident) that sustains conflict. Problems of regulation derive from inherent ambiguities in the data about risk that permit conflicting technical interpretation. Scientists' disagreement and their public disputes reveal their limited ability to predict the environmental and health effects of certain technologies. This is particularly upsetting to those responsible for making decisions. For how are regulatory authorities and legislators to make judgments when faced not only with diverse political interests but also with conflicting evidence and polarized expert opinion? How can decisions be made about the acceptability of risk given the confusion among facts, their interpretation and their subjective evaluation?

A related feature of recent technological disputes is the vague and shifting boundary between the technical and political discourse. Questions of risk can be defined in technical or political terms; even if scientific consensus exists, finding an acceptable level of risk requires weighing costs and benefits and balancing competing priorities.

Finally, concerns about risk are often translated into questions about political authority and the legitimacy of decision-making procedures. This reflects a far broader characteristic of modern industrial societies — the declining trust in decision-making authority and expertise. The credibility of scientific expertise as a reliable basis for decision-making is suspect: public bureaucracies, accustomed to relying on expertise, find that their technical assessment of risk bears little relationship to public attitudes. Other factors, including the acceptability of the decision-making process, intervene to affect the public assessment of controversial projects.

These three features of disputes over science and technology have shaped efforts to resolve them — to find means to establish rules of evidence, to assess the adequacy and competence of information, and above all, to assure the procedural conditions that will assure public acceptance of specific projects and restore confidence in decision-making authorities. This paper will first describe some of these efforts in the United States and Western Europe, and then analyze their limits in the context of the expectations of opposition groups.

EXPERIMENTS IN CONFLICT RESOLUTION

Experiments in conflict resolution follow from the assumption that appropriate procedures to evaluate risk will lead to public acceptance of controversial technologies. However, ideas about what procedures are appropriate vary considerably depending on how the problem of risk is defined. Is it mainly a problem arising from insufficient technical evidence? Then the goal is one of ascertaining "scientific truth" and this leads to a structure based on scientific advice. However, where controversy is defined rather in terms of the public's acceptance of risk, a more participatory or consultative system is developed to advise decision-makers about the terms of public acceptability. Often the problem of acceptability is defined in terms of information; it is assumed that people fear risks because they are poorly informed. The task is then to educate the public. But if concern about risk if defined as a political problem of trust in expertise, then participatory procedures seek to involve the public in defining the salient information as a basis for greater influence. These four definitions of the problem of risk underlay differences in the procedures created to resolve conflict, and we have organized our discussion accordingly. Indeed some experiments are primarily intended to advise decision-makers about scientific opinion and about public concerns. (See Table 1) Others are intended to inform the public and are dissociated from the decision-making process. (See Table 2) Some of these advisory and information procedures are organized around an elite, that is, scientists and officials from large representative associations or governmental institutions. Others are more participatory, including civil servants and a broader public constituency.

Advisory Models: Elitist — Among the more well-publicized and controversial proposals to resolve technological disputes in the United States is the "Science Court", a quasi-judicial procedure in which scientists with different views on issues such as nuclear safety or the effect of high-voltage transmission lines would argue before "impartial scientific judges" [1]. In this proposed forum, debate would be limited to questions of fact: judges are to give opinions only on factual matters, leaving social value questions for the political arena. But the opinion of the Science Court is expected to have enough authority to provide a basis for policy decisions. Democratic control of technology, claim Science Court proponents, follows from establishing "the truth among conflicting claims of scientists". This concept has seeded a crop of similar proposals for a "Technical Review Board" [2], a "Technological Magistrature" [3] and a new profession of "Certified Public Scientists" [4]. All these proposed institutions would provide neutral judgments to help settle disputes. They assume that scientists, through adversary procedures, can reach agreement on specific questions of risk, and that this will lead to public consensus. Similar positivist assumptions motivated the experimental "energy campaign" in Austria where the government set up structured public debates among scientists with opposing positions on the government's nuclear policy [5]. The purpose, however, was not to verify points of agreement, but rather to highlight the controversial dimensions of the nuclear program and to clarify areas of persistent disagreement.

TABLE 1

Consensus Procedures: Advisory Models

Type	Who Participates	General Intention
Elitist		
1. *Based on Scientific Authority* Science Court (U.S.)	Scientists	An adversary procedure to establish a consensus on "scientific truth".
Energy Campaign (Austria)	Scientists	An adversary procedure to establish consensus on which problems remain controversial within scientific community.
2. *Based on Consensus Among Elite* Advisory Commissions of Science Ministry (Germany)	Scientists; Representatives of Major Institutions	A comprehensive discussion of a project or problem area.
Regional Concertation (France)	Local Dignitaries State Officials Engineers	To reach local agreement in order to implement national polices which affect the local environment.
Royal Commissions (England)	Experts and Officials	Comprehensive discussion to achieve agreement on a controversial issue.
Participatory Public Inquiries Windscale Inquiry (England) Berger Commission (Canada) Policy Intentions (Holland)	Scientists Interested Public	Comprehensive discussion of a project or broad problem area to raise public awareness.

TABLE 1

Consensus Procedures: Advisory Models (cont.)

Type	Who Participates	General Intention
Complaint Investigations Declaration d'Utilite Publique (France) Atomlaw (Germany)	Civil Servants Interested Public Scientists	To formulate objections and claims based on individual or collective rights and to correct decisions.
Citizen Advisory Groups (U.S.)	Community Groups	To allow local community to influence a specific decision affecting local interests.
Environmental Mediation (U.S.)	Environmental Groups Project Developers Third Party Mediator	To allow face-to-face confrontation over specific project and to settle disputes.
Referenda (U.S., Austria, Switzerland, Holland)	Voting Public	To arbitrate a controversial decision by direct vote on an issue.

TABLE 2

Consensus Procedures: Information Models

Type	Who Participates	General Intention
Elitist Conseil d'Information sur l'Energie Electronucleàire (France)	Scientists High Officials Reps. of Ecology Associations	Oversee all information by public agencies. Recommend what should be publicized.
Ad Hoc Groups: Groupe de Bellerive (International) Reflection Groups (Holland) Council for Science and Society (England)	Scientists Enlightened Power Elites	To enlighten decision-making and to generate discussion and publicize information about policy.
Participatory Study Circles (Sweden) Burgerdialog (Germany)	Large voluntary associations and their constituency. Scientists Public at Large.	Broad debate over a problem area to raise public awareness.

Other efforts to enhance the public acceptability of decisions involve, in addition to scientists, the leaders of major associations and state and regional officials. France developed a system of "regional concertation" to facilitate industrial siting decisions [6]. For example, in 1976, seeking to win acceptance of nuclear siting plans, Electricité de France (EDF) circulated dossiers providing technical data on 34 potential nuclear sites to her regional assemblies. They were to examine the dossiers and reach an informed agreement about the acceptability of the sites.

In the German political context, consensus on controversial projects is sought by including leaders of key institutions in advisory committees; in 1975, for example, the Ministry of Science and Technology employed 927 consultant experts from research, industry, unions and other interest groups. But 80% of these experts represented scientific and industrial interests, suggesting that harmony within the sector is perceived as the crucial component of political consensus [7].

Advisory Models: Participatory — The declining influence of the citizen in an expertise-based society is a pervasive concern in many disputes, and most procedural reforms try to involve the citizen more directly in the formulation of policy. This is the rationale behind public inquiries, advisory councils and mediation procedures.

Public inquiries serve as a forum for comprehensive discussion of specific projects and as a channel for the expression of a range of opinions. But the spectrum of opinions normally considered by an inquiry varies. At one extreme, the Canadian Berger Commission, created to assess the Mackenzie Valley Pipeline controversy, was extraordinarily open to nontechnical participation [8]. Intervenor groups were given financial support to develop their case. The Commission heard everyone from fishermen to legislators and considered political testimony as important as technical information. However, such an open procedure is unusual. In most commissions the agenda is limited, dominated by scientists and pervaded by the belief that factual evidence should carry more weight than subjective concerns.

This was the case in the Windscale public inquiry into the British plan for THORP, a thermal-oxide reprocessing plant [9]. Hundreds of individual organizations testified at this 100-day inquiry which was to assess the conflicting arguments in order to evaluate the "facts" of government policy. With no source of funding, objectors had great difficulties in developing a coherent and coordinated position to counter the arguments of British Nuclear Fuels Ltd. Judge Parker, left to interpret conflicting evidence, recommended construction of the reprocessing plant.

In the Netherlands, the government organized an elaborate public inquiry system on the principle that the public must be consulted on all decisions affecting the environment. Government plans are preceded by the publication of "policy intentions" dealing with political and philosophical questions: the objectives of growth, the goals of particular projects and their likely impacts [10]. These are widely distributed for public comment. A representative advisory group analyzes the reactions and the appropriate Ministry must answer the criticism and justify the policy. The entire dossier developed through this process serves as a basis for a parliamentary decision.

Other inquiries are essentially inquests into public objections and complaints about specific projects. This is the purpose of the French Declaration d'Utilite Publique (DUP) and certain provisions of the German Atom Law. To build a nuclear power plant in France, Electricite' de France (EDF) must undertake an inquiry through a "DUP" procedure [11]. Originally developed as a channel for objections to the practice of eminent domain, the DUP has recently become a forum for the expression of concerns about environmental risk and personal safety. In the DUP process, all people living within 5 km from a proposed nuclear site have access to a technical dossier for 6-8 weeks, during which time they can voice their objections. The prefect appoints an investigating commissioner, usually a local dignitary or a retired civil servant, who collects and evaluates the complaints and EDF's response. Using this evidence he recommends whether the project should be given public utility status. To date, no commissioner has ever denied an EDF application.

Under the German Atom Law, electricity companies wishing to build a nuclear plant must apply to the Land administration for a construction permit and hold a public hearing. Documents are available for public inspection for a month, and anyone affected by a project is entitled to object; the courts have accepted claims from people within 100 km from the plant site [12]. But the hearing is restricted to discussion of the nuclear safety; other environmental issues and economic and social concerns are considered irrelevant, a restriction that often limits meaningful lay participation.

Several kinds of participatory procedures have developed in the United States in response to controversies. In 1976 a citizen review board formed in Cambridge, Mass. to advise the Cambridge City Council on a policy for allowing recombinant DNA research in the city. This was organized on the following principle:

> Decisions regarding the appropriate course between the risks and benefits of potentially dangerous scientific inquiry must not be adjudicated within the inner circles of the scientific establishment . . . a lay citizens group can face a technical scientific matter of general and deep public concern, educate itself appropriately to the task and reach a fair decision [13].

For four months, the review board discussed complex questions about the risks of DNA research and, with the modification requiring local monitoring, approved the research under the Federal guidelines set down by the National Institutes of Health. A similar citizen review structure has been proposed by a study group at the Oak Ridge National Laboratory to resolve nuclear siting disputes.

Also in the United States, a number of conflict-resolving procedures utilize techniques of social research to provide advice to decision-makers about actual public preferences in controversial areas. In 1970, the Forest Service employed an instrument called "code-involve" to uncover public attitudes about its proposed use of DDT to protect Douglas Firs. Code-involve is a computerized content analysis system designed to transfer information from very diverse sources (editorials, letters, petitions) into a condensed form useful for policy review. This technique can tap a wider spectrum of attitudes than may emerge through other procedures such as public hearings, and theoretically allows the problems to be defined by the public. However, like opinion polls and similar techniques to uncover public preferences, code-involve is a surrogate for direct participation. Moreover, since there are no provisions to generate a public reaction, it is only useful in cases where significant debate has already generated large quantities of controversial written comment.

Mediation is the latest proposed elixir for environmental disputes in the United States. This process is based on voluntary participation by the contesting parties, who meet face to face in discussions with a third party to facilitate the arbitration of disputes [14]. Mediation procedures have helped to reach settlements in several environmental disputes, for example over highway routing and strip mining policies.

Finally, referenda are a growing feature of the political landscape. In the United States thousands of referenda take place each year on questions ranging from local property taxes to airport expansion or nuclear power plant development. Twenty-one states have petition procedures for placing issues on the ballot; twenty-three states permit direct legislation through voting, and thirty-eight permit voters to review laws passed by the legislature. Thus referenda allow citizens to initiate legislation or to repeal existing laws. Significantly, technological decisions are increasingly appearing on ballots in Europe as well as the United States; Austria, Switzerland, and Holland have all held referenda on nuclear power. While posing significant problems of representation, adequate information and co-optation, the growing popularity of the referendum suggests the interest in dealing with controversial technical questions in a more participatory framework.

Information Models: Elitist — Access to information about controversial projects is a necessary precondition of conflict resolution, but information can serve several ends. Some hope that access to information will enhance trust in administrative decisions; others look to greater access in order to influence such decisions. Information is usually controlled by an "inner circle" of scientists and officials, but some experiments have been organized to involve broad segments of the public in defining salient issues.

In France, public information about nuclear power prior to 1974 consisted mostly of promotional material; EDF distributed comic strips to school children and glossy brochures to adults. As coun023groups responded with a "war of words" it was clear that promotional material increased polarization; so in 1975 EDF cut back its public relations campaign and established a "Groupe d'Information Nucleaire" and a documents center for nuclear information. Then in 1977 the government established a Conseil d'Information sur l'Energie Electro-nucleaire made up of scientists, high officials and representatives from ecology groups. Its purpose is oversight: it reviews the information on nuclear power disseminated by the government to the public, evaluates its quality and completeness and recommends what should be publicized.

Another way to expand information is through establishing *ad hoc* groups that can rise above the polarization characteristic of so many disputes. Most such groups are not officially linked to government; they seek a membership that will satisfy both public opinion and governmental authorities and an image as responsible and respected sources of public information. For example, the Groupe de Bellerive, an international council "for reflection and evaluation", formed out of concern about the violent opposition to nuclear power in Europe. Its purpose is to bring together "minds both enlightened and recognized as such (scientists but also jurists, technicians but also philosophers, economists but also the politically aware, leading bureaucrats but also those elected by the people), capable of analyzing complex problems and forming independent judgments."

The British Council for Science and Society, organized in 1973, assesses new developments in science and technology for their potential social impacts. It includes scientists, lawyers and philosophers who meet to consider contemporary problems and to stimulate informed public discussion.

Similarly, *ad hoc* reflection groups often meet in the Netherlands to discuss controversial policy issues. In 1974, for example, a reflection group of scientists, parliamentarians, industrialists and journalists examined the social and economic dimensions of the nuclear program, and called for a five year reflection period to rethink the basis of national planning. With no formal decision-making authority, all such groups exercise influence mainly by trying to create an informed citizenry and attracting the attention of the media.

Information Models: Participatory — In 1974, disturbed by the anti-nuclear movement, the Swedish government initiated an experiment in public education in the field of energy. Using a system of study groups managed by the principal popular organizations and political parties, the government financed a program to inform broad segments of the public about energy and nuclear power [15]. The program involved some 8,000 study circles each with about ten members who met together to discuss those energy-related questions they felt to be most important. It was fully expected that greater information would create more favorable attitudes toward government policy. Yet reports from these groups suggest continued uncertainty and ambivalence — a lack of consensus that was evident when the Center Party successfully used its anti-nuclear position to displace the Social Democrats in the 1975 elections.

The Burgerdialog in Germany, first organized in 1974, represents a similar effort to involve broad sectors of the public in an information program. Organizations such as churches, unions, and adult education groups are funded to organize discussion groups and meetings which include speakers both for and against nuclear power. The goal is "to strengthen confidence in the ability of the democratic process to function, especially in the controversy over nuclear energy, and to restore confidence wherever it may be undermined" [16]. This information effort is distinct from the decision-making process, intended less to ascertain public opinion than to inform citizens about the necessity for nuclear energy and convince them that risk is minimal. The "dialogue" has thus frequently become a monologue, and it has clearly failed to create consensus over nuclear policy.

PROBLEMS OF PROCEDURAL ACCEPTABILITY

We have described a variety of procedures intended to avoid or to resolve the controversies that obstruct so many decisions about science and technology. Most of these procedures rest on a traditional "welfare model," in which risks are defined as *problems* to be dealt with, mainly by experts. It is assumed that if a problem is solved by a respected group of elites or the best available scientific opinion, this will enhance the legitimacy of public authorities. And it is assumed that if the public is adequately informed, this will result in consensus. These hopes are often dashed: conflict and mistrust persist, and the procedures themselves are often debunked. Neither public participation nor enlightened representation appear to systematically assure the acceptability of controversial technologies. For protest groups see the issue of risk not as a problem to be solved, but as a controversial question requiring dialogue and negotiation.

What then must one do to enhance legitimacy? What kind of procedures would be acceptable to critical groups? To explore these questions we turn to five questions that are frequently asked by opposition groups.

How are the boundaries of the problem defined? Who participates in the experiment? Who conducts the procedure? What is the distribution of technical expertise? Is there really a choice?

Definition of the Problem — A first principle of negotiation over controversial policies is that there be mutual recognition of the real source of conflict. Is opposition to a technology really based on concern about risk or is this a surrogate for more fundamental social concerns? Too often highly political issues are defined as technical; questions about the impact of a technology on community values are translated into arguments about the degree of risk involved. It is somehow assumed that agreement about technical issues will help to resolve questions of political choice.

This definitional bias is most obvious in experiments such as the Science Court which, in trying to differentiate facts from values, ignore the value interpretation that enters even into the collection of data. In the social and institutional context of science and technology today, it is highly anachronistic to seek resolution of conflicts through scientific expertise. Moreover, even if technical consensus could be established, this may have little affect on public attitudes. Technical consensus may narrow the range of choices, but procedures that bypass underlying value concerns will have little effect on the resolution of disputes.

Yet commissions and inquiries also tend to give disproportionate weight to technical evidence over subjective concerns. The wide publicity given to the Canadian Berger Commission hearings suggests that their careful balancing of technical and nontechnical criteria was relatively unusual. For example, the German hearings required by the Atom law restrict discussion to the technical dimensions of nuclear risk, avoiding the economic and social issues that are often the primary concern of the nuclear opposition. Even the broadly participatory citizens review in Cambridge was limited to an agenda set by academic scientists that defined the DNA issue narrowly in terms of short range health hazards and adequate safety measures. Questions of long range risks or of ethics were put aside.

Similar definitional problems impede the efforts to open public access to information. How does one select the materials to be released to the public? Information on sensitive topics such as the extent of police controls over nuclear facilities or evacuation plans in case of accidents are often excluded from circulation despite, or perhaps because of, their political importance as a basis for nuclear opposition.

Who Participates? — A second principle of negotiation is that participation must include appropriate interests. In most procedures two criteria are used to establish the right of participation: "affected interests" and "representativeness". The interests affected by a project are often defined by geographic proximity to a project, but this varies. The German nuclear inquiries have no geographic limits to participation in public hearings; the French restrict participation to 5 km radius from a proposed site. The limits that are placed on public involvement may predetermine the assessment of risk, as those living near the site of a controversial facility may use different criteria to evaluate the technology than those further away. For example, the employment and economic benefits expected by those living near a nuclear site may clash with the more abstract environmental concerns of those in adjacent communities. Conversely neighbors of a nuclear facility may feel that they unjustly bear the risks of a project intended to benefit a wider region.

The notion of who is representative and therefore entitled to participate in decision-making procedures also varies. In Sweden, study circles were broadly participative, but organized by a relatively small group of trusted representative elites, the leaders of major associations and political parties. In France, with its comparatively limited tradition of voluntary association, representatives are often defined as civil servants or local elected officials. In the United States participation is often more direct, as in the Cambridge Citizens Review Board.

In some inquiries the question of who should be involved is avoided by the use of a public referendum. Local governments in France, for example, have often organized referenda on nuclear power plant siting. These, however, serve only as a source of information; they are neither decisive nor do they necessarily create consensus, for they fail to account for the intensity of opposition from small but actively critical groups.

Problems in defining appropriate participation also limit the usefulness of mediation procedures. Mediation works best when two major protagonists share a minimum common interest that will lead to a mutually satisfactory compromise. In technological controversies, antagonist groups are not necessarily well defined, nor do they necessarily share the values that will compel a compromise.

Who Runs the Show? — A major source of criticism of consensus procedures focuses on their management — the choice of commissioners and of various supervising and consulting agencies. In Germany, the civil servants in charge of public hearings often serve at the same time on the administrative boards of the firms applying for permits. For example, the Minister of Economics, politically responsible for the nuclear inquiry procedure in Wyhl, was also the acting vice-chairman of the board of directors in the electric utility. The large state-run Institute for Reactor Safety consulted in the German licensing process had formed out of the first generation of enthusiastic German nuclear experts. Nuclear critics feel such groups are biased and prone to underestimate information that comes from outside the nuclear establishment.

In France, the commissioners who evaluate the public inquiry procedures are also suspect. Often retired civil servants, they lack the technical competence to judge the details of EDF's studies. Yet they maintain a monopoly of information in the DUP nuclear siting procedures.

Distribution of Expertise — Because disputes are so often translated into technical terms, resolution of conflicts requires a reasonable distribution of expertise. Indeed, expertise is a crucial political resource, and if parties in conflict are to have any sense of political efficacy, they must have access to technical advice. In major inquiries, such as Windscale, the lack of technical resources among intervenor groups seriously undermined their ability to present an effective counter-argument. The French program for regional concertation has been criticized as a "phantom consultation" because regional councils lacked the expertise to evaluate EDF technical dossiers on the nuclear sites. Thus they could only respond in terms of traditional political alignments.

Inadequate distribution of expertise allows control of information. The Conseil d'Information in France can only respond to the information made available by public authorities. It cannot generate its own data nor insist that information be released. Thus while leaders of ecology associations agree that the Conseil is an important concept, they believe it is "a bluff" — simply another way to subdue opposition.

In the German nuclear inquiries, only those documents that are part of the formal application to construct a plant are officially open to the public. Internal administrative evaluations are available on request but only on the discretion of the administration.

Control over information and its distribution is an important procedural consideration, for the selection of the technical data available for discussion may predetermine final decisions. Aware of the political implications of expertise, *ad hoc* "science for the people" groups have organized in many countries. They include Science for the People in the U.S., the Wetenschapswinkels or "science shops" in the Netherlands, the Health Hazards groups in England and Groupe de Scientifique d'Information sur Electronucleaire (GSIEN) in France. They distribute expertise to environmental and anti-nuclear groups lacking ability to generate the information required to challenge policies affecting their interests.

Is There Really a Choice? — In 1977 a public inquiry for a nuclear plant opened at Le Pellerin where local resistance had long been evident. The mayors of 7 out of 12 communes in the region had refused to use their offices for the inquiry, and documents had been stolen and burned in two city halls. The prefect was forced to open an "annex to the city hall" under police protection [17]. The population agreed to boycott the official inquiry; of the few people who did participate, 95 approved the project, 750 opposed it. 80-95% of the population in each of several municipalities signed anti-nuclear petitions; one had 30,000 signatures. But the outcome of the inquiry was as unexpected as the whole procedure: the inquiry commissioners declared themselves incompetent to judge the issue, but nevertheless concluded in favor of the project.

This is an extreme example, but it suggests the low tolerance for disagreement in the public inquiry procedure. Most are simply structured discussions over predetermined policy with few real options. The financial and administrative investments involved in specific technologies are simply too profound to allow a real margin of choice. Thus, when the first opposition to the nuclear plant appeared in Wyhl, Germany, the Prime Minister of the Land said, prior to the licensing procedures: "There can be no doubt that Wyhl will be constructed" [18]. In the case of a plant in Esensham, Germany, regional officials acknowledged that secret negotiations with the nuclear industry had taken place for more than a year prior to the official application for a construction license. And in several cases, Electricite de France began preparatory work on construction before the end of inquiry procedures.

The limits of choice are evident in the very specific and short term questions approached by most forums, except for those "reflection groups" specifically organized to consider the long term effects of technology. Of all the advisory models only the Dutch "policy intentions" seek to incorporate public opinion at an early stage when policy objectives are first articulated. Most are directed more towards co-opting public support than changing decisions, more towards seeking informed consent than expanding democratic choice. Determination to implement preconceived decisions leads officials to ignore, to debunk or simply to be unaware of opposition. And this results in the transfer of conflict from the hearings to the courts, and often to the streets.

CONCLUSIONS

Governments in most Western countries share a common political problem — how to reconcile technological systems with social value — how to develop sufficient consensus about controversial technologies to permit continued growth — how to clarify citizen interests and produce the political support necessary to make authoritative and acceptable decisions. The inadequacy of existing institutions to deal with this problem has inspired many experimental procedures designed to achieve consensus through better distribution of information and greater opportunities for public input into the decision-making process. These experiments open up a range of possiblities for institutional change and suggest a set of criteria for successful negotiation. But we wish to emphasize that they provide no systematic solutions.

Comparative policy studies often approach common problems in seeking a "best solution" that can be transferred to other contexts. Our analysis is not to be interpreted in this way, for we find that the structure of these experiments — the definition of "appropriate" procedures — varies, reflecting national political styles and expectations about the role of government, about participation and about who

represents the collective will. The approach to solving conflicts in a political context of consensus and compromise will differ from that in an adversary culture: Holland, with its tradition of cleavage reflecting religious and regional differences, has had long experience in accomodating competing interests. It can be expected to develop quite different procedures than Sweden. The French tradition of open ideological factions within a centralized political and administrative structure calls for different conflict-resolving procedures than Germany where consensus through internal negotiation within major party and union organization is expected. The structure of experiments and the assumptions about who should participate in programs seeking public involvement reflect basic political differences and cannot be simply transferred without considerable adjustment. Indeed, transferring means of conflict resolution can pose problems not unlike those of technology transfer.

What can be generalized is not the structure of the experiments but the conditions that will allow dissenting groups to articulate effectively with administrative agencies: a careful definition of the agenda that gives due weight to social and political concerns, the appropriate involvement of affected interests, an unbiased management, a fair distribution of expertise, and a real margin of choice. In fact, these conditions are not likely to produce consensus, but they may reduce mistrust and hostility towards political and administrative institutions in order to at least allow detente. Our conclusion, however, is that detente is a more appropriate and realistic goal. Resolving conflicts is not necessarily always desirable. And seeking consensus is like wanting 100% risk-free technology — hardly a feasible objective. One of the more important effects of recent disputes is an awareness that decisions about technological risk are not simply matters of sufficient technical evidence and adequate information. They embody highly controversial political and social values, requiring institutions and procedures that will allow an open and balance dialogue and enhance the constructive sense of collective responsibility necessary for legitimate and acceptable decisions.

REFERENCES

1. *Task Force of the Presidential Advisory Group on Anticipated Advances in Science and Technology, "The Science Court Experiment," Science, 193: 653, August 20, 1976.*
2. *See discussion in B. Ackerman, The Uncertain Search for Environmental Quality, New York, The Free Press, 156 ff, 1974.*
3. *G. Bugliarelli, "A Technological Magistrature," Bulletin of the Atomic Scientists, 34-37, January 1978.*
4. *J. C. Glick, "Reflections and Speculations on the Regulation of Molecular Genetic Research," and M. Lappe and R. Morison, "Ethical and Scientific Issues Posed by Human Uses of Molecular Genetics," Annals of the New York Academy of Sciences, 265: 189-90, 1976.*
5. *D. Nelkin and M. Pollak, "The Politics of Participation and the Nuclear Debate in Sweden, the Netherlands and Austria," Public Policy, 25: 333-357, Summer 1977.*
6. *Jean-Philippe Colson, Le Nucleaire sans les Francais, Paris, Maspero, 114 ff, 1977.*
7. *Bericht der Kommission fur Wirtschaftlichen und Sozialen Wandel, Bonn, 473, 1976.*

8. D. J. Gamble, "The Berger Inquiry," *Science, 199:* 946-51, March 3, 1978.
9. B. Wynne, "Nuclear Debate at the Crossroads," *New Scientist,* 349-360, August 3, 1978.
10. Nelkin and Pollak, *op. cit.*
11. Colson, *op. cit.*
12. S. Nagel and K. von Moltke, "Citizen Participation in Planning Decisions of Public Authorities," *National Report for Germany, EEC Participation Project, 31 ff.*.
13. J. Sullivan, City Manager, letter to the City Council of Cambridge, August 6, 1976; Cambridge Experimentation Review Board, *Guidelines for the Use of Recombinant DNA Molecule Technology in the City of Cambridge,* submitted to the Commissioner of Health and Hospitals, December 21, 1976.
14. See for example a proposal by the American Arbitration Association, in D. B. Strauss, "Mediating Environmental, Energy and Economic Tradeoffs," *AAAS Symposium on Environmental Mediation Cases,* Denver, Colorado, February 20-25, 1977. See also K. R. Hammond and L. Adelman, "Science, Values, and Human Judgment," *Science, 194:* 389-396, October 27, 1976.
15. Nelkin and Pollak, *op. cit.*
16. International Atomic Energy Agency Document, 1976. See also D. Nelkin, "Technological Decisions and Democracy," *SAGE Publications, 59 ff,* 1978.
17. La Gazette Nucléaire, 17: 8 ff.
18. H. H. Wustenhagen, *Burger Gegen Kernkraftwerke,* Reinbek, Rowohlt, 61, 1975.

DISCUSSION

W. C. Clark *(Int'l. Inst. for Applied System Anal.)*

So we all agree the world is a nasty place. What sort of mediating structures, what sort of public participation structures would you recommend we begin experimenting with to do a little bit better if not to solve the problems.

D. Nelkin

I don't know. I'm not too good at designing particular experiments. I'm rather concerned about the underlying notions. I think if we start designing structures basically intended to manipulate public acceptability then we're bound to fail. I think we have got to think conceptually in terms of opening up to broader kinds of social concerns and to provide some real choice. Otherwise it's a cynical exercise.

D. B. Straus *(American Arbitration Association)*

I've got to congratulate you on the very broad overview, but I must make one criticism. It reminds me that four or five years ago I had a chance to go to China and I heard at that time a description of the United States by a Chinese who had just made a very short visit. I feel, a little bit, that your visit in the mediation field has been a little bit like that, and superficial. Constructively, let me say this, the questions you raise are indeed the questions that are being addressed today. They are not being swept under the rug as you suggest and there are some very serious efforts being made to get into these questions that you are raising. They are not being swept under the rug and there have been some successful mediations.

D. Nelkin

True, but in a twenty-three minute address one cannot get in depth into ten different procedures.

D. Nelkin

W. D. Rowe *(The American University)*

People tend to use nuclear as a model, yet. It seems to me, and I think that Paul Slovic has shown, that nuclear energy is unique in that the issue is so polarized that it may not be a good model for other things. In fact, using it as a model may be to the detriment of setting some of the other problems. Have you any opinion on that?

D. Nelkin

Yes, I think that nuclear power has become a symbolic issue. The fact that so many people have focused in on it is because it reflects much more than the question of risks; it has become essentially a metaphor for other concerns, for example about centralization, about the role of government, the role of regulation. It's a useful model not because it's representative, but because it's symbolic.

M. A. Schneiderman *(National Cancer Institute)*

Dorothy, by the way of the question implied by Clark's question and the things that you've looked at and the activities that you've looked at. Some of them appeared to have been successful, some of them have appeared to resolve the conflict or let people move ahead with what they're doing without the residue of anger and animosity. Have you been able to classify those, or are there some that do work? Are there some that have worked to suggest things that might work in the future?

D. Nelkin

I am on a panel at the Hastings Center on "Closure of Disputes" and one of the things that's quite clear is that disputes don't really close. For some reason or another the DNA dispute has temporarily abated but it hasn't gone away. I think a lot of the concern about recombinant DNA research is now displaced by concern about genetic manipulation and socio-biology. The procedures don't really close disputes because I think that the concerns about risks lie in deeper issues. That's why I commented on the symbolic nature of the nuclear debate. I think the notion of mechanisms to close disputes or of expecting procedures to "work" is not too useful because anxieties simply get displaced in a conflict over something else.

S. Drescher *(GM Financial Staff)*

Between the first two talks this morning both of you paint a very bad picture and I wonder really what we get here. The first talk told me that people who are generally lay people, form subjective judgments which are generally not necessarily correct from a technical point of view as to the true probabilities of hazard and risk. Now, you tell me that if you try to inform people in order to eliminate the conflict, people are not really interested in the risk but it is the value judgments that they make which concerns them. So in effect what you're saying is that even if we were able to disseminate all the correct information we're really wasting our time because we're really fighting just value judgments and philosophy which are different conflicts than risks. So the problem of risk is not the problem, the problem is differing values and the polarization of society.

D. Nelkin

But we've dealt with this in other areas. The fact is that we make decisions on political issues all the time. I would argue, and I think Paul would agree, that the subjective nature of risks and the fact that public perceptions include all sorts of other subjective concerns is important. The theme of this particular panel is decision making in a democratic society. My feeling is that these perceptions exist and they've got to be dealt with in their own right, not just dismissed.

S. Drescher

I'm not arguing for dismissal, I'm trying to point out that maybe we are going around as "experts" doing the wrong thing. I don't know if we can take the information that we've developed and pass it on to the public because they're not willing to accept it, either because they're resistant to the change or their value judgments are such that they don't care whether or not what you have to say is true. They'd still rather make their own estimates.

D. Nelkin

Let me be unfair, and turn the question back to you. Are you absolutely sure that all the research that's going on about risk is asking the important questions or should there be more diversity in the expertise? The fact is that there are disputes among experts. The question of distribution becomes very important and this is part of what we're driving at.

G. Patton *(American Petroleum Institute)*

In your report you mentioned that there was a Canadian commission and that support was provided to public interest groups to help develop the case. I'm not sure you are aware that this issue of financial support to public interest groups is a current issue here in this country. I wonder what your views on it are?

D. Nelkin

I think it's necessary. Given the role of expertise in our society, expertise is a political resource, and distribution of expertise therefore becomes rather important. Therefore the whole notion of intervener funding is something which I support.

G. Patton

How would you control that system? How do we identify who's to be the public spokesman? Do you have any problems with that?

D. Nelkin

I think it necessarily has to be organized groups. How many people in the public even vote for President in the United States. If in fact there was some sense that one could get involved in decision making, maybe we would have a less apathetic public. Maybe more people would get involved in elections and would take a greater role if they thought they had any real choice.

G. Patton

I'm going to turn that around on you and say I'm pleased that some part of this public is not so up in arms as to be either rioting or perhaps even voting if they don't see the need. My concern is that as you go to public intervener financing you just give additional voice to the very activists who I presume vote and do a lot more than that.

D. Nelkin

Well, I basically disagree. The fact that so few people vote is a bad thing. And anything that will give people a greater say in their society and decreases the political alienation that comes from impotence, is a good thing. But that's a basic political judgment.

G. Patton

That's a question in which I would like to know a positive alternative instead of negative ones.

F. E. Burke *(University of Waterloo)*

I have two comments that call for very few replys. One is that you mentioned the World Commission as an isolated Canadian example and I think it's fair to say it is one of a whole trend of Canadian methods of inquiry that have in the last 15 years emerged from an ossified traditional mechanism into a much more lively one. I won't give an example of the lists, I'm sure that you have a list available. Secondly, in this Canadian practice one of the most respected experts available who was called in front of one of these particular inquiries, which incidentally resulted in over 10,000 pages of homework published, was called six times and said that on the third occasion he was faced with the issue of conflict of interest that you have mentioned. He said that when it comes to experts, there is no one who doesn't have a conflict of interest, because if a person is competent, then he is busy with something, and therefore he has an interest in which he is in conflict. And if he is not working on something then he's not competent.

R. J. Tobin *(State University of New York)*

I have two comments. One concerns your comment about apathy among the American electorate. I don't feel that at all. It may be the case that people are very satisfied with the political system and don't feel an obligation to participate. Certainly, many governments here in the United States don't encourage participation. That is to say it's very difficult to register to vote, in most states it's more difficult to register than to actually vote. If governments were interested in encouraging participation here in the United States, it would require registration, it would fine people for not voting and we could do other things that take place in other countries.

The other thing in terms of participation, G. Patton asked about who should be funded. Well the decision has already been made to fund every corporation that wants to become involved in politics. That's to say in terms of tax write-offs. Perhaps it's worthwhile to put an equal amount of money in the public groups which would in turn be comparable to the tax write-offs the corporations get. If I were to testify at a public hearing I would have to drive my car, have to pay for meals, lodging and things like that. I can't write that off. A corporation fellow does the same thing and he writes that off. The government subsidizes their participation, it does not subsidize public participation.

D. Nelkin

Yes, I would agree with you.

G. Patton

Well, I got into the middle of this, and I hope not to waste your time with it, but I was asked to be on a panel last week where EPA listens to the public to rate their views on the adequacy of environmental protection afforded to the DOE research program on energy. This started me thinking, and after all, the government is the instrument to represent the public; that's why there's a Congress. That's why you can vote for your congressman, if you wish. That's why the agencies are responsible to the President. And I get concerned when you say that the answer is simply to spend more money, having more people come and debate the issue. I saw no end to the process and I did bring it up because it's a very unresolved issue. It's nice to say that you should be funded so that you can voice your opinion on something. But when you sit in one of those panels and you listen to people say things like they can't get a geothermal plant put into a place in New Mexico because the locals object to it and you find out that geothermal is probably about the cleanest thing going, you ask who is representing the national interest?

J. L. Holt *(Exxon Company, U.S.A.)*

In the context of the lady's question yesterday from the health commission in regard to practical need, if consensus is as unavailable as zero risk and you maintain that it is, if real issues somehow don't go away but conflict is resolved sufficiently and decisions some how get made whether efficient or unefficient, what criteria do you want us to use other than consensus for measuring the success of our institutions of procedures? I guess I'm confused.

D. Nelkin

This relates in part to what you were saying. There is an extraordinary decline in confidence in the representative system. There is a wide feeling that it's simply not functioning and therefore you have a groping for alternative measures of public involvement. People seek to become a part of the system and if there's no other way, this takes place as protest. I don't regard consensus as an objective. But one should seek to develop a system that people feel enough a part of that they will accept decision even though it's not particularly in their interest. There must be some sense of collective responsibility. This is necessary to resolve the energy problem. At this point you have sufficient alienation and sufficient mistrust that nobody even believes there's an energy problem. And consensus isn't the goal. It's establishing institutions that people have some confidence in.

THE ROLE OF LAW IN DETERMINING ACCEPTABILITY OF RISK

Harold P. Green

The George Washington University
Washington, D.C.

ABSTRACT

Legal criteria and procedures for determining societal acceptability of risk are established by governmental institutions performing legislative (i.e., deliberate formulation of rules with prospective application) and/or judicial (i.e., resolution of conflict on a case-by-case basis) functions. In both cases the law-making process reflects both the subjective preferences of the human beings who administer these institutions and their sense of societal attitudes.

Laws, once formulated, are in no sense immutable. The words used in formulation of the laws are inexorably subject to interpretation, and interpretation necessarily reflects ethical, cultural, and political considerations. Moreover, the laws — whether made by legislature, court, or administrative agency — are always subject to override by political effort culminating in legislative action. Thus courts and administrative bodies are constrained in deciding "how safe is safe enough" by their sense of what the body politic desires or will tolerate.

It is futile to expect that societal determinations as to acceptability of risk can be made in an entirely rational, objective manner that will reflect a "correct" balancing of benefits and risks. Indeed, societal "acceptability" is an ephemeral thing that frequently changes from day-to-day. The most for which we can hope is that law provides procedures through which decision-makers will have access to all relevant data — hard as well as soft — so that they can perform their assigned function of making determinations that reflect their sense of what their constituency wants.

INTRODUCTION

Public policy decisions involving determinations of acceptability of risk are made in two contexts. The first involves decisions whether the government should conduct or support some activity that in itself would involve risk, as well as benefit, to individuals. The second involves decisions whether government should restrict the freedom of individuals or entities to engage in activities that involve risk, or should penalize those who engage in such activities and who, in fact, have caused injury. It is this second category with which this paper is concerned. It should be observed in this connection that the imposition of a penalty is a form of restriction, since it makes the activity more costly and hence less attractive.

My topic is the role of law in determining acceptability of risk. The first point I must make is that the role of law is pervasive, at least as I define law to include decisions and decision-making by courts, legislatures, and administrative agencies with respect to particular activities involving risk. There can be no public-policy decision restricting or penalizing any activity without the involvement of law.

THE COMMON LAW

A useful starting point is to consider the manner in which the common law handles the acceptability of risk. The common law consists of principles derived from the body of precedents that evolves from case-by-case resolution of claims presented to the courts. Judges do not consciously or deliberately make law, but merely decide cases before them; but as their decisions are rendered in particular cases, these decisions become precedents for decisions by judges in subsequent cases. In a sense, therefore, the common law reflects, and is the repository for, the wisdom of the ages.

The area of the common law most relevant to acceptability of risk is the law of torts, which deals primarily with the liability of defendants whose conduct causes injury to plaintiffs. Not every injury to another gives rise to liability, but only those injuries caused by the fault — primarily negligence — of the defendant. Liability is imposed to compensate the innocent plaintiff for her injuries, and, by example, to deter dangerous conduct by others in the future. Negligence is "conduct which falls below the standard established by law for the protection of others against unreasonable risk of harm" [1]. "Risk" is the "chance of harm" [2]. Since it is only conduct that creates an "unreasonable risk of harm" to others that gives rise to liability, such conduct, although not "unacceptable" in the sense that society prohibits it, is "unacceptable" in the sense that society disfavors it by penalizing the actor if the conduct results in harm to others.

It is, moreover, only "unreasonable" risk that is unacceptable in this sense. This means that the actor would be liable only if a fictitious "reasonable" person applying a mythical community standard would recognize that the conduct involves unreasonable risk to others [3]. The risk is "unreasonable" if it is of "such magnitude as to outweigh what the law regards as the utility of the act or of the particular manner in which it is done" [4]. The magnitude of the risk is a function of the social value the law attaches to the interests imperiled, the probabilities that the conduct will cause injury, the magnitude of the likely injury, and the number of persons who may be injured [5]. The utility of the actor's conduct is a function of the social value the law attaches to the interest to be advanced or protected by the conduct, the probability that this interest will be advanced or protected by the conduct, and the availability of other less dangerous alternatives to advance or protect the interest [6].

It is the legal, and not the popular, opinion that is controlling with respect to utility of the conduct, although courts frequently reexamine and correct the law's view of utility when such view is inconsistent with public conviction [7].

These formulations involve a kind of risk-benefit assessment to determine the reasonableness of the defendant's conduct. Although the principles discussed above do not preclude the use of techniques for measuring and assessing relevant factors in objective, quantified terms, they do not contemplate or require such techniques, and, indeed, assessments based on purported objective quantifications are rare. Thus, for example, the operation of railroads and utilities, the conduct of commerce, and the free use of highways are all regarded, without objective evidence, as having sufficient utility to justify acceptance of some injuries to the public. Indeed, the law's reliance on such concepts as "reasonable", "social value", and "opinion" clearly evidences a *qualitative* or common sense, rather than a quantitative, approach.

Let me offer two examples of this. In a 1934 decision, the Wisconsin Supreme Court suggested that a careful driver should not be liable to a pedestrian splashed with muddy water on a rainy day, because the benefit of allowing people to travel under such circumstances clearly outweighs the probable injury to pedestrians. Such a conclusion made no pretense at quantification of the competing considerations, but was derived from the wisdom and experience of the court and the traditions of the people, as perceived by the court. Similarly, one of the most eminent federal judges, Learned Hand, in a case involving liability for injury resulting from the breaking loose of a vessel from its pier, reduced the question to algebraic terms with three variables:

P - the probability that the vessel will break away

L - the gravity of the injury

B - the burden of adequate precautions,

so that liability would depend on whether $B > PL$. Nevertheless, Judge Hand gave content to the algebraic terms in a purely non-quantitative analysis that included consideration of custom in the industry [9].

In the interest of completeness, it should be mentioned in passing that the common law recognizes strict liability, i.e., liability without fault, in cases involving "abnormal risk" [10]; and in some cases extraordinarily hazardous or unpleasant conduct may be prohibited by injunction [11]. The same general approach to assessment of risk, and balancing of utility against risk, as is discussed above with respect to negligence is applicable in these special situations.

LEGISLATION

The perceived inadequacies of the common law in protecting society against hazardous activities frequently lead to legislative action resulting in statutes that merely set standards of general applicability, that also prescribe penalties for violation of the standards, or that prescribe a regulatory pattern to be administered by a regulatory agency. Such statutes are based on some kind of determination as to the acceptability of risk that is made by legislatures. Where a legislature enacts a statute dealing with protection of people against risks, such legislation, unlike a judicial decision, is a positive, deliberate making of law intended to control conduct in the future.

Sometimes the statute flatly prohibits the conduct of an activity regarded as hazardous. For example, statutes may prohibit the sale, use, or possession of certain items such as fireworks, heroin, or pornographic materials, reflecting a legislative judgment that the risks associated with the activity are of such magnitude that the activity should be banned notwithstanding the magnitude of any possible offsetting benefits. In most such cases, the activity in fact is perceived as having very little, if any, social utility; but there are some cases in which the activity is flatly prohibited even though substantial offsetting benefits are present. A leading example of this is the Delaney Amendment [12] which prohibits the use of any chemical in food where the chemical is known to cause cancer when ingested by animals.

The important point with respect to these examples is that the legislature has made a determination that the activity in question involves a risk that is unacceptable from the standpoint of society, and should therefore be prohibited. The evidence on the basis of which such a determination is made is invariably more anecdotal than objective or empirical, and there usually is no effort on the part of the legislature to consider either risk or benefit in quantitative terms. There is, moreover, no satisfactory rationale to explain why the legislature prohibits some activities involving risk and not others. It appears anomalous, to say the least, that a statute might prohibit sale or possession of marijuana or pornography, but not of liquor and tobacco.

In most situations that involve risk as well as compensating benefits, the resulting statute reflects the striking of a risk-benefit balance. For example, although the use of motor vehicles involves great risks, the legislatures have recognized the importance of minimizing these risks while at the same time preserving the benefits of the technology. This balance has led to numerous statutes permitting the use of motor vehicles subject to various conditions intended to reduce risk: e.g., age requirements for operators' licenses; requirements for lights, windshield wipers, and other safety equipment; speed prohibitions; and traffic rules all designed to make the use of motor vehicles safer.

Again, it should be noted that, although legislative consideration of such safety requirements to make the risk (more) acceptable may involve some inputs that reflect some attempt at scientific analysis or quantification (e.g., how many lives would be saved if the speed-limit were reduced from 65 to 55), the resultant statute usually is more a product of subjective and political factors. Each legislator has her own experience and attitudes relating to pleasure and pain, advantage and disadvantage, the speed at which a car should be driven, the perils of cancer, diabetes, and obesity, etc., and it is such subjective views, coupled with the attitudes of constituents who also have such views and the pressure of special interest groups, that shape the manner in which the legislator will vote.

We are all prepared, I believe, to accept as a fact of life in a democracy, that our legislatures make decisions and enact statutes on the basis of emotional and political factors that result in a seemingly irrational fabric of law. What we seem less willing to accept, however, is that a regulatory statute almost always reflects an amalgam of, and compromise among, competing or conflicting considerations, so that the regulatory scheme cannot be regarded as optimum from anyone's point of view. Sophisticated observers of the regulatory process know that when, for example, Congress enacts a coal mine safety statute, even though the rhetoric of stated congressional goals may be extremely tough, the operative provisions of the legislation may in reality be quite soft.

REGULATORY AGENCIES

But even if we are willing to recognize the vagaries of the democratic political processes in legislative bodies, we seem to expect much more from our regulatory agencies. After all, the personnel of these agencies are much more expert to begin with, and their functions are focused on a relatively small piece of the vastly broader ambit of the legislature's concerns. They can, therefore, bring to their regulatory function a body of knowledge, experience, and expertise that simply could not be brought to the object of that regulation by the much more diffuse and generalized legislative process.

The remainder of this paper will, therefore, be devoted to a discussion of the role of the law in determination of acceptability of risk by regulatory agencies.

At the outset the relevance of law to the activities of regulatory agencies should be stressed. To begin with, regulatory agencies are creatures of law. They come into existence with characteristics established by statute, and they may exercise only such authority and perform only such functions as the legislature decrees by statute. Moreover, the agency's procedures, jurisdiction, functions, and authority are subject to revision by the legislature's enactment of a new statute at any time; and, indeed, any action taken by the agency is subject to legislative correction or reversal. For this reason, regulatory agencies are often regarded as "creatures of the legislature" performing a legislative function delegated by the legislature. It is useful to keep in mind that the legislature itself has the theoretical power to accomplish by statute much, perhaps most, of what a regulatory agency is created to do. As a matter of fact, in the early history of the United State most such functions were, in fact, performed by the legislature. Regulatory agencies are a fairly recent development as the increasing complexity of economic life has overcome the ability of the legislatures to cope directly.

Regulatory agencies usually have the power to prescribe rules and regulations establishing substantive requirements with respect to the activities subject to their regulation. In performing such a "quasi-legislative" function, the regulatory agency, in effect, makes much the same kind of law that the legislature otherwise would. As it promulgates rules and regulations to protect the public against hazards, the agency makes determinations as to acceptablity of risk as it decides which activities are to be prohibited, which are to be permitted, and what conditions will be imposed with respect to the latter category. Such rules and regula-

References pp. 266-267.

tions, although not of the same dignity as statutes, have the "force and effect of law."

In making rules and regulations, the agency interprets the statutory provisions governing its existence and functions. These statutory provisions, as well as the agency's own rules and regulations, are further interpreted and elucidated in particular agency decisions as the agency applies the statute and rules to individual cases over which it has jurisdiction. This body of agency decisions becomes in turn a body of precedent — much like judicial decisions — and a source of law.

As noted above, the discretion of the agency is limited by the power of the legislature to correct and reverse its particular actions and to revise its basic charter. Inevitably, some interests will be pleased and some will be displeased by particular agency rules, regulations, and decisions. Those that are displeased have recourse to the legislature, and recourse to the courts is also available.

A person aggrieved by an action of an administrative agency may seek judicial review. Although the standards for judicial review vary, there are a few generally accepted factors. The courts will reverse the agency's actions if it is clearly inconsistent with governing statutory law, if it is arbitrary or capricious, or if it was taken otherwise than in accordance with prescribed procedures. Judicial decisions in cases involving review of agency actions also involve interpretation of the applicable law and become legal precedents governing future cases. In connection with the role of the courts, it should be noted that the legislature not infrequently avoids coming to grips with "hot potatoes" by deliberately leaving such questions unresolved in the statute, so that they will ultimately have to be dealt with by the courts. On the other hand, just as agency actions are subject to correction and reversal by the legislature, so is the law made by the courts. In the last analysis, the law is made by the legislature to the extent the legislature desires to exercise the power.

This complex process is fully applicable, in various ways, to myriad situations in which regulatory determinations are made as to acceptability of risk. Such determinations are made by numerous federal and state agencies. Within the federal system, it can be said that no two regulatory schemes or administrative agencies concerned with such determinations have even remotely similar statutory foundations. The practical effect of this is that the nature and level of acceptable risk is not, and cannot be, universal, but varies and is dependent in each case upon the particular activity that gives rise to the risk, the particular statute that purports to regulate the activity, and the particular agency assigned to implement the statute.

The starting point for regulatory determinations of acceptable risk is always the statutory provisions that authorize the agency to make such determinations. Each such statute provides some insight or direction, explicit or implicit, as to the acceptable levels of risks — not objectively or absolutely, but solely for purposes of application of that statute. But language is often (probably always and intrinsically) imprecise and ambiguous, and statutes are often (probably usually) less than adequately drafted. Part of the process of statutory interpretation is, therefore, resort to the legislative history of the statute.

Examples of Safety Regulations — Let us look at several examples illustrating the typical approach to safety regulations.

Chemical Food Additives — The Chemical Food Additives Amendments to the Food, Drug and Cosmetics Act [13] were enacted in 1958. Under this legislation, a chemical (except one "generally recognized as safe") is "deemed to be unsafe" and may not be added to food until it has been pretested and falls within an FDA regulation prescribing the conditions under which it may be "safely used" [14]. Although the statute does not attempt to define "safely used," the report of the Senate Committee on Labor and Public Welfare [15] defines "safety" to require "proof of a reasonable certainty that no harm will result from the proposed use of an addition."

Of course, considerable ambiguity and latitude in interpretation remain even with the gloss thrown on "safely used" by the committee's explanation. What does seem clear is that an additive may be used only if its proposed use is shown with reasonable certainty to cause no harm. One can, however, debate the meaning of "reasonable certainty," and "used;" and the word "harm" embraces a broad spectrum of possible, probable, and/or actual consequences of use. The statutory language can be further interpreted by reference to the more specific language of the so-called Delaney Amendment flatly prohibiting use of any chemical additive "found to induce cancer when ingested by man or animal" [16]. As is generally known, this provision has been interpreted to prohibit the use of any chemical found to cause cancer in animals even where enormously large quantities (relative to quantities likely to be ingested by man) have been fed to animals. We can infer, therefore, that a somewhat more relaxed standard applies to noncarcinogenic chemicals.

Of importance equal to that of the interpretation of the specific concept of safety found in the Act and the legislative history is the basic procedure laid out in the statute. An additive may not be used until FDA, in effect, licenses its use. An applicant for the promulgation of an FDA rule permitting use of the additive has the burden of satisfying the FDA that it may be "safely used;" and the FDA has no burden to show, so long as its decision is not arbitrary or capricious, that harm will in fact result. It can almost be said, therefore, that the statute reflects a bias *against* additives and is not concerned about a loss of the benefits incident to their use.

Flammable Fabrics — A second example is the Flammable Fabrics Act of 1970 [17] which banned the movement in interstate commerce of fabrics that do not meet flammability standards prescribed by the Secretary of Commerce. The touchstone for determining acceptability of risk was the statutory requirement that each such standard be based on findings that the standard

"is needed to adequately protect the public against unreasonable risk of the occurrence of fire . . ., is reasonable, technologically practicable, and appropriate . . ." [18].

This standard of acceptable risk is obviously less severe than that in the Chemical Food Additives legislation. The latter bars *any harm*, while the Flammable Fabrics Act bars only unreasonable risk of harm. Moreover, whereas the former prohibits the treatment of food with chemicals unless adequate safety is shown, the latter *requires* the treatment of fabrics with chemicals to achieve adequate safety, but only if each treatment is "reasonable, technologically practicable, and appropriate . . ." Thus, although economic considerations are not relevant in the food additive area, they must be taken into account in the flammable fabric area.

Here again, a consideration of the locus of burden of proof is instructive. The Flammable Fabric Act is in no sense a licensing statute. No permission or license is required to ship flammable fabrics in interstate commerce, but one is subject to criminal prosecution (in which the burden of proof rests with the government) if her products do not meet the flammability standards. Moreover, the government has the burden to show that its flammability standards are "reasonable, technologically practicable, and appropriate." One would expect therefore that the standard of acceptable risk is much less severe with respect to fabrics than in the case of food additives.

Toxic Substances — A third useful example is the Toxic Substances Control Act of 1976 (TOSCA) [19] which authorizes EPA to regulate any chemical upon a finding of "unreasonable risk of injury to health or the environment" [20]. Here, some degree of risk is clearly acceptable, since it is only unreasonable risk that triggers regulation. The statute seems to contemplate some degree of EPA tolerance since it specifies that the regulation impose the "least burdensome requirements" necessary to protect against "unreasonable risk."

The "unreasonable risk" criterion is identical to that found in the Flammable Fabrics Act, but the report of the House Commerce Committee [21] provides a rather explicit definition of "unreasonable risk" for TOSCA purposes. It states that the determination of "unreasonable risk" involves

". . . balancing the probabilities that harm will occur and the magnitude and severity of that harm against the effect of proposed regulatory action on the availability to society of the benefits of the substance or mixture, taking into account the availability of substitutes for the substance or mixture which do not require regulation, and other adverse effects which such proposed action may have on society."

The report explicity states that a "formal benefit-cost analysis under which a monetary value is assigned to the risks . . . and to the cost of society" is not required.

All three of these examples reflect the typical approach to safety regulation. All three statutes were enacted after several years of legislative effort — at least six years in the cases of both chemical food additives and toxic substances. In all three, moreover, the legislative history includes ample documentation — a parading of horribles — of the evils to the public that existed under the status quo. In each case also, the industry that was to be regulated fought the legislation. Typically, as is shown primarily in the flammable fabrics and toxic substance examples, the ultimate legislation reflects a compromise of sorts in that the industry is often able to obtain ameliorating statutory language, or language in a committee report, designed to curb the zeal of the regulatory agency and to require it to be "reasonable."

Nevertheless, in each of the three cases, although the statutes and legislative histories clearly contemplate the applicability of varying interpretations of acceptable levels of risks, the regulatory agency had a clear mandate to prevent and to correct evil. The basic presumption underlying all three statutes, and the consequent regulatory programs, was that elimination of harm to society was more important than societal enjoyment of the full benefits flowing from the subject matter of the regulation. Nevertheless, within the framework of this presumption, there are some interesting distinctions to be drawn among the three cases.

Some Distinctions — In the case of chemical food additives, the presumption is that an additive is unsafe until it has been shown to be safe. The burden of persuasion rests with the industry, and under the statute FDA has virtually unrestrained discretion to decide the conditions under which use of an additive will be permitted. Under TOSCA, the burden is likewise with the industry, but statutory language operates somewhat to constrain EPA and to limit its discretion. On the other hand, under the Flammable Fabrics Act, a fabric may be used unless and until its use is prohibited by regulation. The initiative is taken by the agency and not by the industry, and the agency has the burden of finding that the regulation is necessary to achieve adequate safety, and that it is "reasonable, technologically practicable, and appropriate."

Finally, it should be observed that the considerations outlined above are subject to judicial review and, perhaps of greater importance, to constant monitoring by Congress — or at least by those members who have an interest from the standpoint of either public health and safety or of industry. The statute, once enacted, is not the end of the law-making process. The words used in the statute are subject to interpretation through the agency's application of the statute on a case-by-case basis and through judicial decisions. The agency's action are subject to continuing legislative oversight, and an interpretation out of line with the views of an interested and influential member of Congress is likely to produce some form of counter influence — a legislative hearing, a public rebuke, a criticism, an effort to cut the agency's funds, etc. The agency's discretion is thus constrained not only by the language of the statute, but also by the necessity to keep agency decisions within the ambit of political reality — a necessity resulting from the desire of agency personnel to protect their egos, keep their jobs, and advance in the bureaucracy or the world of politics. Ultimately, of course, the constraint is that Congress may respond to what it regards as incorrect or improper regulation by enactment of corrective legislation.

The above typical examples may usefully be compared with the case of nuclear power, which is atypical. Under the Atomic Energy Act of 1954 [22], there is evidenced an obsessive concern with the "health and safety of the public." That phrase, or close equivalents, was used in the original Atomic Energy Act of 1954 no fewer than 30 times, and numerous additional references have entered the statute through amendments since 1954. Nevertheless, the Act contains only the most general criterion as to safety; i.e., it provides that no license for a nuclear power plant may be issued if the NRC determines that issuance of the license would be "inimical to the health and safety of the public" [23]. This seems to require a higher threshold of acceptable safety than the criteria "safely used" or "unreasonable risk" as discussed above, since "inimical" connotes (Random House Dictionary) "adverse in tendency or effect" as opposed to actual hazard or harm. The NRC's rules provide another criterion of acceptable safety which, it has been said, is inferred from the language of the statute: [24] in the case of a construction permit, "reasonable assurance that . . . the proposed facility can be constructed and operated at the proposed location without undue risk to the health and safety of the public" [25] or, in the case of an operating license, "reasonable assurance that the activities . . . can be conducted without endangering the health and safety of the public" [26]. "Without undue risk" is probably a more relaxed standard than that found in the Chemical Food Additives statute, but it is probably close to the equivalent of the "unreasonable risk" criteria of the Flammable Fabrics and Toxic Substances statutes. Indeed "undue" seems to invite a balancing of risks against benefits. On the other hand, "without endangering" seems to be at least as severe a criterion as "safely used" in the Chemical Food Additives statute, and approaches the "inimical" criterion in severity.

Thus, the Atomic Energy Act has three separate criteria of acceptable safety that, at least from the semantic standpoint, have differing value connotations. In actual practice, however, the AEC and NRC have treated these criteria as essentially identical [27], and this approach has some judicial sanction [28]. The equivalence of these three criteria is less a product of analysis than of political circumstance.

The Atomic Energy Act of 1954 stands in sharp contrast to the three other statutes discussed above and, indeed, to any other statutory scheme for regulation in the interest of health and safety. Unlike the typical regulatory statute that is some years in gestation, the elapsed time from conception to enactment of the Atomic Energy Act of 1954 was less than two years. Moreover, the Act was in no sense the product of perceived abuse of the public interest crying for a remedy. There was no parading of horribles and no record of injuries suffered. Indeed, despite the apparent obsession, as noted above, with the health and safety of the public, there is scarcely a word in the entire legislative history as to the nature or magnitude of the hazards against which the health and safety provisions were directed. There was no nuclear power in existence when the 1954 Act established the regulatory scheme for nuclear power. What the Act did was to establish a framework for dealing with risks that were expected to arise when a nuclear power technology did come into existence. Moreover, the basic purpose of the 1954 Act was really more promotional than regulatory; it was to establish a framework for bringing a nuclear power industry into existence. The regulatory provisions of the Act were intended to ensure that the industry would be appropriately regulated when the time was ripe, but it was clear that the regulation was to be sufficiently benign as not to strangle the embryonic and infant technology. For example, the statutory provision under which all nuclear power plants were licensed prior to amendment of the Act in 1970 directed the Commission to impose the "minimum amount" of the regulation [29].

Just as there was no discussion whatsoever of the risks of nuclear power anywhere in the legislative history, so in the first 15 years or so of experience under the 1954 Act there was no significant Congressional concern about nuclear safety. The absence of Congressional concern, plus the strong pro-nuclear bias of the Joint Committee on Atomic Energy, for many years the only group in Congress with any particular interest in or knowledge about atomic energy, gave the Atomic Energy Commission in effect carte blanche to structure a regulatory program that would not discourage the rapid development and growth of nuclear power. The Atomic Energy Commission's role was more that of the big brother than the cop. Acceptability of risk was measured largely in terms of the extent to which industry was capable of reducing the risk without jeopardizing an economic and financial environment conducive to continuing development of the technology.

This is not to say that the Atomic Energy Commission was lax or permissive in determining acceptable levels of safety, or that the health and safety of the public was impaired. All that is suggested is that the Atomic Energy Commission took more of a political than a legal approach to the question of acceptable safety, and did not evidence the spirit of vigorous single-minded safety regulation that characterizes the typical agency. Whereas the typical regulatory agency errs, if at all, on the side of more stringent safety requirements, the Atomic Energy Commission bent over backwards to avoid any error in the direction of excessive safety stringency.

This approach developed a body of legal precedent that was doctrinally weak. When the Energy Reorganization Act of 1974 [30] split the Atomic Energy Commission into two new agencies, the Energy Research and Development Administration and the Nuclear Regulatory Commission, taking great pains to strip the latter of any promotional mandate or authority, this body of precedent remained in place as legal precedent and was used by NRC as the basis for its own safety determinations. Thus, the essentially political approach that AEC took to balance its promotional and regulatory functions, as this approach was embedded in its regulations and licensing decisions, became the precedent on which NRC built its supposedly non-promotional regulatory program.

CONCLUSION

In considering the relevance of law to the determination of acceptable levels of risk, it is well to understand more generally the role of law in a democratic society such as the United States. All societal actions take place within a framework of laws that, not necessarily with clarity, establish the acceptable ambit of individual action and define the possible consequences if the limits are exceeded. The basic philosophy underlying the legal system is that every individual should be free to act as she perceives to be in her self-interest, subject only to limits established to protect the interests of others. The basic function of the legal framework is, therefore, to resolve the conflict that inevitably arises as individuals pursue their respective self-interests.

The laws that exist at any given moment — whether made by legislature, administrative agency, or courts — reflect societal values that have reached expression through a complex socioeconomic-political process. They are not

References pp. 266-267.

immutable, but are always subject to change. They may be changed by legislative action, by executive implementation, by judicial application and interpretation, and/or by a combination of these, again reflecting the values that emerge from the socioeconomic-political process.

Unfortunately, the nature and function of the legal system is not always well understood by many, even sophisticated, Americans, perhaps as a consequence of oversell by the legal profession itself. Law is too often viewed as a means for implementing justice based on truth, and truth is regarded as an absolute, immutable, objective concept. In reality, however, the function of law is to do justice on the basis of optimum resolution of conflict, and for this purpose truth is a variable concept. The law may accept something as true for one purpose or at one time, but not for another purpose or at another time.

Acceptability of risk, whatever the context of public-policy decision-making, is intrinsically a socioeconomic-political question. Although the elements of the exercise may be defined in scientific, objective terms as Judge Hand did in the judicial opinion discussed above, the content of these terms cannot be reduced to meaningful numbers that have more than ephemeral value in the decision-making process. There is no doubt that our polity could benefit from improved means for determining acceptability of risk, but any such improvement can, as a practical matter, come about in our democratic system only through an overall betterment of the political process and the public whose process it is.

REFERENCES

1. *Restatement (Second) of Torts   282, 1965.*
2. *Id. § 282 (g).*
3. *Id. § 283.*
4. *Id. § 291.*
5. *Id. § 293.*
6. *Id. § 292.*
7. *Id. § 291 (d).*
8. *Osborne v. Montgomery, 203 Wis. 223, 234 N.W. 372 (1931).*
9. *United States v. Carroll Towing Co., 159 F. 2d 169 (2d Cir. 1947).*
10. *Restatement, supra note 1, § 519, et seq.*
11. *Id.   933, et seq.*
12. *21 United States Code § 348(c)(A) (1976).*
13. *Public Law No. 85-929, 72 Stat. 784 (1958).*
14. *21 United States Code 348(a) (1976).*
15. *Senate Report No. 85-2422, 85th Congress, 2d Session (1958).*
16. *21 United States Code § 348(c)(A) (1976).*
17. *Public Law No. 90-189, 81 Stat. 568 (1967).*
18. *15 United States Code § 1193(b) (1976).*
19. *Public Law No. 94-469, 90 Stat. 2003 (1976).*
20. *15 United States Code § 2603(a) (1976).*
21. *House of Representatives Report No. 94-1341, 94th Congress, 2d Session (1976).*
22. *Public Law No. 83-703, 68 Stat. 919 (1954), 42 United States Code 2011, et seq. (1976).*
23. *42 United States Code § 2133(d) and 2134(d) (1976).*
24. *Power Reactor Development Co. v. International Union, 367 U.S. 396, 407 (1961); Maine Yankee Atomic Power Co., 6 AEC 1003, 1004 at n.4 (1973).*

25. *10 CFR   50.35(a)(4).*
26. *10 CFR   50.57(a)(3).*
27. *Maine Yankee Atomic Power Co., 6 AEC 1003, 1009 (1973).*
28. *Citizens for Safe Power v. NRC, 524 F.2d 1291, 1298 n.12 (1975).*
29. *42 U.S.C.   2134(b).*
30. *Public Law No. 93-438, 88 Stat 233 (1974), 42 United States Code   5801 et seq. (1976).*

DISCUSSION

H. P. Green

D. Okrent *(University of California-Los Angeles)*

Harold, I wonder if the qualitative nature of the decisions made in the courts in the past reflects the nature of the arguments that were given to the court and if a series of cases were to be given to the courts presented in, what I'll loosely call, quantitative terms, that in fact you might build up a common law that was more quantitative in nature?

H. P. Green

I'm inclined to doubt it. First of all, if we're talking about the regulatory context where the main issue lies, it is no function of the courts to decide whether the agency's decision was right or wrong. It is the function of the court only to satisfy itself that there was substantial evidence in support of the agency's decision and that the proper procedures were used. Therefore, the courts themselves really have no role in hanging numbers on risks or benefits for this purpose. With respect to the common context, I can visualize that in a whip lash injury case the plaintiff and the defendant might present some very sophisticated quantitative evidence to the courts. The problem with that, however, is that we rely very heavily on our jury system and the jury may not react positively to a party who burdened it with complicated information of that kind.

D. Okrent

In the regulatory context, suppose parties to a decision chose to try to interpret the work "unreasonable" as presented in the act and gave an argument as to whether the agency's ruling was in conformity using numbers and this became an important part of the court's decision making process on whether the fact the agency was pursuing the intent or spirit of the act.

H. P. Green

I find it difficult to believe that an appellate court would ever seriously entertain that kind of argument. As far as evidence is concerned, all that the appellate court is interested in is whether there is substantial evidence in support of the agency's decision. And if there is substantial evidence in support of the agency's decision, no matter how much more substantial the evidence may be on the other side, under the Administrative Procedure Act the court may not disturb that opinion. That's at the federal level; it may be different in state courts.

W. C. Clark *(Int'l. Inst. for Applied Systems Anal.)*

Dr. Lave commented the other day on his opinion that the legal system had been relatively ineffective at bringing about this optimal conflict resolution which you set for it. Would you comment on your opinion of his opinion and particularly upon whether there exists a transition within the legal profession of reviewing the efficacy of the legal procedures for achieving what they set out to do?

H. P. Green

Well, I would say, first of all, that since I, as Bill Rowe indicated, am a graduate of the University of Chicago in Economics and a graduate of the University of Chicago Law School, I align myself with the right wing school of economists to some extent, and therefore I would have to express some dissatisfaction with the existence of government regulations and the process of government regulation. So I don't have any serious quarrel with Ron Howard and Lester Lave's condemnation of the system, if you strip away from it *ad hominem* comments about lawyers. My purpose in this talk really was not to describe utopia, but to describe the system as it now exists. I don't know anybody — laymen or lawyers — who would regard the system as ideal. On the other hand, the kind of reform that was suggested yesterday afternoon, in my opinion, is a very long way off and probably cannot be implemented politically, where it has to be implemented, without a really massive change in public attitude. When you get Milton Friedman elected President of the United States, you might then begin to think about whether or not this kind of reform would be politically feasible.

D. Owen *(SRI, International)*

Let me ask a question about the liability issue that, I think, Ron Howard raised yesterday. Suppose that I bought a Ford product, a small compact car, and in the sales contract it had agreed that a certain value, V_S, be paid in the event of my death due to a design defect. Would that kind of a contract hold up in the courts?

H. P. Green

Yes, I think it would except that the Ford Motor Company might have to get a license to sell insurance.

SESSION IV
DIRECTIONS AND PERSPECTIVES
OF SOCIETAL RISK ASSESSMENT

Session Chairman

Howard Raiffa

Harvard University
Cambridge, Massachusetts

AESTHETICS OF RISK:
CULTURE OR CONTEXT

Michael Thompson

Institute for Policy and Management Research, Bath, Avon, England

ABSTRACT

High standard Himalayan climbing is, quite probably, the riskiest business there is: the chances of being killed are around 1 in 8 or 1 in 10 per expedition. So Himalayan mountaineering provides an ideal (if dangerous) laboratory for the investigation of how and why people come to accept a very high level of risk. By a fortunate coincidence, such discussion of risk as exists in the anthropological literature is centered on the phenomenon of Himalayan trade, and some of these traders (the Sherpas) are now heavily involved in mountaineering.

The conventional anthropological theory is that individuals are guided in their choice between risk-avoiding and risk-accepting strategies by their world views - their culture. A more radical hypothesis accepts this but goes on to suggest that both chosen strategy and culture are, in turn, closely related to the social context that an individual finds himself in. Since an individual's social context can be changed, either by his own efforts or by the actions of his fellows, it follows that his culture and his chosen strategy may also change. On this hypothesis, it is a waste of time trying to discover which is the correct (or best) strategy. Strategies are not right or wrong; they are appropriate or inappropriate. At this point, the anthropology of risk begins to acquire practical implications. But just what these implications are is not (as yet) too clear.

1. Complex industrial societies are likely to generate a wide variety of social contexts and, at the same time, the social policies that such societies adopt are likely to alter the distribution of those contexts - increasing the number of individuals in some and decreasing the numbers in others.
2. Each context will generate its own strategy, its own appropriate pattern of behavior . . . its own rationality. The interesting question then becomes: how do these different rationalities impinge upon one another — what does the risk avoider stand to gain as well as lose, from the activities of the risk accepter, and vise versa?

Could this mean that there is some optimum configuration of social contexts — some particular mix of contradictory rationalities — at which

the welfare of the totality will reach a maximum? If the answer is "yes"
then public policy can take an oblique approach to risk; advocated
policies can be assessed simply according to whether they are likely to
bring the mix of rationalities nearer to, or further from, this optimum.
There will be no need for anyone to say how much a human life is worth.

INTRODUCTION

High standard Himalayan climbing is a risky business. Indeed, it is quite possibly
the riskiest business there is. The fact that it is also an expensive business makes it
difficult to understand why anyone should choose to engage in it and scotches right
at the outset the sort of explanation, favoured by some, that people take risks
because they are poverty-stricken. Whilst it is undoubtedly true that people sometimes do take risks from economic necessity, the fact that climbers often, in effect,
pay heavily to put their lives at risk suggests that economics has not got *that* much
to do with it. And, of course, people sometimes are poverty-stricken because they
are not prepared to take risks. It would be a curious sort of explanation that
required us to hold that poverty was the cause of risk-taking and that the avoidance
of risk-taking was the cause of poverty.

Such an explanation is curious not because it is wrong but because it is
incomplete. It would be perfectly valid if the two states, poverty and its absence,
were cyclically related in such a way that the collapse into poverty took place under
conditions of risk avoidance and the rise out of poverty took place under conditions of risk acceptance. A complete explanation would have to include some sort
of delayed trigger mechanism to provide the switching from one set of conditions
to the other. For instance, it would not be too unreasonable to assume that poverty
leads to desperation and desperation to risk taking. If the risk taking proved successful the poor person would find himself on the up-and-up and might well find
his newly adopted strategy reinforced by it evident success. Only when he was well
and truly removed from poverty might Prudence catch up with him and, by suggesting that he could now afford not to take risks, set him on his downward course.

How, then, do we account for these two regimes: risk taking and risk avoiding:
and for the switching mechanism between them? As it happens, anthropologists
have already described the two regimes but to understand the switching mechanism we have to look at the aesthetics, not the economics, of risk.

ANTHROPOLOGICAL DISCUSSION OF RISK

The anthropological discussion of risk centres upon the fascinating phenomenon
of Himalayan trade.[1] To explain why Himalayan trade exists is a simple matter,
but to explain why some people engage in it and others do not is a problem that is,
as yet, unresolved. The Greater Himalayan Range, running roughly from East to
West, separates two remarkably different regions. To the North is the high, dry,
cold plateau of Tibet sparsely populated by Lamaist Buddhists. To the South are

the low, wet and hot Middle Ranges quite densely populated by Hindus. The largely pastoral people of Tibet have a lot of hairy animals, virtually unlimited deposits of salt but very little grain. The largely agricultural inhabitants of the Middle Ranges produce a lot of grain, particularly rice, on their terraced hillsides but their cattle are not very hairy and they have no deposits of salt. Beyond Tibet lie the great civilisations of Mongolia and China whilst, to the South, a narrow strip of malarial jungle (the Terai) separates the Middle Ranges from the plains of India. The Greater Himalayan Range, though high, is surprisingly narrow and this means that anyone prepared to move into the uninhabited high valleys below the even higher passes into Tibet would be onto a very good thing. He would be in a position to generate wealth as effectively as a hydroelectric station that happened to have a massive head of water in both directions would be able to generate electricity. He could become the profitable channel through which wool and salt would flow down from Tibet whilst grain of various sorts flowed in the opposite direction. Once these flows were established he could expect the varied and sophisticated products of China and Mongolia to the North and of India to the South to be sucked in by the strong trade currents he had set in motion.

As if this were not enough, the various cross-breeds between the Tibetan yak and the Indian cow are highly prized in Tibet: they yield more milk and they are more tractable than the pure-bred yak whilst their hardiness is scarcely less. Indian cattle would perish in Tibet whilst Tibetan yak cannot survive in the thick air of the Middle Ranges. Thus the high valleys provide the only possible meeting ground and, in consequence, anyone moving into these valleys will be able to breed and export these desirable but fortunately, for him, infertile beasts.

But the price he will have to pay in taking up this tempting position is the acceptance of a high level of risk. He will have to live in a place the natural resources of which are probably not sufficient to support him and this means that trade will be vital, not just to his prosperity, but to his very existence. And trade across passes of up to twenty thousand feet or more is a risky business. A sudden storm can wipe out not only his entire stock but him as well. If he gets safely across he may be robbed by Tibetan bandits, he may find that political changes (such as the Chinese occupation of Tibet) have made it difficult or impossible for him to sell his goods, or he may find the market in these goods has been flooded, or that for some other unforseeable reason there is no longer any demand for them. And all this time that he is away he may be worried about the arrangements back home: the ploughing of the fields, the planting and harvesting of the crops, the husbanding of the yaks, the health of those members of his family who have remained behind to see to all these tasks. And then he has the long and dangerous journey back!

There are two possibilities concerning the occupying of this tempting but risky middleman position. Either the Buddhists could move down or the Hindus could move up. In all cases, it would appear, it is the Buddhists who have moved and it is the Hindus who have stayed put. All attempts at explaining why this should be so have foundered in a classic anthropological whirlpool. Do they move or stay put according to whether they are Buddhists or Hindus, or do they become Buddhists or Hindus according to whether they move or stay put?

All the evidence suggests, not that one of these answers is right and that the other is wrong, but that they are both right. Yet, since they are mutually contradictory, they can't both be right *at the same time.* Sometimes, we must assume, one is right: sometimes, the other. If this is the case, then there is still a large piece missing from the explanation; we need to know *when* one answer is right and *when* the other answer is right and, of course, it would be nice to know *why* as well. So let us have a look at risk taking and at the aesthetics of those who take the risks.

AESTHETICS OF RISK TAKING

The real physical risks in Himalayan mountaineering - the avalanches, the frostbite, the verticality, the cerebral and pulmonary oedema, even the leeches and the Nepalese food - would probably be all too apparent to non-mountaineers even in the absence of the books and slide lectures which, with their relentless and emphatic rehearsal of these horrors, are the favoured means by which those climbers who have survived recoup the financial losses incurred in their latest exploit and accumulate something towards the expenses of the next.

But sometimes an expedition will entail, as well as these physical risks, financial risks that in their own way are every bit as great. When Barclays Bank agreed to back the 1975 British Everest Expedition there had already been six attempts by powerful teams, including one led by Chris Bonington who was also to lead our 1975 attempt, all of which had failed by a considerable margin to climb the South West Face: the formidable "last great problem" on Everest. Obviously, Barclays were ill-equipped to assess our chances of success but there were plenty of experienced mountaineers, and many a self-appointed pundit, only too willing to bend their corporate ear. Most were strongly pessimistic but prudently elected not to pronounce too specifically on the likely outcome. But one journalist, Chris Brasher, actually went so far as to quote the odds as fifty to one against.

The full enormity of Barclays' financial risk becomes apparent when you see the same institution that is so reluctant to lend a customer just a few hundred pounds against the ample security of his freehold house, calmly handing out a hundred and fifty thousand pounds, completely unsecured, for a madcap scheme that they know has only a fifty to one chance of succeeding. Of course, Barclays will point out that this is not what they were doing and that the money, in fact, came entirely from that part of their budget allocated to advertising and public relations. But the fact that they were bombarded with letters from incensed customers suggests that the general public has difficulty in visualising the Big Five banks as benign grannies with the cash they have earmarked for various purposes distributed between different tins and vases on their mantleshelves. Rather, they employ a simple input-output model. They see their money going into the bank and they see that same money being dished out to Chris Bonington, his friends and a whole lot of opportunistic Sherpas on the other side of the world.

In the event the prophets were confounded, the South West Face was climbed and Barclays' great gamble paid off in the sense that they have now got back much more than the hundred and fifty thousand pounds that they laid out. I hasten to add that these profits have not disappeared into their coffers but have all been carefully placed in a little tin on Grannie Barclay's mantleshelf and are to be devoted to the encouragement of youthful adventure.

There can be no doubt that Everest climbing involves massive physical and financial risks. The reason why Everest climbing, unlike say air travel, has not got safer with the passage of time is to be found in its uselessness. Once air travel became useful there were powerful economic incentives to increasing the likelihood of a passenger being delivered live to his destination. Technology, management, the selection and training of personnel, even international law, were all bent towards this paramount aim of increasing the probability of a passenger arriving safely and on time in the place where he wished to be. Mountaineering has been mercifully free of such utilitarian constraints . . . until recently.

The first ascent of Everest (twenty three years before ours), coinciding as it did with Queen Elizabeth's coronation, was one of the great imagination-capturers of this century. Almost every child in Britain saw the film: "The Conquest of Everest" . . . the members of the expedition regrouped into lecture teams to visit every corner of the Kingdom. Plain men with simple tastes became Knights of This and Companions of That, and found themselves sustained on a diet of champagne and smoked salmon. "We knocked the bastard off" entered the Oxford Dictionary of Quotations.

Every man destroys the thing he loves: the leader, Brigadier John Hunt, left the army and, as Sir John (later, Lord) Hunt, headed the Duke of Edinburgh's Award Scheme designed to channel the pure spirit of the great achievement to every schoolboy and schoolgirl in the land. Outdoor Pursuits arrived in education and with it came a whole new profession: Outdoor Pursuits Instructors: hideous Frankensteins, half teacher half mountaineer.

One result of all this was that a small number of children, who left to themselves might never have gone near a mountain, died. Mountaineers are an irresponsible lot, teachers are responsible: the faces of the Frankensteins were contorted with anguish. In vain did they hold Official Inquiries, introduce Codes of Safety, initiate Mountain Leadership Certificates and weigh down their charges' rucksacks with devices that would enable them to extricate themselves from every conceivable eventuality. Some children still died.

In desperation they even asked the mountaineers why this should be so. They replied: "Some children die because mountaineering is dangerous." The message was clear: there is no place for mountaineering in education. It was also unacceptable. If mountaineering was removed from education Outdoor Pursuits Instructors would be left with no children to instruct and Mountain Leadership Certification Boards would have no candidates to certify. The solution was simple and obvious: *mountaineering must be made safe.*

In this way, a programme originally inspired by a great achievement is now poised to bring about a situation in which such an achievement will be impossible. Nearly all the Buddhists have been converted to Hinduism: there are very few of us left. Before we become extinct, and before achievements involving a high level of risk become impossible, let me enter a plea for our preservation. I do not ask that we be recognized as yet another oppressed minority and granted the security of a Buddhist sanctuary: to make such a request would be to capitulate and to join the

Hindus in their prescription-ridden and risk-free world. Rather, I would urge that we understand the Hindu-Buddhist cycle, and its switching mechanisms before it finally breaks down. If we understand it we can rebuild it and so retain access to the full range of capabilities that it alone can generate.

The aesthetics of high standard mountaineering are such that a proposed route is only felt to be worthwhile if there is considerable uncertainty as to its outcome. It is for this reason that we wished to climb the South West Face. Advances in equipment and technique, and the familiarity resulting from its many ascents, have rendered the original route by the South Col of little interest to the leading climbers of today (unless it be an attempt with a very small party or without oxygen). To repeat the original route with a large party or with helicopters to ferry loads into the Western Cwm, as happened recently, is simply to do less with more and to render the outcome almost a foregone conclusion. The traditional mountaineering response to this aesthetically repugnant behaviour is ridicule, and I was interested to discover that the Sherpas who accompanied us on the South West Face also entered into this aesthetic framework and disparagingly referred to the line by which Hillary and Tensing first reached the summit as "The Yak Route".

In sharing this little joke, European climbers and Nepalese Sherpas are both revealed as Buddhists poking a little malicious fun at some European Hindus. For a moment, as we chuckle, the mists of cultural difference clear and we see through to the universal mountain that usually they obscure. These mists are formed by our personal processes of risk management. Risks, it turns out, come in several different forms and the way in which we emphasise one and play down another often clouds our understanding of what is actually going on.

As well as physical risk and financial risk there is a third type, intellectual risk. A person takes on intellectual risk when he sets out to provide an adequate explanation for something where previous attempts have failed, and he takes an intellectual risk when he sets out to question the validity of some explanation which most people believe to be perfectly adequate. In taking an intellectual risk a person stands to lose neither his life, nor his fortune, but his credibility. Since knowledge, like air travel, is usually believed to be useful there are strong disincentives to intellectual risk-taking, and anyone who wishes to take such risks would be well-advised to immerse himself in some relatively useless area of knowledge, such as anthropology.

If Everest climbing and anthropology are united to the extent that they are both pretty useless, they are set apart by the very different kinds of risk-taking that each encourages. The picture is further confused by the intrusion of financial considerations. Though neither the Everest climber nor the anthropologist is particularly interested in financial risk both need money to indulge in the sorts of risk-taking that do interest them. When it is not forthcoming the problems they face and the risks they must run are compounded. For example, Don Whillans, in the first batch of mail to arrive at Base Camp after he and Dougal Haston had returned exhausted but triumphant from the summit of Annapurna, received just two letters. One, from the Lord Mayor of Rawtenstall, offered him the freedom of his native borough, the other, from a different room in the same Town Hall, informed him that if he did not pay his rent arrears he would be evicted from his council

house.

An example of intellectual risk-taking in the face of financial difficulties every bit as severe as those besetting Whillans is provided by the application, by Professors Hoyle and Wickramasinghe, to the British Science Research Council for funds to investigate their hypothesis that life originated in outer space and arrived, and is still arriving, on earth by meteorite. Although the credentials of these two gentlemen are quite impressive, and although eight thousand pounds is a quite small price to pay for evidence that might upset widely held beliefs about evolution, the application was rejected. By the simple expedient of saving itself eight thousand pounds the Science Research Council stood a good chance of preventing the evidence being gathered, thereby avoiding the risk that the amount of certainty in a wide field of very useful knowledge might be suddenly and dramatically reduced. At the same time it could avoid the risk, inherent in approving the application, of itself losing credibility and becoming, in the eyes of its faithful, the Science Fiction Research Council.

Now intellectual risk-taking is not usually much in evidence on Himalayan expeditions. A climber *knows* he wants to climb Everest and his main concern is to try to do it in as aesthetically pleasing a way as possible. If he were all the time asking himself *why* he wanted to climb Everest he would probably not get far beyond Base Camp, and might well fall down a crevasse if he did. Mallory's famous reply: "Because it's there": was not an answer to the question: "Why climb Everest?": it was a way of stopping people asking it for long enough for him to have a stab at doing it. The charm of the Sherpas' little "Yak Route" joke is that, by momentarily clearing those mists, it encourages me to take a large and exciting intellectual risk. In anthropological terms I want to try to formulate a general theory of risk. In everyday terms I would like to have a go at answering that perennial question: "Why climb Everest?"

It is perhaps only appropriate that that ugly pyramidical lump which happens to terminate in the highest point on earth should act as a focus not just for physical risk but for financial and intellectual risk as well.

Aesthetics, of course, have always been recognised as an important part of mountaineering. The aesthetic form may change, from the stiff upper-lips of the pre-war Everest climbers, through Smythe's "spirit of the hills", Winthrop-Young's craftsmanship and Whillan's job-of-work-to-be-done, to Bonington's shameless exposure of his inner states, but, whether it be the aesthetic of reticence, nature mysticism, esoteric skill, hard graft, or of letting it all hang out, there can be no doubt that more than just economic considerations motivate the mountaineer.

Yet, curiously, such aesthetic niceties are not assumed to extend to the Sherpa who throughout the history of Himalayan mountaineering have carried the Sahib loads, and sometimes the Sahibs themselves, up their chosen peaks. True, virtually every expedition book is full of praise for the Sherpas' fortitude, courage, cheerfulness and dependability, but at the same time there is always something rather stereotyped about this handsome expression of credit and inevitably the reader finds the Mingmas, the Dorjes, the Pasangs and the Kanchas merging into a succession of indistinguishable brown, hairless, smiling faces. The same conver

anonymity permits their formidable achievements to be condensed into a statistical table of loads carried and altitudes reached somewhere among the appendices on food, health, logistics, communications and the like, towards the end of the book. Despite the numerous best-selling accounts of expeditions, and despite the real bonds of affection and admiration that link many a Western climber and his Nepalese counterpart, the Sherpa still remains the Cheshire cat of mountaineering literature: little more than a big smile at the opposite end of the arm to the Sahib's pre-dawn mug of tea. The basic assumption is that the Sahib climbs Everest because it is there whilst the Sherpa climbs it for the money.

It is a convenient assumption. If climbing is mostly about aesthetics and if the Sherpas are concerned only with economics, then their contribution to any mountaineering achievement can be equated with that of, say, Barclays Bank. You need money to climb Everest so the argument runs, and you need Sherpas to climb Everest, but both are simply the pre-conditions for Himalayan mountaineering: neither has anything to do with its essence, with what climbing is *really* about.

In other words, our personal risk management leads us to expand the aesthetic scope of our own actions and to contrast that of the Sherpas: East is East, we say, and West is West and cultural difference explains all. But, of course, all that this appeal to cultural difference does is set a limit to what we are prepared to explain. Sharing a joke with Sherpas overrides these distortions produced by our risk management and reveals that the frames of our aesthetic scopes are identical. Suddenly, appeals to cultural difference are of no avail: we can no longer call upon Kipling to bail us out when the intellectual risks become too great.

Probably the greatest achievement of anthropology has been to shatter the convenient assumption that, in the same sort of situation people will tend to do the same sort of thing, and no sooner does an economist, a psychologist, a sociologist or a political scientist produce some elegant universal model of some aspect of human behaviour than an anthropologist will jump up to spoil his fun by adding the carping codicil: "in our culture". Anthropologists have become so carried away by their spoilsport success that they have almost completely lost sight of the really interesting, and difficult, question which is: "Granted that different people in the same sort of situation may do different things, *why* do they do the different things that they do?" This is the question that a general theory of risk will have to answer. It will have to offer some satisfactory reason why Mallory wanted to climb Everest, and why Mingma wants to climb Everest, and it will have to give some satisfactory reason why all sorts of other people *don't* want to climb Everest.

RISK TAKING AND RISK AVOIDING

Since climbing Everest is both a voluntary and a risky business, an explanation of why some people accept risks and others avoid them will go a long way towards answering these questions. The answer to the secondary question, "Why climb (or avoid climbing) Everest, in particular?", is simply that Everest climbing is very risky and very useless. The risk taker, anxious to expand the pure aesthetic scope of his preferred style, could not ask for a more perfect objective. There is nothing

peculiar, or hard to understand, in his choosing Everest. On the contrary, if Everest did not exist it would probably be necessary for him to invent it. The risk avoider is positively repelled by these very same properties that the risk taker finds so attractive. For him, risk of any kind is nasty and a useless *and* voluntary risk is just about the nastiest thing there could be. So the whole explanation hinges upon these two personal styles: risk taking and risk avoiding.

For every proverb and catchphrase there is, it would seem, a contradictory counterpart: "Look before you leap" versus "He who hesitates is lost", "There's safety in numbers" versus "Only a dead fish swims with the current". If we were to collect these contradictory pairs and line them up with one another we could sketch out the world views: the sorts of predictive frameworks: that the risk taker and the risk avoider use in choosing, and justifying, the very different courses of action that each follows. The risk taker's world view corresponds to that of the adventurous Himalayan trader: the Buddhist. The risk avoider's world view corresponds to that of the cautious stay-at-home cultivator: the Hindu. One grants credibility to one set of proverbs, the other to the opposing set. Once equipped with these very different perceptions of the world it is hardly surprising to find that, when confronted with uncertainty, they operate very different strategies.

The Buddhist is an optimist: his response to uncertainty is positive. He acts boldly, but not foolhardily, in the hope of reaping rich rewards. The Hindu is a pessimist: his response to uncertainty is negative. He prefers not to act for fear that one thing may lead to another. He subscribes to a "domino theory" that insists on the connectedness of everything. The Buddhist operates on a "one-off theory" that allows him to disregard those possible consequences that lie outside his immediate concern. He goes in for risk narrowing: "Spot on he wins, way out he loses". The Hindu goes in for risk spreading: "A trouble shared is a trouble halved". Why should one be led to adopt the risk narrowing strategy, the other the risk-spreading strategy?

First of all, it is not because one is a Buddhist and the other is a Hindu. "Buddhist" and "Hindu" are simply convenient labels to identify a person's commitment to one or other set of proverbs. No, the answer to why a person accords credibility to one set of proverbs rather than the other is quite ridiculously simple, and has nothing to do with cultural difference.

The Hindu adopts a risk sharing strategy, and subscribes to the pessimistic all-embracing world view that justifies such a strategy, because he has someone to share with. The Buddhist adopts a risk narrowing strategy, and subscribes to the optimistic piecemeal world view that justifies such a strategy, because he has no one to share with.

Social context is enormously persuasive. If there is no-one there to share your risks with you, you cannot go in for risk sharing and, conversely, only a mug would take a huge personal risk knowing that, if he was successful, he would have to share the rewards among all his cautious risk shunning fellows. Of course, a risk avoider may in certain circumstances, be prepared to take certain risks: those that are not for personal gain but for the survival, the glory or the honour of the group. These altruistic acts can sometimes be so risky as to be suicidal: the team spirt of Horatius and his comrades and the selfless heroism of the Kamikaze pilot typify the risks a risk avoider may accept: "Never volunteer for anything, except certain death". Paradoxically, the risk acceptor has a strong aversion to these sorts of risks. As a perceptive mountaineer, Tom Patey, once put it: "He should never underestimate the importance of staying alive".

I can foresee two possible objections to this devastatingly simple answer to the question "Why climb Everest?" The first is that, if my theory states that individualism encourages risk taking and collectivism encourages risk avoidance, then, in equating Everest climbing with individualism, I have got it all wrong because Everest climbing is a collectivist activity wholly dependent for its success upon superlative teamwork and upon highly motivated and skilled individuals selflessly surrendering their personal ambitions (to be *the* man on top of Everest) to the common cause (getting *a* man on the top of Everest). Now, though this heroic picture is often what those who do not go on Everest expeditions choose to see and, indeed, is often what those who do go on them are happy to paint for their armchair public, Everest expeditions are not like that at all.

High standard mountaineers are extraordinarily individualistic and are only prepared to join together to form an expedition in the first place because they know that, regrettably, they can't get to the top of Everest unaided. The steady progress of an Everest assault can only be maintained by continually changing the lead climbers: those who have done their stint returning to Base Camp for a rest whilst others, still fresh, take their place. Going down to Base Camp takes one or two days, coming back to the sharp end is at the rate of one camp per day. It is most fortunate that the total amount of time it will take to climb the mountain is unknowable; if the members of an expedition knew just how many days lay between the first pair of climbers setting out from Base Camp and the placing of a man on the summit, they could quite easily work out which position in the sequence of lead climbers they should take up in order to end up on the summit. Having myself watched and been involved in these complex dynamics, I am quite certain that, if the optimum position happened to be that of the first pair to set out from Base Camp, all the members of the expedition would hurl themselves simultaneously at the mountain and that, if this was not the optimum position, not a soul would stir from his sleeping bag.

Of course, as the climb proceeds so this degree of uncertainty decreases and towards the end it becomes quite predictable, either that the summit will not be reached, or that it will be reached in a certain number of days. But, by this time, it is usually too late for the individualists to be able to do much about it. If they are resting at Base Camp they have had it: if they are moving up towards the top camp they are in luck. Of course, those in the top camp, sensing victory, could if they were selfish enough refuse to return to Base Camp and this, indeed, is often what happens. On one brilliantly successful expedition Base Camp contained only the Sherpa cook and a Yorkshireman with a wooden leg: all the other members were crammed into the top camp and the last one to drag himself up to this over-populated spot received for his pains, not a steaming mug of tea, but a punch on the nose.

But, usually, events move too quickly and many of the members are stranded without any real hope of getting to the top camp in time. It is at about this moment that they turn their hands to good works — selflessly ferrying essential supplies to the higher camps . . . manning the Base Camp radio (in their sleeping bags) in case the summit party should suddenly come on the air . . . getting a good sweet brew ready for the returning victors. The leader for his part plans increasingly unrealistic second, third, fourth and even fifth summit parties in a pretence of fair shares for all, and the Sherpas, sensing the end of the expedition, lose their upward momentum and divert their energies to the stripping of the camps and the accumulation of staggering loads of personal booty.

The second possible objection is that, if the answer to this perennial question really is so simple, how come somebody hasn't come up with it sooner? The reason is twofold. First, people are usually very strongly committed to their view of how the world is, and this commitment is partly maintained by denying the validity of other ways of seeing the world. Second, though the world view to which they are committed depends upon their social context, they are unaware of this dependence. As far as they are concerned, their world view is not some artificial construct, it is an accurate factual account of how the world is. One might say that the collectivised context is to the Hindu what the earth is to the worm and that the individualised context is to the Buddhist what the air is to the bird. Each is as firmly and unquestioningly committed to his world view as are the worm and the bird to their respective eye-views.

M. Thompson

But there is a crucial difference. For worms and birds the environments are natural: for risk avoiders and risk takers the environments are social. No matter what actions worms and birds take they will never find themselves living in one another's media but, for the risk avoider and the risk taker, there exist the possibilities equivalent to the worm sprouting wings and the bird slithering into the soil. Social contexts can change, either as a result of the actions of the individuals who constitute the totality, or as a result of external natural or social pressures. This means that, whilst wormhood and birdhood are two clearly separate and unrelated states, risk avoiders can be transformed into risk takers and risk takers into risk avoiders. Yet the path to risk acceptance is not the reverse of the path to risk avoidance: there is a cyclical relationship between the two states: a Hindu-Buddhist cycle.

Though people are convinced that the view they have of the world is something natural — something self-evident — they have in fact worked very hard to make it appear like that. The Hindu in his collectivised context *learns* to avoid risk taking. Since he learns from his mistakes, as well as by getting it right, he will have built up a considerable profit and loss account by the time he is firmly locked onto the set of proverbs that makes sense of his world. In the same way, the Buddhist will build up his distinctive pattern of investments that commits him firmly to the other set of proverbs. The consequence of all this work — all this aesthetic investment — is that, in order to let go of the Hindu world view and acquire a firm grasp on the Buddhist world view, you have to dismantle one investment structure and build another. This means that, if you went right round the cycle from Hindu to Buddhist to Hindu again, you would, on the first leg, cling to your risk avoiding strategy long after you had passed the mid-way point between collectivised and individualised contexts and, on the return leg, you would similarly over-persist with your risk accepting strategy.

Taking someone from Hinduism to Buddhism and back again is rather like that school experiment in physics in which you start off with an iron bar magnetised north-south, demagnetise it, magnetise it south-north, demagnetise it and then magnetise it north-south again. Such cycles can be depicted by a graph called a "hysteresis" in which the area enclosed by the paths between the two magnetised states provides a measure of the work done in going round the cycle.

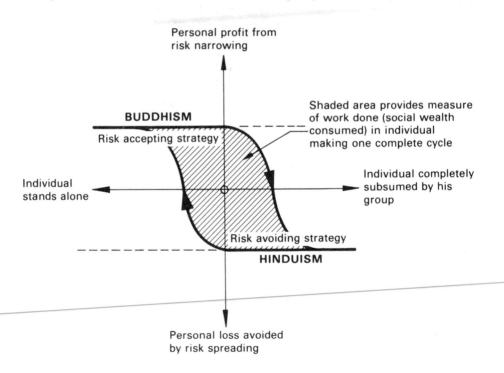

Fig. 1.

Apart from providing some sort of answer to the question "Why climb Everest?", what does this little diagram tell us that we don't already know?

It tells us:

1. That changing people's social contexts is a costly business.
2. That, though risk taking and risk avoiding strategies are contradictory, little is to be gained and a great deal could well be lost by insisting that one is right and the other is wrong. Rather, each is appropriate in a particular kind of social context.
3. Though each is convinced that his view of the world is the right one, Buddhist and Hindu each stand to gain, as well as lose, from the activities of the other. This means that in all probability there will be, in the distribution of Hindu and Buddhist contexts, some optimum arrangement at which the gains minus the losses for the totality reaches a maximum.
4. Since a modern industrial society inevitably generates both sorts of social context, and since the social policies that such societies implement inevitably result in the transformation of some individual's contexts, there exists the possibility of evaluating these policies according to whether they will bring us nearer to, or take us further from, this optimal.

Having put forward these few suggestions, I am assailed by awful premonitions. Could it be that I am about to destroy the thing I love: have I got something *useful* out of Everest climbing?

WITCHES, FLOODS, AND WONDER DRUGS: HISTORICAL PERSPECTIVES ON RISK MANAGEMENT

William C. Clark

International Institute for Applied Systems Analysis
Laxenburg, Austria
and
Institute of Resource Ecology
University of British Columbia
Vancouver VGT 1W5, Canada

ABSTRACT

Risk is a people problem, and people have been contending with it for a very long time indeed. I extract some lessons from this historical record and explore their implications for current and future practice of risk management.

Socially relevant risk is not uncertainty of outcome, or violence of event, or toxicity of substance, or anything of the sort. Rather, it is a perceived inability to cope satisfactorily with the world around us. Improving our ability to cope is essentially a management problem: a problem of identifying and carrying out the actions which will change the rules of the game so that the game becomes more to our liking.

To cope better is to better understand the nature of risks and how they develop. It is naive and destructive to pretend that such understanding can carry with it the certainties and completeness of traditional science. Risk management lies in the realm of trans-science, of ill-structured problems, of messes. In analyzing risk messes, the central need is to evaluate, order, and structure inevitably incomplete and conflicting knowledge so that the management acts can be chosen with the best possible understanding of current knowledge, its limitations, and its implications. This requires an undertaking in policy analysis, rather than science.

One product of such analyses is a better conceptualization of "feasibility" in risk management. Past and present efforts have too often and too uncritically equated the feasible with the desirable. Results have been both frustrating and wasteful.

Another is an emphasis on the design of resilient or "soft-fail" coping strategies. The essential issue is not optimality or efficiency, but robustness to the unknowns on which actual coping performance is contingent.

The most important lesson of both experience and analysis is that societies' abilities to cope with the unknown depend on the flexibility of their institutions and individuals, and on their capability to experiment freely with alternative forms of adaptation to the risks which threaten them.

Neither the witch hunting hysterics nor the mindlessly rigid regulations characterizing so much of our present chapter in the history of risk management say much for our ability to learn from the past.

FEAR, RISK, AND HISTORICAL PERSPECTIVE

At the center of the risk problem are people and their fears. Fears of loss, fears of injury, and — most of all — fears of the unknown. How we cope with those fears affects both the material well-being, and the spiritual character of the individuals and societies we become.

Students of risk therefore face a number of difficulties in addressing their subject matter, and surely one of the most serious is that of establishing a useful perspective. The fears of risk are our fears, the people making and taking risks are ourselves and our neighbors. When we intellectualize ourselves away from these ambiguities, our work becomes sterile, our subjects ciphers. When we tackle them directly, our involvement makes critical interpretation impossible and broader interpretations irrelevant. Unable to see inside the problem, we trivialize it. Unable to see outside the problem, we become part of it.

I suspect that work on contemporary risks will always contain some elements of this contradiction. To better appreciate the problem we are in, to better orient our directions for the future, a backward look into the history of societal risk assessment therefore seems in order. With the perspective of time, we should be able to perceive some of the pitfalls and opportunities of risk management which our intimate involvement in the contemporary scene denies us.

Unfortunately, what must have been a truly monumental environmental impact assessment on the Seven Days of Creation has been lost. But societal risk assessment nonetheless has a history as long as man's efforts to explain, manipulate, and cope with his fears and the unknown. Much of this is still accessible to us, and should have some lessons to teach. That at least is a possibility, and one that has not yet been explored.

For several years now I have been trying to convince some competent historians to undertake a retrospective study of societal risk, all to no avail. The present historical "essay" is the result, and I emphasize that I use the term in its original sense of an amateur's first attempt. That attempt has been fun, and has provided me some interesting perspectives on our present risk dilemmas. I hope its transgressions of historical fact will sufficiently outrage professionals that their rebuttals can begin the serious study which I seriously do believe is needed.

In the next three sections I review what seem to me three particularly instructive episodes in the history of social risk assessment. The European witch craze of the 16th and 17th centuries is treated in the first section, some North American experience in renewable resource management in the second section, and medical drug regulation in the third. The historical perspective derived from these three studies is then used to shed some light on the unasked question of "What are we arguing about?" in the contemporary risk debate. Finally, I look forward to the prospects for adaptive design of risk management policy.

WITCH HUNTS: ON THE SOCIAL PSYCHOLOGY OF RISK*

For several centuries spanning the Renaissance and Reformation, societal risk assessment meant witch hunting. Contemporary accounts record wheat inexplicably rotting in the fields, sheep dying of unknown causes, vineyards smitten with unseasonable frost, human disease and impotence on the rise. In other words, a litany of life's sorrows not very different from those which concern us today.

The institutionalized expertise of that earlier time resided with the Church. Then, as now, the experts were called upon to provide explanation of the unknown and to mitigate its undesirable consequences. Rather than seek particular sources of particular evils, rather than acknowledge their own limitations and ignorance, these experts assigned the generic name of "witchcraft" to the phenomenology of the unknown. Having a name, they proceeded to found a new professional interest dedicated to its investigation and control.

As the true magnitude of the witch problem became more apparent, the Church enlisted the Inquisition, an applied institution specifically designed to address pressing social concerns. The Inquisition became the growth industry of the day, offering exciting work, rapid advancement, and wide recognition to its professional and technical workers. Its creative and energetic efforts to create a witch-free world unearthed dangers in the most unlikely places; the rates of witch identification, assessment and evaluation soared. By the dawn of the Enlightenment, witches had been virtually eliminated from Europe and North America. Crop failures, disease, and general misfortune had not. And more than half a million people had been burned at the stake, largely "for crimes they committed in someone else's dreams" [2, p. 221].

How did the expert institutions of the day come to wreak such havoc on the people they sought to protect? Answers to this question provide some useful perspectives for our present attempts to assess societal risk.

Witches and witchcraft can be traced back to the very beginnings of history. For centuries, people had found "witches" a convenient label for their fears of the unknown, an adequate explanation for the inevitable misfortunes which befell one's crops, health, and happiness. For centuries, the Church adopted a skeptical and largely academic approach to these explanations, preaching the difference between fact and fantasy, and placing witches squarely in the latter category. Witchcraft, if it existed at all, was an illusion sent by the devil. These illusions were

---

*This section draws from Trevor-Roper [1], Harris [2], Duerr [3], and Summer's translation of the Malleus [4].

frowned upon and even prohibited by law. But individual witches and witch mongers were not sought out by the Church and, if brought to its attention, were merely advised of their errors and urged to desist. Persistent individuals were simply excommunicated. The social structure, represented by the Church, declared itself no longer interested in or responsible for the welfare of those individuals. If the miscreants chose to ignore responsible advice, their subsequent fate was their own business, and the devil's. Witches remained an individualized risk, requiring individual responses by individual members of the Church and lay communities.

Wildavsky [5] has spoken of the "watershed" which is passed when such individualized issues are collectivized under unified social policies. Institor and Sprenger's *Malleus Maleficarum (The Hammer of the Witches*, published in 1486) was for the witch hunting business that collective consciousness watershed which Kefauver's *Hearings* would later be for drug regulators, and Rachel Carson's *Silent Spring* for environmentalists. Massive in its scope and evidence, impelling in its argument, the *Malleus* showed that witches did in fact exist, with real power for evil. As agents of Satan, they were a heresy — a dangerous sin in need of eradication. Not individuals, but society as a whole was in peril as long as witches remained at large. Pope Innocent VIII credited this argument, and his bull *Summis desiderantes affectibus* authorized full application of the Inquisition, including torture, to the eradication of witches. The witch risk, to use another of Wildavsky's [5] terms, had been "socialized". Collective action by the central authority was henceforth required, and any action taken against a particular individual was justified in the name of the common good. In the case of the witch hunts, this "common good" justified the carbonization of five hundred thousand individuals, the infliction of untold suffering, and the generation of a climate of fear and distrust — all in the name of the most elite and educated institution of the day.

Modern risk assessors do not incinerate their fellow citizens. Furthermore, they seek to insure against milder forms of witch hunting by a scientific approach to gathering and evaluating evidence on risk issues. But the history of witch hunting suggests that what we say we are doing or wish to be doing in contemporary risk assessment may be far removed from what actually occurs. Again, the historical perspective may help us to recognize some of these discrepancies, and to provide a basis for their rectification.

The institutionalized efforts of the Church to control witches can be seen, in retrospect, to have led to witch proliferation. Early preaching against witchcraft and its evils almost certainly put the idea of witches into many a head which never would have imagined such things if left to its own devices. The harder the Inquisition looked, the bigger its staff, the stronger its motivation, the more witches it discovered. Similar trends have been documented in the modern literature on hazard and have long posed difficulties for those seeking to document the crime prevention effectiveness of larger policy forces [e.g. 6, Fig. I-1]. A general question therefore arises concerning the causal relationship between assessment and risk: Which is driving which? A strong case can be made for the notion that search effort *creates* the thing being sought. Since the resulting higher discovery rate of witch risks obviously justifies more search effort, the whole process becomes self-contained and self-amplifying, with no prospect of natural limitation based on some externally determined "objective" frequency of witch risks in the environment.

The way we ask questions, and the kinds of evidence we admit in our attempts to answer them, are of the utmost importance here. The Inquisition asked "Are you a witch?" and proceeded to examine the evidence to see if you were. Today, whatever we title our symposia, we ask "Is this a risk?" and proceed accordingly. In neither case is there any conceivable empirical observation which could logically force an answer "No!". In neither case is there a "stopping rule" which can logically terminate the investigation short of a revelation of guilt.

In witch hunting, accusation was tantamount to conviction. Acquittal was arbitrary, dependent on the flagging zeal of the prosecutor. It was always reversible if new evidence appeared. You couldn't win, and you could only leave the game by losing. The Inquisition's principal tool for identifying witches was torture. The accused was asked if she was a witch. If she said no, what else would you expect of a witch? So she was tortured until she confessed the truth. The Inquisitors justified ever more stringent tortures on the grounds that it would be prohibitively dangerous for a real witch to escape detection. Of course an innocent person would never confess to being a witch (a heretic with no prospects of salvation) under mere physical suffering. The few who lived through such tests were likely to spend the rest of their lives as physical or mental cripples. Most found it easier to give up and burn.

And today? What is not a risk with a parts-per-trillion test can always be exposed to a parts-per-billion examination. If rats cope with the heaviest dose of a chemical that can be soaked into their food and water, you can always gavage them. Or try mice or rabbits. Again, the only stopping rule is discovery of the sought-for effect, or exhaustion of the investigator (or his funds). Many of the risk assessment procedures used today are logically indistinguishable from those used by the Inquisition. The absence of "stopping rules" means that both fail to meet Popper's [7] "demarcation" criterion for true science. Since neither is advancing falsifiable propositions, neither is capable of producing anything more than propaganda in support of its own prejudices.*

Modern science's defense against self-delusion relies upon a spirit of open and critical inquiry. This, though hardly infallible, ostensibly uncovers errors and thereby proceeds towards objective truth. Once again, however, exactly the same high principle failed in actual practice during the heyday of the Inquisition. Within the 16th century Church, hardly a voice was raised against the witch hunt, while those outside defended the accused only at great personal risk. After all, argued the *Malleus*, with such crop losses, child mortality, marital infidelity, and general aches and pains as exist today, "Who is so dense as to maintain . . . that all their witchcraft and injuries are phantastic and imaginary, when the contrary is evident to the senses of everybody?" Who, indeed? Only those in league with the devil.

---

*On "propaganda" in this context, see Feyerabend [8]. It is worth noting that both witch hunts and risk assessments also fail to meet Kuhn's weaker "puzzle solving" criterion of science [see 9].

And so the few philosophers and physicians who did speak up against the hunts were themselves harassed, excommunicated, and in many instances burned along with the witches.

Today, anyone querying the zeal of the risk assessors is accused at least of callousness, in words almost identical to those used by the *Malleus* five hundred years ago. The accused's league with the devil against society is taken for granted. Persecution in the press, courts, and hearing rooms is unremitting, and even the weak rules of evidence advanced by the "science" of risk assessment are swept away in the heat of the chase (see section on medical drug regulation below). This is not to say that risks don't exist, or that assessors are venal. It is to insist that skeptical, open inquiry remains theory rather than practice in the majority of today's risk debates. That those debates are so often little more than self-deluding recitations of personal faith should not be surprising.

A last insight into our modern treatment of risk evidence comes from the historical demise of witch hunting as a profession. In 1610, after a century of witch hunting, the exceptional Inquisitor Alonso Salazar y Frias carried out an extensive analysis of witch burnings at Logrono, Navarre. He showed that most of the original accusations had been false, that torture had created witches where none existed, and that there was not a single case of actual witchcraft to show for all the preaching, hunting, and burning which had been carried out in the name of the Church. He did not rule on whether witches existed. He did order that the Spanish Inquisition no longer use torture under any circumstances, and that accusations no longer be considered unless supported by independent evidence. The number of witches brought to trial declined precipitously.

In modern terms, y Frias had instituted a grand jury condition between accusation and trial. Further, he had introduced rules of evidence which recognized the perverse and essentially meaningless forms which unstructured "facts" could take. Neither of these reforms has yet been introduced into major streams of the contemporary risk debate. And very few retrospective studies of the sort carried out by y Frias have yet been conducted by the modern risk assessment community.* When we realize that y Frias' study occurred only after a century of active witch hunting, and that the practice was not completely stamped out until more than a centurey later, the prospect of rationalizing contemporary risk assessment seems distant indeed.

It is all very well to note the psychological and evidential problems which led the Church to protect its fold by burning goodly numbers of them. But witch hunts continued as an organized political activity for over two hundred years, and it requires a certain credulity to pass off such persistence as a product of excess zeal and logical error. We may be forgiven for joining the lawyers in asking "*Cui bono?*": Who benefited from this complex, expensive, and destructive undertaking?

The invaluable studies of Lawless [10] focus on cases where a serious risk existed, but was recognized late. Missing are the complementary studies of legislated restrictions where no risk existed. I discuss some retrospective looks in the matter of drug regulation below.

Anthropologist Marvin Harris has pointed out that to believe that the main purpose of the witch hunters was the annihilation of witches is to accept uncritically the lifestyle consciousness professed by the witch hunting community. Looking at "its earthly results rather than its heavenly intentions" [2, p. 237], the witch hunt supports a rather different interpretation. Whether individual witch hunters sincerely believed in what they were doing is not the point. As with risk assessment today, what actually happens may be radically different from what people think is happening. The benefit of our historical perspective on the witch phenomenon is that, with hindsight, we can see that difference, and try to learn from it.

To begin with, there was certainly an element of opportunistic careerism in the Inquisition, and there is almost certainly an element of opportunistic careerism in the present risk assessment movement. However small this element, it is clear that it can do a lot of damage to the world that the profession is trying to protect, and can bring the profession into disrepute in the process. The same reform Inquisitor Alonso Salazar y Frias who restructured the rational side of the witch hunt was evidently a worldly man as well. Besides instituting grand jury hearings and rules of evidence, he revoked the law that property of a convicted witch could be confiscated by the Church. Again, the rate of witchcraft accusations plummeted. It is interesting to speculate on what might constitute a similar perturbation experiment for today's risk assessors.

A second point illuminating the witch hunt phenomenon is that virtually no members of the clergy or aristocracy were accused, much less executed.* At best, the profession was evidently incapable of coping with findings which refracted on itself. In fact, it reacted like a powerful elite which finds its own ox is about to be gored. One assumes that heretics accusing these privileged elites were promptly identified as the devil's agents. Anyone who has followed the recent debates over the risk of recombinant DNA research** will recognize that things haven't changed much, and can imagine the outrage with which the Church must have reacted to accusations upon its own house. The same episode justifies a certain skepticism regarding the presumption that today's science community is willing to pursue its risk assessment activities into areas striking close to home.

A third historical issue is less firmly established but, for purposes of understanding the present risk enterprise, much more significant. Harris continues his analysis of the witch hunts with an argument that they functioned directly to increase the power of the elite institutions which conducted them, and simultaneously directed discontent against those institutions into relatively nonthreatening channels:

> The poor came to believe that they were being victimized by witches and devils instead of princes and popes . . . Against the people's phantom

---

*H.C.E. Midlefort (*Witch Hunting in Southwestern Germany,* Stanford Univ. Press, Stanford, 1972) shows that of 1258 executions between 1562 and 1684, 82% were female. Three members of the nobility were accused, and none were executed.

**For a review, see P.B. Hutt. Research on Recombinant DNA: The Regulatory Issues. South. Calif. Law Rev., 51: (6) 1435-1450, 1978.

References pp. 311-313.

enemies, Church and State mounted a bold campaign. The authorities were unstinting in their efforts to ward off this evil, and the rich and poor alike could be thankful for the energy and bravery displayed in the battle. The practical significance of the witch mania therefore was that it shifted responsibility for the crisis of late medieval society from both Church and State to imaginary demons in human form . . . Not only were the Church and State exonerated, but they were made indispensable. The clergy and nobility emerged as the great protectors of mankind against an enemy who was omnipresent but difficult to detect [2, pp. 237-238].

Valid or not, there is an obvious modern parallel to this interpretation of the witch craze. Science has been under growing attack in recent years for a variety of ills ranging from wasting the tax dollar, to pompus arrogance, to greedily destroying our environment for short-term personal gain. The science establishment has recognized this, and governments are now funding grand programs on "research applied to national needs". Individual scientists, with all the good will in the world, speak of the need for "critical science" focussed on just such needs. If professional interests such as risk assessment continue on their present course, it will not be long before science can display its difficult and unstinting efforts to ward off evil, its indispensability as "great protectors of mankind against an enemy who is omnipresent but difficult to detect". This scenario does not require venality, but only self-interest and self-delusion. For that reason alone, it merits our attention.

To read too directly from the witch hunts of the 16th and 17th centuries to the risk assessments of the present would be to fall into the trap of historical determinism. To declare without further ado that "It can't happen here" would be to display naivete of another sort. At a minimum, as Trevor-Roper [1] has argued, the existence of the witch craze in the midst of the Renaissance is "a standing warning to those who would simplify the stages of human progress."

The "new professional interest of risk assessment" is not necessarily a progressive step. Neither its professed rational-scientific foundations nor its concern for collectively redressing ills of the human condition are enough, in themselves, to make it so. Both the potential for bettering human life and the potential for witch hunting are latent in contemporary risk concerns. Our pressing task is to learn how we can cultivate one aspect of this Janus-faced creature while suppressing the other. In the next section I consider some more recent risk assessment experience which illuminates further aspects of this problem.

RESOURCE MANAGEMENT: ON THE FUNCTION OF UNCERTAINTY*

Some particularly useful insights on the basic nature of risk phenomena can be drawn from a consideration of man's attempts to manage environmental

---

*This section is based on the work of my colleagues at the Institute of Resource Ecology, University of British Columbia [11, 12], and on Ian Burton, Gilbert White, and Robert Kates's studies on man's relationships with environmental hazards [13, 14, 15].

resources. The dual character of "risk" is again apparent. The river that brings water, irrigation, and transport also brings floods. The plants and animals with which we cohabit the earth provide us with oxygen, food, labor, and a variety of more subtle benefits. Under other circumstances, they may compete with us as "pests", attack us as diseases, or inconsiderately disappear under the various demands we place on them.

There is nothing witch-like or imaginary about the risks encountered through our relations with such resources. Failure to cope leads not to the ambiguities of future purgatory, but to the definite and immediate consequences of drowning, starvation, and consumption.

Anthropological studies have shown that pre-industrial "folk" societies adjust to such environmental risks largely through modifications in human behavior. From an external perspective, these adaptations often appear mystical and arational. On closer examination, they often exhibit the notable virtues of being effective a good deal of the time, of being flexible and easily adapted, of requiring action only at the individual or small group level, and of imposing little stress on the environmental system as a whole [15, p. 982; see also 16, 17]. Modern industrial societies have tended to pursue an opposite course of adaptation, controlling and reducing the variability of nature by means of large, long duration, capital intensive "engineering" projects. These have indubitably succeeded in achieving many of their short-term goals. But a look at the historical record shows that many of those gains have been bought in exchange for expensive and unanticipated long-term consequences. We have begun to discover that variability and uncertainty are in fact important "structural" factors, responsible in large part for the way our environmental and resource systems work. In general, they cannot be removed or reduced without precipitating major changes in those "workings". In particular, the control of small, frequent fluctuations has resulted time and again in a growing vulnerability to rare but large perturbations. Consider some particular examples.

Throughout the middle part of this century, the United States devoted unprecedented expenditures to the control of river flooding. By 1960, however, it was clear that the country's increasing flood control efforts were supporting an ever *rising* level of flood loss and damage [18]. As might have been expected, there followed a great deal of acrimonious debate amongst the flood-protection industries, Congress, flood victims, and sundry academics. The facts were denied, explained away, attributed to extraneous factors, and so on. But gradually there grew a body of evidence showing that the early technological view of flood risk protection had been seriously incomplete. People, together with their reactions to perceived flood frequency, had been left out of the picture. When empoundment and levee construction made former flood plains less prone to flooding, people reacted by moving into areas which now appeared "safe enough". Good control of normal river fluctuations was indeed achieved, and the previously farmed lands became more and more densely settled, their flooding history a more and more distant reality. When an exceptional flood eventually did exceed capacity of the flood works — or, much more rarely, when those works failed under less than their designed tolerances — the floods which resulted caused unprecedented damage [14, 18]. Only now are comprehensive strategies, incorporating the human element, beginning to be devised. And these, almost without exception, are emphasizing mitigation and flexibility of response rather than the old litany of flood "control" [13].

A related phenomenon is documented repeatedly in the pest control literature. For example, in the period 1947-1974, agricultural use of insecticides in the United States increased over ten-fold. Over the same period, the rate of crop loss to insect pests *rose* by a factor of two [19]. In American corn production, while the acreage treated for pests has risen from 1% to 52%, crop losses have *risen* nearly four-fold [19]. Nor can we simply stop using insecticides and hope things will go back to their original variable but endurable state of affairs. Though this might be possible in some theoretical long run, the short-term implications for farmers and food supply would be devastating. We are, sadly but simply, hooked on a risk control policy which gives us little, but which we can no longer do without. The broad result of our efforts to control pest risks has been to increase not only immediate damages, but also vulnerability to future surprises. There is no simple explanation for these seemingly perverse relationships. In some individual crops the results have been better. In others the losses are in part due to changes in tillage and usage practice which accompanied the increases of insecticides. But in many well-studied cases, it is clear that man's crude efforts to eliminate natural variability in the resource system are directly responsible for the ensuing debacle.

One such case is documented by Canadian studies of the spruce budworm [20]. Under natural conditions, this normally rare insect erupts into epidemics at intervals of 30 or more years, defoliating and killing a good proportion of the older coniferous forest as it does so. This forest destruction eliminates the insects' habitat, the outbreak collapses, and a healthy, young forest grows back in its wake. But these temporal uncertainties make efficient commercial utilization of the forest impossible, and insecticides were applied to control an incipient eruption in the 1950's. By preserving the forest in a mature condition — by reducing its variability — this policy also preserved the biological conditions which precipitated the eruption in the first place. Today, under the relatively unvarying conditions of insecticide "control", budworms have spread throughout the entire province of New Brunswick where they persist at intermediate to high densities. Continuous, expensive applications of insecticides are required to prevent an epidemic. The forest and forest industry are more vulnerable, and at greater risk, than ever before, should the control policy fail or be abandoned. The most intensive analyses of this dilemma have been unable to design remedial policies with any but the most painful withdrawal symptoms.

Not surprisingly, a number of parallels to these pest-control histories can be found in man's efforts to control human disease [11, 21]. The case of poliomyelitis provides an especially illuminating example. It seems that prior to the 20th century severe cases of polio were rare. Minor infections were probably contracted by most children, producing immunity but few obvious symptoms in most. By the 19th century, however, improved living conditions and public health measures — in part introduced to combat cholera and other "unsanitary" diseases — had begun to isolate the more well-to-do segments of society from their traditional childhood exposures to various diseases. The reduced frequency of contact meant that these "diseases of cleanliness" were often first encountered in adult life, with violent or fatal results.

This growing toll of polio cases, perversely focussed on the most meticulously hygienic classes, fuelled the successful search for a vaccine. By the late 1950's, the incidence of polio was again extremely low in the United States and an organized scheme of inoculations was reaching a very large proportion of the school-age population. Once again, however, there is some suggestion that this "control" of uncertainty and fluctuation may well have increased vulnerability to large-scale disaster. Today, polio has become for many a threat of the distant past. Public health officials are finding it harder and harder to guarantee that significant proportions of the population are not missed by immunization and booster campaigns. It has been suggested that the growing complacency over the "non-risk" of polio may well be leading us to conditions which could support a major epidemic. The same is true for a variety of other diseases [see 11, 21].

Obviously, public health and vaccination campaigns have done a great deal of good, and will continue to do so in the future. But the reduced frequency of disease brought by vaccination programs is invariably accompanied by increased risks of other sorts. Such alternative risk structures — not the simplistic myths of natural exposure versus ultimate eradication — should be the focus of policy discussions and analysis in each particular disease case. Such explicit weighing of realistic alternatives is not particularly in evidence. German measles vaccinations for children are a case in point. Here, in exchange for protecting small children against a relatively mild illness, we leave adults susceptible to disastrous and debilitating attacks. It seems virtually certain that a broader perspective would encourage the disease in childhood, and vaccinate only adults who have missed natural exposure in their youth. As elsewhere, however, a simplistic and counterproductive predilection for "control" per se has so far prevailed.

The unpleasant surprises historically associated with efforts to manage pest and disease risks might be passed off as special consequences of introducing exotic substances into complex biological systems. But precisely the same sorts of unanticipated results have been encountered in apparently straightforward efforts to reduce the risk of forest fire in America's National Parks [2]. Once again, initial efforts were successful, leading to adoption of the policy throughout the park system. Only later did it become clear that brush and scrub unnaturally accumulated in the absence of small periodic fires were providing fuel for conflagrations of a size and intensity never before experienced. Again, "withdrawal" from the initially successful risk-reduction policy has been delicate and expensive in the extreme [11].

A final example of the relationships among uncertainty, risk, and resource management concerns the role of genetic variation. Studies in evolutionary biology have shown that variable environments give rise to populations with substantial genetic differences in traits relevant to the populations' survival. One genetic type will be slightly better adapted to one type of environmental condition, one to another, and so on. As a result, over a wide range of environmental conditions, disturbances, and surprises, some members of the population will do relatively well. The occasional variation in the environment shifts the balance and prevents one form from replacing all the others. It is true, however, that if environmental conditions can be kept constant, one form highly adapted to those conditions will usually do better than a mix of forms. In agriculture, this situation has led to the breeding and distribution of genetically pure crop strains supremely well adapted to the controlled (low risk) conditions of water and nutrient availability which modern farms can provide.

A sobering lesson in the risks of such strategies was delivered to American corn producers in 1964 [23]. Huge tracts of land were by then planted in a single genetic strain of high yield corn. When a disease arose to which this particular strain was not resistant, a very large proportion of the crop was lost. Disaster was averted because some resistant strains were still available and could be used to replace the susceptible one. As a result of this and similar surprises, much more attention is now being devoted to the development and preservation of mixed genetic stocks in agriculture. The lower short-term yields obtainable from such approaches is judged an acceptable price to pay for the increased ability to cope with the unexpected.*

Looking across these diverse examples of resource management experience, several common themes stand out. In each case, uncertainty or variability in the natural system was initially viewed as a source of risk/hazard. Without exception, it was assumed that removal of the variability would be an unmitigated good, resulting in reduced risk and improved performance of the resource system.

Initial successes led to optimism that the proposed management policy would be an effective one. But they also led to changes in the system itself. In each case, the existence of variability and uncertainty turned out to have played an important role in establishing and maintaining key relationships among the system components. With that variability removed, relationships shifted to accommodate the new reality: people settled the unflooded flood-plain, budworms spread through the undefoliated forest, brush accumulated on the unburned understory, and so on. As a result, the decreased frequency of variation in the system was accompanied by increased vulnerability to and cost of variation when it finally broke loose from managerial controls. Management efforts had changed the kinds of risks encountered, but not the fact of risk. And more often than not, management shifted the risk structure from one sort people were accustomed to dealing with to one they had never before experienced.

Failures and surprises of the sort described here have been instrumental in sensitizing managers to the internal role played by variability in resource systems. Detailed investigation have begun to tease out the mechanisms involved in this sensitivity, and to let us make use of it in our policy designs [e.g. 11]. But if we have learned something about the different structures which variation and uncertainty can take, our ignorance still remains more substantial than our knowledge. It is now clear that we are unlikely to reduce unpleasant surprises in resource management merely by increasing knowledge or imposing crude "controls".

---

*Unfortunately, it seems that we need to learn this lesson anew for each resource system. Present efforts to enhance the production of salmonid fish stocks in the Pacific Northwest seem likely to select for dangerously narrow genetic stocks throughout the system (K.H. Loftus, " Science for Canada's fisheries rehabilitation needs," J. Fish. Res. Board Can. 33: 1822-1857, 1976.).

Rather, we must learn to design resource management schemes so that they can better cope with the failures which are guaranteed by our ignorance and the inherent variability of resource systems. This need for designing "soft-failure", uncertainty-tolerant management policies is receiving growing attention in areas beyond resource management [24]. Coupled with a concern for increased institutional flexibility, it forms the core of an approach to adaptive management which my colleagues and I at IIASA and the University of British Columbia have been exploring over the last few years [12]. At the end of this paper, I will discuss some of the implications of this adaptive management notion for societal assessment of complex, incompletely known risk systems. First, however, I wish to consider one more set of historical lessons, this time taken from a field in which solid scientific knowledge of risk is at its most complete.

DRUG SAFETY: THE LIMITS OF REGULATION

The history of drug development and regulation shows the risk assessment profession at its best. The products in question are medicines designed from the beginning to combat specific ills of man and to improve directly his health and well-being. In return for their favors, medicines themselves pose risks, but of a very special kind.

First, exposure to the risk is limited to those seeking the related benefits. Second, the risk is undertaken in close consultation with a professional trained to help his patient balance personal risks, benefits, and alternatives in particular circumstances. Third, the nature of the risk itself has been carefully investigated, evaluated, and described through rigorous and sophisticated experimental investigations.

Each of these features of medical drugs should make their assessment and regulation easy relative to other risk situations. In fact, people dealing with nuclear, toxic chemical, or even traffic risks would almost certainly be thankful if even one of the properties listed above pertained to their systems. Looking at the history of drug safety efforts over the last several decades, we might therefore expect to learn something about the best that can be hoped for from risk assessments in other less mature and tractable fields.

This task is facilitated by the National Academy of Sciences' sponsorship a few years ago of a symposium with the famililar title "How Safe is Safe?". That symposium reviewed experience in the design of policy on drug development and regulation [25]. Papers were presented by a variety of senior drug regulators, producers, and consumers. With the recorded discussion, these papers provide a lively review of the current debate on drug safety issues. In so doing they raise serious questions regarding the limits of risk management. I review some of these below.

The basic procedures for risk-benefit assessments of medical drugs are well established. Preliminary screening makes use of extensive information and experience on similar products. Promising candidates move on to limited trials in lower animals, intensive evaluations in higher animal forms, and finally to closely supervised clinical trials on volunteer human subjects [26]. Real differences of opinion arise not regarding the logic of this basic plan but on the judgmental issue of how much, and what kind of, assurance is needed before drugs are approved for human consumption.

If a drug is approved with minimal testing to make it quickly available to those who need it, people may be the guinea pigs who reveal unanticipated side effects: The specter of thalidomide is never far in the background when more rapid licensing procedures are debated. On the other hand, efforts to approach zero risk through exhaustive pre-release testing are extremely expensive and time-consuming. New drugs are delayed in reaching those who need them, and marginal* drugs may not be developed at all.

Dilemmas of this sort exist in most risk management situations. The drug case is virtually unique, however, in that different solutions have been adopted in different countries, providing a prospect for empirical comparisons of regulatory efficacy. The two most thoroughly exercised and extreme solutions are those adopted by the United States and United Kingdom. The U. S. emphasizes intensive pre-market testing to mitigate the risk of unanticipated side-effects, while the U. K. promotes prompter release, relying heavily on an extensive system of post-marketing monitoring.

The explicit comparisons which have been carried out between these two approaches are in no sense definitive or free from methodological problems. With some unanimity, however, they conclude that the U. K. practice better advances the public interest [26, 27, 28]. American regulatory caution is argued to be needlessly expensive, stifling of new product development, and not superior in its ability to assure drug safety. In particular, the stringent safety testing procedures instituted in the United States following the Kefauver hearings and thalidomide episode of the early 1960's, are demonstrated to have been a mistake in classic risk-benefit terms [9]. The clear and vociferously stated conclusion of such studies is that some more rational form of regulation, including less expensive and time-consuming assessment procedures, is long overdue for the U. S. drug industry.

But while American drug regulators and risk assessors are being condemned as overly conservative by collective social welfare studies of the sort cited above, powerful, articulate, and convincing consumer groups are simultaneously attacking them for "caving in to industry" and neglecting their responsibility to assure the public's safety [e.g. 30]. Advocates of this position cite the regulators' failures to detect risks which "could have been" detected, and their ambiguous reactions to ambiguous evidence as proof that the public safety is too important to be left to even the best of safety experts. The beleaguered regulators have accepted consumer representation on their drug review panels, without anyone being sure just what those representatives are supposed to represent. Congress has responded to the political importance of drug safety by almost continual intervention in and reorganization of FDA. Significantly, however, Congress' direct attempts at

---

*I use "marginal" here in the economic sense of low market potential, low profitability products. Some of these may be literally a matter of life and death to the few who need them, and pharmaceutical concerns do market some "public interest" drugs on which they will never make a profit. But one cost of regulation will always be to make some such marginal drugs not worth developing by even the most public spirited of concerns.

"representative" safety regulation have resembled nothing so much as Keystone Cops scenarios (e.g. DES, saccharine). And Congress has failed repeatedly to meet FDA's own requests for an unambiguous legislative mandate specifying what balance of risks and benefits *does* constitute the public good, how this is to be democratically determined and achieved.

What emerges from the "How Safe is Safe?" debate in the drug field is that, for better or worse, public safety is now and is likely to remain a primarily political issue. Scientific data and economic analyses — even of the inordinately high quality encountered in the drug field — are simply not going to be the central issue in even the most technical of risk decisions.

This is not to say that science, data, and rigorous analysis are irrelevant to actual decisions in drug licensing. Nor does it suggest that carefully reasoned risk assessments do not have a role to play in other fields, even when these are destined to deal in even greater ambiguities of "objective" analysis than do drug safety trials. It does suggest, and strongly, that the would-be "professional interest of risk assessment" must reconsider its basic goals, and reassess its own potential for real contribution to the public interest.

One direction which such a reconsideration might profitably explore is suggested by Joshua Lederberg's "systems analytic" contribution to the drug safety symposium cited earlier [30]. He argues that contemporary safety testing procedures, even in the drug field, often resemble catechismal obstacle courses. These procedures undoubtedly do make it very time-consuming and expensive to introduce new products or proposals, but rarely has any effort been made to determine whether they actually do catch the hazards to which they ostensibly are a response. Some of the drug screening evaluations already cited in this section suggest that they often do not [e.g. 29]. Furthermore, the cases I discussed in the earlier section on resource management suggest that simplistic or intuitively plausible "safety" measures may frequently increase total risk.

Lederberg concludes that we must come to treat the issues of drug regulation and management as problems of experimental design. Instead of routine adherence to large scale screening experiments on mice, or bizarre attempts to determine cancer "causing" dosages of some agent, he calls for "creative investigations that look for problems on the basis of some theoretical rationale" [31, p. 80]. It is the development of such rationales, rather than of arcane methodological treatments for eventually irrelevant data, which constitutes the central scientific challenge of contemporary risk assessment.

At a more prosaic level, Lederberg's call for an experimental design approach in drug safety regulation can be extended to the way in which we make use of experience and information that we already possess. The comparative evaluations of regulatory performance referred to earlier are valuable attempts to advance the public interest. On closer examination, however, they offer little actual policy guidance. Virtually no regulatory activity in any field has ever been shown to have a clean bill of health when subject to essentially economic evaluation [32]. To conclude from such analyses that we need "less regulation", or "deregulation", may not be wrong, but neither is it particularly instructive. The "don't regulate" vs. "do regulate" choice is a sterile and artificial one. To begin creating effective policies of risk management, we must surely begin to view these issues at a finer level of resolution. We need carefully designed studies to show what *kinds* of risks our present testing procedures can catch, and which kinds of risks they let slip by. Armed with such knowledge, we could begin to determine the kinds of tasks which various post-marketing monitoring schemes can perform effectively, and the kinds of situations where intensive pre-release investigation is justified. Only when we begin to blend the results of such studies in the careful design of integrated risk management strategies will we be able to move much beyond the present unsatisfactory state of regulation by polemic.

Finally, appropriate blends of risk assessment tactics are not likely to emerge from even the most sophisticated contemplation. We will have to learn to make efficient diagnostic use of the different empirical experiences emerging in different countries under different regulatory approaches. This brings us almost full circle to the notion of "adaptive risk management" already suggested by historical experience in resource management. In the final sections of this paper, I shall attempt to close that and other circles suggested by this survey of historical perspectives, and to suggest some general directions for future work in risk assessment.

WHAT ARE WE ARGUING ABOUT?

The various attitudes towards the unknown suggested in my historical reviews were captured nearly a hundred years ago by Frank Richard Stockton in his studies on the ancient myth of The Lady or the Tiger?*[33]:

> The young man could open either door he pleased. If he opened the one, there came out of it a hungry tiger, the fiercest and most cruel that could be procured, which would immediately tear him to pieces. But if he opened the other door, there came forth from it a lady; the most suitable to his years and station that His Majesty could select among his fair subjects. So I leave it to you, which door to open?
>
> The first man refused to take the chance. He lived safe and died chaste.
>
> The second man hired risk assessment consultants. He collected all the available data on lady and tiger populations. He brought in sophisticated technology to listen for growling and detect the faintest whiff of perfume. He completed checklists. He developed a utility function and assessed his risk averseness. Finally, sensing that in a few more years he would be in no condition to enjoy the lady anyway, he opened the optimal door. And was eaten by a low probability tiger.
>
> The third man took a course in tiger taming. He opened a door at random and was eaten by the lady.

*Stockton's initial translation of 1884 has been questioned on several grounds, but remains the most complete version available. I have used his work for the first paragraph quoted here, but employ some of the more credible alternatives for its variant endings, following the reasoning I developed in an earlier study of the myth [34].

To interpret, we respond to the unknown by trying to retreat from it, or trying to comprehend it, or trying to control it. The first approach is evident in longings for return to a simpler "risk-free" life that never was. The second is reflected in the fantasy of synoptic risk assessment: Measure the requisite probabilities and trade-offs, calculate the social risk-benefit function, and the common good will have been defined. The third approach is in the tradition of professional engineering. It serves us well, but as engineers have been among the first to point out [35], it has met its match and more in the complex social risk problems it increasingly is called upon to address.

All of these traditions have one thing in common. They set themselves in opposition to the unknown and try to overcome or control it, thereby hoping to establish a more predictable and less frightening world. The history of risk management shows the inadequacy of this approach. The unknown is not a wrinkle to be ironed out of the social fabric. The analysis that predicts the tiger will always be surprised by a carnivorous lady. Our ignorance will always remain greater than our knowledge.

Fortunately, none of this need present really serious obstacles to effective coping with the unknown. There is an alternative tradition of coping which, though virtually absent from the contemporary risk debate, has nonetheless long been a practical mainstay of successful coping in man and beast. This tradition accepts the inevitability of incomplete knowledge, seeks to accommodate rather than control the unknown, and thereby aims to coexist with and prosper from surprise. In this tradition, the "risk problem" is not uncertainty of outcome, or violence of event, or toxicity of substance, or anything of the kind. Instead, it is the challenge of coping confidently, effectively, and creatively with the surprising world around us. The fundamental question is not how to calculate, control, or even reduce risk. It is how to increase our risk-taking abilities.

Nowhere is this distinction clearer than in the questions of medical drug safety which I reviewed above. By any imaginable criteria, the complex, biologically active compounds generated by modern pharmaceutical concerns are risky things indeed. The sheer volume of production is frightening enough. Add the high proportion of that production that comes into contact with humans and you have a situation bound to dispatch a modern risk assessor for his injunctions and press agents. In the medical drug case, however, the existence of a professional managerial framework within which these dangerous chemicals are characterized *and* administered *and* monitored makes them into risks we can afford to take, thereby improving our health and well-being. Any narrow attempt to create a world free from the very real risks posed by such chemicals would entail obviously unacceptable consequences. Moreover, since many drugs are now most valuable in roles for which they were not originally envisioned, any preemptive risk-benefit accounting would produce similarly unfortunate results. In contrast, improvements in our ability to take risks — in our knowledge of how the drugs confer their risks *and* benefits, in doctors' and patients' understanding of the trade-offs involved, in the monitoring and diagnosis of unanticipated (positive and negative) drug reactions — all increase our capacity to cope with disease and improve our health.

References pp. 311-313.

A similar emphasis on increased risk-taking abilities, rather than decreased risk per se, emerges as a strategy for coping with the unknown in a number of pragmatic fields which have tried the alternatives. Portfolio designers long ago recognized the fallacies of "risk-free" earning strategies [36]. Boehm-Bawerk's Law, based on one of the most rigorous theorems in economics, states that existing means of production can yield greater economic performance only through greater uncertainty — i.e. through taking greater risks. Strategic corporate planning has been defined by one of its most successful advocates as creating the "capacity to take a greater risk, for this is the only way to improve entrepreneurial performance" [37]. Most biologists, myself included, would concur with W. H. Auden's poem "Unpredictable but Providential", wishing only that we had put the central experience of our discipline so well:

> . . . for the animate, to last was to mean to change,
> existing both for one's own sake and that of all others,
> forever in jeopardy . . .
> As a rule it was the fittest who perished, the misfits,
> forced by failure to emigrate to new unsettled niches, who
> altered their structure and prospered . . .

Rene Dubos states the biologists' conclusion more bluntly [38]: "Willingness to take risks is a condition of biological success." This point is critical to our understanding of human risk-taking. Willingness to take risks, together with knowledge of risk-taking consequences, determines our ability to cope with the unknown. Confidence is as important as understanding if we are to shape the future in a rational way. The real challenge for the "new professional interest in risk" is to contribute to both.

In seeking to meet this challenge, it is reasonable to begin with the paradox of contemporary social risk history: The more we learn about risk the less confident we seem to be of our risk-taking abilities. Hence we have the spectacle of an American society which has a greater life expectancy, higher level of material welfare, and more knowledge than ever before, frightening itself into virtual catatonia, unable to mobilize the risk-taking efforts necessary for coping with the unknown. A "new professional interest in risk" which cannot bring itself to address, much less explain, such a central fact of its subject is hardly a thing to inspire confidence.

Nobody knows what makes one individual or society believe in itself while another heads for the bunkers.* After all, Columbus was venturing into the void at the same time Institor and Sprenger were inciting witch hunters to new heights of paranoia. It seems virtually certain, however, that risk assessors' sincere knowledge-seeking efforts to identify potential dangers can undermine the very confidence which would be necessary for creatively coping with those dangers.

The proverbial "little knowledge" is both a dangerous and a frightening thing. We have already seen the workings of this in the witch hunts of the Renaissance. Authoritative and, let us presume, sincere preachers preached valiantly the dangers of witches and of the devil's incredibly subtle and cunning ways of infiltrating society. In so doing, they amplified society's latent fear of the

---

*For some provoking thoughts on the subject, see Gardner [39].

unknown, undermined its confidence, cohesion, and common sense, and thereby contributed to the public hysteria which later fuelled the excesses of the Inquisition. Today, authoritative and, let us presume, sincere scientists preach valiantly the dangers of risks and their incredibly subtle and cunning ways of infiltrating society. The Chief Counsel of the Food and Drug Administration admits "we often regulate more out of fear of the unknown than out of respect and appreciation of the known" [40, p. 123]. Society's attitudes towards risks such as cancer and nuclear reactors are not readily distinguishable from its earlier fears of the evil eye.

This is not to say that today's society does not face real risks, nor to deny the real accomplishment of risk management professionals in dealing with those risks. It is to insist that the dual character of the coping problem — the need for knowing *and* willing — is virtually ignored in contemporary literature on the risk problem.* Preoccupied with the knowledge aspects of early warnings and assessments, we are caught inside the risk problem and become part of it. Unable to see the relationship between our knowledge-seeking work and the fear of the unknown it may engender, our contributions to the real problem of enhancing society's risk-taking and coping abilities are correspondingly dissipated or flatly counterproductive.

The challenge of establishing a rational perspective from which to view risk problems and our interventions in them is, however, greater than merely coming to understand the relationship between fear and knowledge.

Alvin Weinberg provided the glimmer of such a perspective in his concept of "trans-science", first enunciated in a discussion of "How safe is safe enough?" for nuclear reactors [43]:

> Attempts to deal with social problems through the procedures of science hang on the answers to questions which can be asked of science and yet which cannot be answered by science. I propose the term *trans-scientific* for these questions. . . Insofar as public policy depends on trans-scientific rather than scientific issues, the role of the scientists in contributing to the promulgation of such policy must be different than is his role when the issues can be unambiguously answered by science. . . When what we do transcends science and when it impinges on the public, we have no choice but to welcome the public — even encourage the public — to participate in the debate. Scientists have no monopoly on wisdom where this kind of trans-science is involved; they shall have to accommodate the will of the public and its representatives.

What is this "different role" required of the risk scientist? How is he to promote scientific knowledge without spreading social fear? How is the "will of the public" to be accommodated in risk problems? Neither Weinberg nor anyone else has proposed definitive answers to these questions, but in the last several years several lines of inquiry have been opened.

---

*The problem is not unique to risk studies. Lindblom [41] distinguishes between the knowledge-based preferences which inform economic theory, and the will-based volitions which, together with preferences, inform political choice. In a penetrating essay on the nature of useful knowledge [42], he characterizes these two complementary forms of social evaluation as "thinking through" and "acting out".

In reviewing these, it seems to me that there are two distinct if related issues at stake. One concerns the incompleteness of scientific understanding which can be brought to bear on risk questions. The other involves the conflicts of individual wills, values, and freedoms which bear on those questions.

TOWARDS THE ADAPTIVE DESIGN OF RISK MANAGEMENT POLICY

Let us first consider the problem of incompleteness. In mature academic science, the incompleteness and fallibility of knowledge should cause no fundamental difficulties. Theories are held tentatively, contingent on new evidence. Contrary evidence and new interpretations are granted equal access to the debate. Independent experiment and peer review provide checks and balances against error and unscrupulous behavior. Of course the ideal standard is often bent or broken in practice, but in the long run, in the majority of cases, good science does seem to replace bad.*

This is not the case, however, in what historian Jerome Ravetz has called the less developed or "immature" sciences, especially when those sciences are applied to social problems [45]. In such circumstances a variety of factors conspire to suppress tentative outlooks and to seize on incompleteness as an excuse for polarization. The result is bad science, leading to unnecessary public alarm, unjustified and ineffective regulations, and an unwillingness to undertake the risk-taking ventures necessary for coping with the unknown [46].

In part, the phenomenon can be explained in terms of a breakdown of quality control within the scientific discipline. The relative absence of established facts or criteria of competence tends to make peer review ineffective. Add the pull of a socially relevant, "public interest" discipline, and there is a real danger that the field will experience "an accretion of cranks and congenital rebels whose reforming zeal is not matched by their scientific skill" [45, p. 427]. Where recognition and grant money both accrue to those making the first, loudest, and most frightening noises, where accusations of corruption, cowardice, or insensitivity are the most likely rewards of the careful skeptic, then the "great confidence game" portrayed by Ravetz cannot be far off.

The fault, however, does not all lie with science. Harvey Brooks has pointed out

> . . . an interesting parallel between the scientist's desire to establish priority for a discovery or invention, and the politician's search for new issues on which he can make a name for himself. . . The potential alliance between individual politicians and scientists, though often beneficial, can also be dangerous because neither side is subject to the normal checks and balances of peer groups. Once a politician has staked out scientific territory for himself, his colleagues tend to stay away. At the same time the scientisit is speaking in a forum to which opposing scientific views are not more or less automatically accorded equal access. The politician is free to select his own

---

*Unfortunately, it also often displaces the merely different. For a particularly readable and disturbing account of Inquisitorial intolerance in modern science, see Feyerabend [44].

experts to develop an issue in the way that has maximum political utility to him. Truth may be only incidental [47, p. 259].*

Even the most conscientious risk scientist, trying to present a balanced view of a complex and uncertain issue, is likely to find his argument caricatured and polarized in such a process. Brooks continues:

> Scientists inexperienced in the political arena, and flattered by the unaccustomed attentions of men of power, are often inveigled into stating their conclusions with a confidence not warranted by the evidence, and . . . not subject to the same sort of prompt corrective processes that they would be if confined within the scientific community [47, p. 259]**

While these and other problems of the incomplete scientific knowledge in risk matters are widely recognized, most responses have essentially called for a resolution through better science. This misses the central issue completely. Thus we have the 1976 Bellagio Conference on Science and Technology calling for the scientific community to "evolve and sustain new standards of scientific rigor appropriate to research in support of early warnings and policy decisions" [49, p. 33]. Or, for those with less faith in their fellow scientists, there are the science court proposals for what amounts to super-peer review [50]. In both cases, the underlying assumption seems to be that rigorous science, or rigorously reviewed science, would not be subject to the incompleteness, polarization, and exploitation that characterizes risk science today. With all respect to good intentions, the historical experience of risk management*** makes this assumption hard to accept.

An alternative, or perhaps complementary, response to the incompleteness dilemmas of trans-science is provided by the growing craft of policy analysis. In a recent *Science* editorial on the subject, M. Granger Morgan argues

> Good policy analysis recognizes that physical truth may be poorly or incompletely known. Its objective is to evaluate, order, and structure incomplete knowledge so as to allow decisions to be made with as complete an understanding as possible of the current state of knowledge, its limitations, and its implications [51, p. 971].

Policy analysis of the sort Morgan describes is just beginning to emerge from its uninspiring past as a branch of applied mathematics. There are indications, nonetheless, that it does indeed offer a realistic and rational perspective from which Weinberg's trans-scientist can shape his "different role" in the social risk debate.****

---

*Brooks draws the latter part of his suggestion from the studies of Nelkin [48].

**The difficulties encountered in scientists' statements to the media are similar in kind and origin, and even more sensational in outcome.

***I would argue, for example, that risk management of medical drugs already has both the "rigorous standards" of Bellagio and the super-peer review of the science courts. The debate is none the less acrimonious.

****The remainder of this section draws heavily on the concepts of policy analysis as developed by Majone [52], Wildavsky [53], and Lindblom [54]. For another more formal view see Quade [55].

In some ways, this role seems likely to take on more the character of a jurist than a traditional positivist scientist. Policy analysis recognizes, above all else, that "the data" are always insufficient to dictate unambiguous conclusions. Rather, particular data are generated, selected, and inserted into an argument as evidence in support of a particular conclusion [52]. The listener may be persuaded by the force of the argument and the strength of its evidence. But there is no suggestion that data of themselves are either necessary or sufficient to a given conclusion. The debate therefore shifts away from a preoccupation with "facts" and their "proof". It turns instead to the careful development of rules for the admissibility of legitimate evidence, and for the form of legitimate argument. Such rules are known to be fallible — the guilty can be acquitted and vice versa — but fallibility is accepted as an inevitable consequence of our lack of omniscience. On the other hand, careful attention to developing mutually agreed-upon rules of evidence can create that essential willingness to proceed in the face of fallibility. It has done this for drug regulation, and our health is the better for it. Attention to rules of evidence can also assure against the wilder tyrannies of self-evident "fact" where no effective peer review exists. It did this in Alonso y Frias' reforms of the Inquisition's torture and indictment procedures. Perhaps most important, formal rules of evidence constitute formal hypotheses on how we can best cope with the unknown. Viewed in this manner, they invite us to use our continuing experience in risk management to evaluate our present rules, and to suggest improvements in them. We therefore can learn from both our successes and our failures and hope for some cumulative improvement in our risk-taking, surprise-coping abilities. Contemporary risk management's inability to effect such cumulative improvements, its insistence on re-fighting all the old battles with each new risk issue, is one of the most discouraging aspects of its exclusively "fact"-focussed approach.

The notion of learning from error is central to modern policy analysis, as it is to those pragmatic coping strategies of man and beast which I outlined earlier. The litany goes something like this: If knowledge is incomplete, if the future is uncertain, then mistakes and surprise are inevitable. The categorical imperative is to recognize such mistakes, to learn from them, and to modify future actions accordingly.*

In this view of life, rationality becomes a retrospective but still respectable concept. Since actual performance is contingent on facts unknown, futures unborn, and choices we ourselves have yet to make, the "rational" is evident only in retrospect. It is what turned out to be adapted to the conditions that occurred, and turned out to be adopted by the powers that were [52].

The problem of rational management is therefore to design self-evaluating policies which adapt themselves to the developing situation and, in so doing, cultivate the will necessary for their adoption and continuing pursuit. Such a reconstructive concept of rationality is central to evolutionary (as opposed to teleological) thinking in a number of fields [52]. Its appropriateness as a guide to action has been argued in terms of social psychology [56], economic theory [57], and the purest of scientific endeavor [58]. Furthermore, as Majone points out,

---

*It has been said that a fool makes many mistakes, while a damn fool makes only one. Over and over again.

This explanation makes sense of behavior frequently observed among policy makers — such as incrementalism, adaptive adjustments, imitation, and "rationalizations" — which must appear to be irrational and/or dishonest in the prevailing models of policymaking [52, p. 215].

Those "prevailing models", unfortunately, are the ones which inform a good proportion of contemporary risk management activity. The synoptic planners, cost-benefit analysts, and regulatory bureaucrats seem wedded to prospective, knowledge-presuming notions of "rationality" in policy making: "Optimal" or "best possible" decisions and decision-rules are derived on the basis of available information, and implemented by virtue of their rationality (social optimality, expert consensus). Subsequent performance can be taken for granted, provided always that compliance with the rational rules is rigorously enforced.

If this sounds too extreme a caricature of present practice, try any other one that comes to mind. Michael Crozier, in his classic study of *The Bureaucratic Phenomenon*, defines bureaucracy as "an organization that cannot correct its behavior by learning from its mistakes" [59, pp. 186-187]. Regulation by such bureaucracies has become almost synonymous with risk management in America today. From the policy analysis perspective, with its insistence on an adaptive, "error-embracing"* response to the unknown, it therefore comes as no surprise that risk management is in trouble. More constructively, policy analysis suggests that effective, rational coping behavior may depend more than anything on our ability to design flexible, adaptive management institutions:** Institutions which can respond to and learn from the inevitable surprises awaiting us. Institutions which can mobilize public will in risk-taking enterprises. Institutions which can improve our ability to cope with the unknown.

Explicit policy analysis focussed on design of alternative institutional structures for risk management has barely begun. The sterile debates over "regulate" versus "don't", "threshold" versus "not", and the like have so far occupied center stage and most of the wings [32]. Some of the notable exceptions include Michael's [60] and Thompson's (this symposium) studies from the human behavior perspective, the comparative institutional studies I referred to in the discussion on medical drug regulation, plus those of Nelkin reported elsewhere in this symposium, the explicit policy analyses of Majone [61, 62, 63], and the applied work being done under the several banners of "mediation" [64, 65]. These suggest a productive future for policy analyses of risk problems and their institutional settings, if only the debate can be turned in the constructive directions they have suggested.

What that future will be like I am not so silly as to suggest in a paper emphasizing uncertainty and surprise. My personal favorite for attention concerns the "scale" of our risk management institutions and arrangements. There is a strong tendency

---

*The term is from Michael's [60] insightful study of the human aspects of Learning to Plan and Planning to Learn.

**Significantly, this was also the overriding need identified by the previously mentioned Bellagio Conference on Science and Technology; See [24].

today for every fear, every unknown to be met by mandatory regulation at a national or even supernatural scale.

This approach might possibly be justified in a world where one socially optimal regulation could be computed in advance, or where the externalities of local risk-taking decisions would be truly national in scope and unbearable in effect. It might, in general, be justified if everyone wanted it. But in most cases of risk management, not one of these conditions is met.

An opposite extreme, less well explored, is a variation on the thousand flowers blooming approach to cultural revival. I suspect a careful policy analysis would show that maximal social learning and political will could be mobilized by designing the scale of particular risk management ventures to fit the character of the risk under consideration. Thus while we might require regional scale regulation in such externality-laden fields as air quality management, we might find that much smaller scales — and therefore more, different learning experiments and less compulsion — would be appropriate in other cases.

Medical drugs, for example, would seem to present the perfect situation for experimenting with much more "local" autonomy in risk management decisions, even down to the level of the individual. My finer fancies imagine the Generally Recognized as Safe (GRAS) list of drugs being complemented by one General Recognized as Uncertain (GRAU). Use of drugs on the GRAU list might be at the user's discretion, with a full description of the known risks and benefits available as advice, but a minimum of absolute constraint. The liabilities issue would be difficult, but could doubtless be resolved with sufficient ingenuity. Appropriate, perhaps, would be voluntary "de-socialization" of the risk in the form of an agreement not to hold the manufacturer accountable or insurance agencies liable for adverse effects. I can imagine circumstances under which I would agree to such conditions, just as I can imagine preferring the de-socialiation of excommunication to the alternative of a witch trial.

The more general point is that to the extent that large-scale monolithic regulations can be avoided, "local" risk-taking preferences can be left to run their course as experiments in risk management. Government can shift from its stressful role as incompetent regulator into a more congenial role as broker of information. Is California's (or San Francisco's, or J. Fred Muggs') approach to Laetrile working better than New York's? In what ways? Who has a third approach? And so on.

Note that the apparent ethical dilemma in fact is less than it seems. If we *really* don't know how to manage a risk, then we're all guinea pigs. The fight over whose expert to believe can be transposed into a contest over whose expert guessed better, and learns faster.

The challenges of helping to design alternative — even competitive — coping strategies and institutions, of evaluating and comparing their actual performances, and of redesigning adaptively in response to what is learned should be enough to satisfy the most ambitious of risk policy analysts. They might even help to make the future of risk management a more satisfactory endeavour than its past and present.

ACKNOWLEDGMENTS

Edith and Joe Clark, and Ulrike and Ray Hilborn provided valuable suggestions and comments on an earlier version of this manuscript. Anna Maria Krebl is responsible for the carnivorous lady. My thanks to all.

REFERENCES

1. H. R. Trevor-Roper, *The European Witch Craze of the 16th and 17th Centuries and Other Essays*, Harper and Row, New York, 1968.
2. M. Harris, *Cows, Pigs, Wars, and Witches*, Vintage, New York, 1974.
3. H. P. Duerr, *Traumzeit: uber die Grenze zwischen Wildnis und Zivilisation*, Syndicat, Frankfurt am Main, 1978.
4. M. Summers, *Malleus Maleficarum* (trans.), London, 1928.
5. A. Wildavsky, "No Risk is the Highest Risk of All," *American Scientist*, 67: 32-37, 1979.
6. R. W. Kates, ed., *Managing Technological Hazard*, Institute of Behavioral Sciences, Boulder, Colorado, 1978.
7. K. Popper, *Conjectures and Refutations*, Routledge Keegan Paul, London, 1963.
8. P. Feyeraband, *Against Method*, New Left Books, London, 1975.
9. T. S. Kuhn, "Logic of Discovery or Psychology of Research?" in I. Lakatos and A. Musgrave, *Criticism and the Growth of Knowledge*, Cambridge University Press, Cambridge, 1-23, 1970.
10. E. W. Lawless, *Technology and Social Shock*, Rutgers University Press, 1977. 1974.
11. C. S. Holling, C. J. Walters, and D. Ludwig, "Surprise in Resource and Environmental Management," unpublished manuscript, 1979.
12. C. S. Holling, ed., *Adaptive Environmental Assessment and Management*, John Wiley & Sons, Chichester, 1978.
13. I. Burton, R. W. Kates, and G. F. White, *The Environment as Hazard*, Oxford Univ. Press, New York, 1978.
14. I. Burton, R. W. Kates, and G. F. White, *The Human Ecology of Extreme Geophysical Events*, University of Toronto, Department of Geography, Natural Hazards Working Paper 1, 1968.
15. R. W. Kates, et al, "Human Impact of the Managua Earthquake," *Science*, 182: 981-989, 1973.
16. R. A. Rappoport, *Pigs for Ancestors*, Yale University Press, New Haven, 1968.
17. M. Sahlins, *Stone Age Economics*, Aldine, Chicago, 1972.
18. G. F. White, et al, *Changes in Urban Occupance of Flood Plains in the United States*, University of Chicago, Department of Geography, Working Paper 57, 1958.
19. D. Pimentel, et al, "Pesticides, Insects in Foods, and Cosmetic Standards," *BioScience*, 27: 178-185, 1977.
20. W. C. Clark, D. D. Jones, and C. S. Holling, "Lessons for Ecological Policy Design: A Case Study of Ecosystem Management," *Ecological Modelling*, 7: 1-53.
21. W. H. McNeill, *Plagues and Peoples*, Anchor Press, Garden City, 1976.
22. C. S. Holling, "Forest Insects, Forest Fires, and Resilience," in *Fire Regimes and Ecosystem Properties*, H. A. Mooney, J. M. Bonnicksen, N. L. Christensen, J. E. Lotan, and W. A. Reiners, eds., USDA Forest Service General Technical Report, Washington, D.C., in press.
23. National Academy of Science (USA), *Genetic Vulnerability of Major Crops*, NAS, Washington, 1972.
14. L. M. Branscomb, "Science in the White House: A New Slant," *Science*, 196: 848-852, 1977.
25. National Academy of Science, *How Safe is Safe? The Design of Policy on Drugs and Food*

Additives, NAS, Washington, D.C., 1974.

26. W. M. Wardell and L. Lasagna, *Regulation and Drug Development*, American Enterprise Institute, Washington, D.C., 1975.

27. W. M. Wardell, "The Drug Lag Revisited," in *Clinical Pharmacology Therapeutics*, 24: 499-524, 1978.

28. L. G. Schifin and J. R. Tayan, "The Drug Lag: An Interpretive Review of the Literature," *International Journal of Health Services*, 7: 359-381, 1977.

29. S. Peltzman, *Regulation of Pharmaceutical Innovation: The 1962 Amendments*, American Enterprise Institute, Washington, D.C., 1974.

30. J. S. Turner, "A Consumer's Viewpoint," in [25] 13-22.

31. J. Lederberg, "A Systems-Analytic Viewpoint," in [25] 66-94.

32. R. G. Noll, "Breaking Out of the Regulatory Dilemma: Alternatives to the Sterile Choice," *Indiana Law Journal*, 51: 686-699, 1976.

33. F. R. Stockton, *The Lady or the Tiger and Other Stories*, Schribner's, New York, 1884.

34. W. C. Clark, "Managing the Unknown," in R. W. Kates, ed., *Managing Technological Hazard*, Inst. Behav. Sci., Colorado, 109-142, 1977.

35. W. Haefele, "Hypotheticality and the New Challenges: The Pathfinder Role of Nuclear Energy," *Minerva*, 10: 303-323, 1974.

36. D. D. Hester and J. Tobin, eds., *Risk Aversion and Portfolio Choice*, Cowles Foundation Monograph 19, New Haven, 1957.

37. P. F. Drucker, *Management*, Harper and Row, New York, 125, 1973.

38. R. Dubos, quoted in W. C. Wescoe, "A Producer's Viewpoint," in [25] 28.

39. J. W. Gardner, *The Recovery of Confidence*, W. W. Norton, New York, 1970.

40. P. B. Hutt, "A Regulator's Viewpoint," in [25] 116-131.

41. C. E. Lindblom, *Politics and Markets*, Basic Books, New York, 1977.

42. C. E. Lindblom and D. K. Cohen, *Usable Knowledge: Social Science and Social Problem Solving*, Yale University Press, New Haven, 1979.

43. A. M. Weinberg, "Science and Trans-science," *Minerva*, 10: 209-222, 1972.

44. P. Feyerabend, *Science in a Free Society*, New Left Books, London, 1978.

45. J. R. Ravetz, *Scientific Knowledge and Its Social Problems*, Oxford University Press, Oxford, 1971.

46. C. Comar, "Bad Science and Social Penalties," *Science*, 200: 1225, 1978.

47. H. Brooks, "Expertise and Politics: Problems and Tensions," *Proc. Amer. Phil. Soc.*, 119: 257-261, 1975.

48. D. Nelkin, "The Political Impact of Technical Expertise," *Social Studies of Science*, 5: 35-54, 1975.

49. L. Branscomb, ed., *Science, Technology, and Society. A Prospective Look*, National Academy of Science, Washington, D. C., 1976.

50. A. Kantrowitz, "The Science Court Experiment: Criticisms and Responses," *Bull. Atom. Sci.*, 33: 43-50, 1977.

51. M. G. Morgan, "Bad Science and Good Policy Analysis," *Science*, 201: 971, 1978.

52. G. Majone, "The Uses of Policy Analysis," *Russell Sage Foundation Annual Report for 1977*, 201-220, 1977.

53. A. Wildavsky, *Speaking Truth to Power: The Art and Craft of Policy Analysis*, Little Brown Co., Boston, 1979.

54. C. E. Lindblom, *The Policy Making Process*, 2nd ed., Prentice-Hall, New York, 1979.

55. E. S. Quade, *Analysis for Public Decisions*, Elsevier, New York, 1975.

56. K. E. Weick, *The Social Psychology of Organizing*, Addison-Wesley, Reading, 1969.

57. A. Alchian, "Uncertainty, Evolution, and Economic Theory," *J. Pol. Econ. 1950*: 211-221, 1950.

58. I. Lakatos, "History of Science and Its Rational Reconstruction," in R. Buck and R. Cohen, eds., *Boston Studies in the Philosophy of Science*, 8: 92-122, 1971.

59. M. Crozier, *The Bureaucratic Phenomenon*, University of Chicago Press, Chicago, 1964.

60. D. N. Michael, *On Learning to Plan — and Planning to Learn*, Jossey-Bass, San Francisco, 1978.

61. G. Majone, "Process and Outcome in Regulatory Decision-Making," Amer. Behav. Sci.,
    22: 561-583, 1979.
62. G. Majone, "Standard Setting and the Theory of Institutional Choice," Policy and Politics,
    5: 35-50, 1977.
63. G. Majone, "Technology Assessment in a Dialectic Key," Int. Inst. Applied Systems
    Analysis PP-77-1, Laxenberg, Austria, 1977.
64. D. B. Straus, "Managing Complexity: A New Look at Environmental Mediation," Envir.
    Sci. Technol., 13: 661-665, 1979.
65. N. E. Abrams and R. S. Berry, "Mediation: A Better Alternative to Science Courts," Bull.
    Atom. Sci., 33: 50-53, 1977.

DISCUSSION

F. E. Burke (University of Waterloo)

Bill, I was most intrigued by your last example of drug control in the U.S. and the U.K. I was wondering if you had some indication of perhaps the worst drug disaster that we know, Thalidomide. Did it occur both in the U.K. as well as here on this continent and, if so, do you have any frequency data? Clearly this is one of those things that slips by, but are there any things that you have learned from it?

W. C. Clark

W. C. Clark

The problem is that trying to learn from the Thalidomide case is like trying to learn from the nuclear case. It's in very many ways an outlier. The facts are that Europe did have a major Thalidomide tragedy and America did not. But it's not clear that America escaped Thalidomide because the regulatory test schemes we had going at the time would have caught it.

F. E. Burke

It did happen here, incidentally.

W. C. Clark

Well, o.k., let's not quibble on the details. (Note added: Thalidomide was never distributed for commercial clinical use in the U.S.). It is not necessary to make the distinction between the U.S. and U.K. schemes on the Thalidomide issue. There's a much richer body of evidence on new drug indications and soon, a series of studies out of AEI (American Enterprise Institute), some work by Peltzman, things coming out of Rochester with Lasagna and his group (see Ref. 26-29 in Clark's paper). But again, these things have not always been balanced reviews of the subject matter. And, unfortunately, there haven't been good studies from the English side to match these with. I would like to see some.

E. Pate *(Massachusetts Institute of Technology)*

I love the story of the lady and the tiger, Bill. The thing that I do not see is the incompatability between risk assessment and this kind of flexible risk management solution that you propose. I think they are very sound. But I don't see why some sort of a priori risk assessment can hurt to the extent it can give you a wider range of solutions.

W. C. Clark

That's a view I'd probably be well advised not to fight with. I'm not speaking against risk assessment per se. I am talking about incomplete approaches to the problem which concentrate 99% on the assessment issue and leave the management or response issue as something we don't worry about. But I've got no quarrel in principal.

W. D. Rowe *(The American University)*

I think you probably hinted that we ought to burn a few risk assessors, scientists, economists . . .

W. C. CLark

and lawyers . . .

W. D. Rowe

and lawyers, too. And the real question I get is "Are we really asking for holistic analyses of the kinds that you could do in ecology when we are beseiged within our management system by institutional barriers?" Institutional barriers are very real, everybody has to operate within his barriers, and nobody can talk across the barriers. Now, it would be very nice to be able to step back and do studies looking from the top down to have some holistic view, but I'm not sure there are people here or anybody available who can do them . . .

W. C. Clark

Certainly there are. I know that Mr. Straus of the American Arbitration Association is in the audience. Some of his organization's work shows that, sure those institutional barriers are there, but it's when we begin focusing our attention on them — breaking down the things that don't let us learn from each others' experience and encouraging some flexible responses — that we sometimes make a little bit of progress. In fact, the most discouraging thing that I've heard at this conference was the little exchange between Dr. Starr and somebody on what we learned from Three Mile Island. That conversation degenerated into an argument on whether someone should have known ahead of time. Well, damn it, if that's what happens every time an agency or an individual says, "Jehoshaphat, we blew it. Let's try to figure out what we can learn from this mistake" — if even an assembly like this conference displays an inquest mentality asking "Who's at fault?" "What were you doing ahead time?" "Who's culpable?", instead of "What can we learn from it?" — then what in heaven's name hope do we have of dealing with the public, Congress, and the press?

W. C. Rowe

Because I'm arguing why I wasn't responsible.

W. C. Clark

Nolo.

E. V. Anderson *(Johnson and Higgins)*

I like your lady and tiger analogy very much. There's another part of the story. He was given advice by the King's daughter and he doesn't know whether she's jealous or really loves him and I think you can draw it on further and further.

W. C. Clark

So do I. I have 75 variant endings to that story and I sort of shuffle them depending them on the audience . . .

C. Starr *(Electric Power Research Institute)*

I found your paper delightful. It has many messages, not just one. I wanted just to mention that the adaptive mechanisms in the development of a technology or of the innovation of a technology in society are historically what has happened throughout the history of technology. If you read any of the literature on technology development, that adaptive mechanism went on. It went on in the development of the automobile as a major sector of our society.

Now the issue that has come up — and I'd like your point of view on this — is that at the time the adaptive mechanism was working, the public was in fact a silent intermediary, if you wish, in that whole process. If you look through the history of the automobile and its development as a safety device, of the use of electricity, the public was essentially an intermediary that didn't have much to say about the mechanisms or much to say about their own involvement. They were guinea pigs, if you wish, by which the process was developed. What has happened now, in the last few decades, is that the public has become, for various reasons, conscious of its role. Now there are groups within the public that say "We do not want to be part of this cycle. We want to be protected from being experimented on." Now, how do you handle that, because that's really where the issues come to the crux. There's absolutely no question that as new things come into our society — new drugs, new technologies — the public is going to be impacted upon. The public now says at the beginning, "We don't want to be part of the cycle of this development because things may happen to us that we can't foresee or anticipate." It's the visibility of their role which is raising the public issue.

W. C. Clark

If I get the answer right do I get a lollipop? There are many responses, or none, or some such thing, to a question like that. Clearly, there are many things happening in the contemporary risk debate. What you're calling public consciousness, or public notions of their role in decisions of this sort, are clearly different now than they were yesterday. That's something we have to contend with and worry about; I don't have any panacea for it.

I think, though, that by focusing our attention on the truly large scale issues — such things as the nuclear issue and fluorocarbons — we tend to use phrases like "The Public" as though they were at least everybody in the United States or maybe everybody in the Western World or some such thing. And that, of course, is where a whole lot of our trouble comes from. It's very hard to obtain compromises with a 200 million person game. Many of the issues referred to in earlier times were resolved in smaller groups. The obviousness of this fact leaves me bewildered by how little it's been taken up by the regulatory community. Here I do have a grudge against the folks whose business is efficient regulaton, pure and simple. Over and over again, in the absence of compelling evidence, I hear them saying, "Here is a social optimum, you cannot possibly do better than this." They inflict global solutions and global (national, statewide) regulations or rules for how we're going to do it. Now this is about as antithetical to the litany I've presented as you can possibly get, and it is so for at least two reasons.

One is given that no one knows the answers, what possible justification is there in saying that my opinion, my knee jerk reaction, as a regulator or a Congressman should be inflicted on all you "public" characters? Legitimate regulatory goals aren't helped by such regulation-for-its-own-sake, and we aggravate an inordinate number of people by shoving our opinions down their throats when it's obvious that we don't have anything more than sort of a first cut guess about what we are doing.

Second, the predilection for global rules and regulations means that we don't get the benefits of experimenting with 25 different ways of trying the regulating. I would like to see us pay more attention to the scale of our regulatory efforts, stepping back and saying "In this particular case we could do best at the level of, say, state-determined regulations, in another case regionally determined regulations, and so on." The regulatory community could then become an information broker. This idea was raised yesterday. We can sit back and we can observe these different regulatory approaches without taking any legislative or constraining authority. We may be able to do a great deal of good by sitting there and saying "Hey, look, this group and this group and that group seem to be doing awfully well with their approaches, while these other approaches are getting dreadful results. At least be aware of this." Certainly there would be circumstances notifying one further regulatory step, saying "We know those four approaches to be so disasterous that they should be banned." But the present monolithic approach just doesn't help anyone but the Inquisitors.

D. McLean *(University of Maryland)*

Bill, it seems that I'm confused by these messages and an analogy that I don't know where to take. I think this is the same question, but I'll put it in two parts. First you are telling us that regulators don't really know and they're promulgating these laws and regulations on the rest of us. Not only do they not know, but as in the case of certain ecological situations that we described at the beginning, they do not know there are all these hidden side effects that will come out. But it seems to me that the very same can be said about industry and other people that the regulators are concerned about so I don't know if we're really worse off having the regulators or better off for having them.

W. C. Clark

In general, neither. The useful question is not whether, but how to regulate.

D. McLean

The other thing is that I don't know what to make of the witchcraft analogy, which is a very interesting case in point. I'm glad you brought it up because witchcraft and alchemy have befuddled philosophers of science recently because they're perfectly respectable according to all the canons of scientific method. It's not a group of individuals that went crazy, it's the institution that went crazy in some way, or at least that's our evaluation as we now look back on it. I don't know how we're supposed to take the moral of such lessons. Of course we were wrong to do all that witch hunting. But on the other hand, to say that there are no witches means that perhaps we should not go searching for all these things. The advice seems to be extremely cynical.

W. C. Clark

No, it's not cynical, it's scared. And it is perversely arguing *for* the role of the professional and of professional ways of operating in this business. I know this always gets misinterpreted — I misinterpret it myself — but I have little use for the dumping on experts and deification of the public which has become popular in the risk debate. If you don't like the way experts manage the complex technical systems we're dealing with now, try your average public demagogue who wants to take that complexity and use it to obfuscate the non-reasons behind his desire to inflict one thing or another on us. Experts aren't saints, but when they're acting in a professionally responsible manner, they have a handle on some of the few hard facts of the situation. They're really our only protection against the nutty witch hunters who are looking for band wagon careers.

D. McLean

I see, I thought the experts were the witch hunters.

W. C. Clark

No, the experts are not, or at least need not be. Witch hunting was not a scientifically responsible operation precisely because it had no rules of evidence which could lead to closure of the debate. An accused witch was a convicted witch because of the methods adopted to find out whether she was. There was no law, rules, or test ever promulgated by the Inquisition which would say "If we observe this evidence you're no longer a witch." Today, a possible risk is an actionable risk as soon as you raise the possibility that it might be, because we have no way of proving non-riskiness or safety.

PROSPECTS FOR CHANGE

Melvin Kranzberg

Georgia Institute of Technology, Atlanta, Georgia

ABSTRACT

There will be change — no question about that. But there are questions about the direction and rate of change, whether or not the change can be controlled, and if so, who will direct it and how.

The problem of controlling changes raises the issue of technological determinism. Is technology autonomous, as Langdon Winner claims? Is it not also possible that there is autonomous social change?

In modern industrial society, neither technological nor social change is autonomous. They are interconnected and interactive. And the directions of both social and technical changes will be determined in large measure by our assessment of societal risks and benefits, which, in the last analysis, will be both cause and effect of changes in our values.

Some technological changes are already under way. Certain of these are the result of advances in our scientific technology, such as developments in computers, communications, and information science. Others derive from a changing public attitude toward the environment, while still others result from our profligate use of energy and materials in the past, and are heightened by international political developments.

How we view those changes — how we choose among the available technologies or which ones we will endeavor to make available — will depend on how they interact with a series of sociopolitical-cultural changes which are also occurring or are about to make themselves felt.

These social changes include a burgeoning world population (affecting the population/resource ratio); transformation in the basic institution of the American family; revolutionary changes in learning, living, and leisure patterns; the growing older of the American population; the emergence of participatory democracy; alterations in the city-suburb interface; and the like. Perceptions of risk will undoubtedly change as new sociocultural and demographic patterns emerge.

Inasmuch as our technology has not yet fully responded to these massive social changes which are already underway, we might be faced with a "technological lag." Despite the present crisis of confidence, there is ample evidence to indicate that American science and technology can respond effectively to these changes while providing new and changing answers to the question of "How safe is safe enough?"

INTRODUCTION — THERE WILL BE CHANGE

Being an historian, I start my story in the dark and distant past, indeed, near the very beginnings of mankind and womankind. And because our knowledge of the very distant past is so dim and hazy, I can only start with a rumor, not a fact. Hence it is rumored that as Adam and Eve were being driven out of the Garden of Eden, Adam turned to Eve and said, "We are living in an age of transition."

Ever since then, man has been living in an age of transition, as the conditions of life and work, travel and play, have been transformed over the centuries — at first slowly, but with growing rapidity as we approached modern times. One of the chief engines for the transformation of man's way of life has been technology.

For most of history technological change occurred at a pace slower than a snail's. For millennia the bulk of mankind lived in rural areas and followed agricultural pursuits. They farmed the same land in the same fashion as their fathers had done; their horizons were limited and life was a constant struggle against the elements, natural disasters, and man-made troubles. Empires rose and fell, political power and institutions disintegrated and reintegrated, and great philosophical and religious systems developed slowly over the centuries — but man's daily life and toil remained relatively unchanged until some two centuries ago.

The Industrial Revolution changed all that. The hearth and home were no longer the center of production and of all family life. Instead, men moved to the cities, and the factory became the productive unit. Men's lives and horizons were broadened. New political and social institutions provided the opportunity for social mobility, while new technology gave geographical mobility to an extent undreamt of in the preindustrial world. As the pace of scientific discovery and technological innovation accelerated, there came concomitant changes in economic, political, and social relationships as well as major transformations in cultural attitudes and institutions.

The rapidity of change is manifest during the past century. This year (1979) we celebrated the centennial of the birth of Albert Einstein, whose probing imagination opened new vistas for science and set the stage for major technological developments now and in the future. In a couple of weeks (October 20, 1979) we will celebrate the hundredth anniversary of Edison's successful demonstration of the first incandescent light bulb. This set the stage for the electrical age and the great electronic manifestations of our own times.

At the turn of this century, man had not yet mastered the art of heavier-than-air flight. Yet the summer of 1979 witnessed the tenth anniversary of man's first landing on the moon, the culmination of the Scientific Revolution of the 17th century combined with the Industrial Revolution of the 18th century, as well as the fulfillment of one of man's most ancient dreams.

It is no wonder that we say we are living in a "technological age." We call it that not because all men are engineers, nor even because all men can understand technology. We call it that because we are aware, as never before in our history, that technology has become a major element in our lives, affecting where we live and work, and how we think and play.

But man has always been living in a technological age inasmuch as much of his life has been bound up in his technology. The chief difference lies in the fact that we are aware as never before of the significance of technology in our lives and institutions, and also because of the acceleration of technological changes — with their impact on society — which has occurred in our times.

The rapid growth of science and technology in our own times thus impinges itself on our consciousness as never before. There are more scientists living today than the total of all those who lived before the world's history [1]. And Philip Handler, President of the National Academy of Sciences, claims that "perhaps 80% of all of science has been learned since the birth of the National Science Foundation, which occurred only some thirty years ago" [2]. In the field of technology the growth has been equally spectacular. Whole new fields of human endeavor have opened up; great advances have been made in transportation, communications, in the use and fabrication of materials, in the development of power sources and resources. Everywhere one turns one can talk about the technical "marvels" or "miracles" which have occurred just within our own century.

THE PROBLEM OF CONTROLLING CHANGE

While the litany of technical triumphs would go unchallenged, there are many who would question whether or not the great technological transformations of our times have been matched by improvements in the human condition. Furthermore, questions are raised about the direction and rate of technological change, whether or not that change can be controlled, and, if so, who will direct it and how.

Is technology, as Langdon Winner has claimed [3], autonomous? Winner's argument is more sophisticated than Jacques Ellul's that technology is out of control and has become an end in itself, rather than a means [4]. While Winner agrees that technology is out of control — the subtitle of his book is *Technics Out-of-Control as a Theme in Political Thought* — he uses a more subtle argument, namely, that we have developed our technology in order to suit our material wants and needs, and these technical artifacts have so shaped our society, including our values, that they have become our masters. Our present technological structure, says Winner, follows rules of its own making, giving no voice to the public — although he fails to detect the contradiction to his view that technology has already shaped the public's will.

NEITHER TECHNOLOGICAL NOR SOCIAL CHANGE
IS AUTONOMOUS

But if we were to accept the argument that technology is autonomous, might we not also raise the question of whether or not social change is also autonomous? Might there not be changes in society which go on independently of technology or irrespective of the technological context?

I reject both these alternatives of autonomous technology and autonomous social change. In modern industrial society, neither technological nor social change is autonomous. They are interconnected and interactive. Hence, in attempting to delineate the interactions between technology and society, I have been forced to formulate Kranzberg's First Law. It reads as follows: Technology is neither good nor bad, nor is it neutral.

By that I mean technology sometimes interacts with society in ways which do not seem inherent in the technology itself or which might have consequences far beyond the original intent of the technology.

Partly this is due to the scale on which we apply our technology, and sometimes it is because a series of technical changes come together to produce synergistic effects which go far beyond the technologies taken singly. Furthermore, these intertwine with political policies, social goals, and economic developments and hence have far-reaching consequences.

The automobile provides a splendid example of all these points. At the turn of this century, the automobile was extolled as the solution to the safety, congestion, and pollution problems posed by horse-drawn transportation. That was a time when in New York City alone horses deposited some 2-1/2 million pounds of manure and 60,000 gallons of urine in one day. The automobile promised relief from these problems, but, as we know, the large-scale use of the automobile brought back these problems in heightened and different form.

The automobile corresponded to a general desire for personalized transportation, and mass production methods — involving a host of technologies — made that possible. At the same time, large-scale use of the automobile demanded a host of auxiliary activities — roads and highways, production and distribution of fuel, and the like. Whole new industries sprang into being, and political and economic considerations came to the forefront as more and more millions of Americans opted for the quick, easy, and at one time inexpensive individualized transportation provided by the automobile.

External sociopolitical considerations also affected America's dependence on the car. Following World War II, the government, filled with good intentions of rewarding the returning veterans, provided the GI Bill of Rights with low-cost VA mortgages. Inasmuch as most Americans still maintained something of the pastoral ideal and wanted to have a private home with some greenery around it, they moved to the suburbs — the only place where land was available. Since the population density in the suburbs could not support mass transportation, the automobile became a necessary adjunct of American middle-class life. Thus a whole series of social choices and technical requirements and economic demands resulted in our organizing our country spatially and economically around the automobile.

A few years ago it was said that America's "love affair" with the automobile was over. That might be so, but the fact is that we have married the automobile. And, as in the case of many marriages, the cost of divorce is simply too high to contemplate.

Indeed, any attempt to interfere with our right to drive, such as a fuel crisis which causes long lines at gasoline stations, brings out the violence which sometimes lies beneath the American character, as witness the behavior in the gas lines in the early summer of 1979. The automobile has become part of the American credo of "life, liberty, and the pursuit of happiness." The American idea of happiness is the liberty to drive one's car in a dense smog through a traffic jam while endangering somebody else's life!

Besides, it isn't quite true that our love affair with the car is over. Anyone with a teen-age son can attest to that fact.

Perhaps Langdon Winner would say that this proves his point, namely, that our technologies have developed in size and complexity to the point where they determine our social needs themselves, so that we are unable to control them. But this postulates an unacceptable dichotomy between technology and society, for it was social need which first established the pattern and which continued to interact with the technology. In brief, technology *is* a part of society. Hence the future directions of both social and technical changes will be determined in large measure by our assessment of societal risks and benefits which, in the last analysis, will be both cause and effect of changes in our values.

As is evident from the need to hold a symposium on the topic, there are many problems associated with societal risk assessment. For one, there is the problem of measuring risks, and as David L. Bazelon has pointed out, "We have learned that even our experts often lack the certain knowledge that would ease our decision-making tasks" [5]. It is difficult to sort out the scientific facts, inferences, and values in risk regulation — and as our ability to measure improves, we learn that there were risks of which we had not previously been aware.

However, the purpose of gauging risks is to improve the "quality of life" — itself a very elusive topic. It is clear that the quality of life involves values, and technology enters into those values. So what must concern us are changes in values which involve changes in society and changes in the technology which is an integral part of our society. It therefore behooves us to look at some of the changes, both technological and social — and the interactions between them — which will affect our future prospects.

TECHNOLOGICAL CHANGES UNDERWAY

Some technological changes are already underway. Certain of these are the result of advances in our scientific technology, such as developments in computers and information science which will affect communication and the ability to store, retrieve, and manipulate vast amounts of information for the resolution of problems which have long confounded the human mind. Others derive from a changing public attitude toward the environment. Thus we have debates about extraction versus recycling technologies, as well as varying energy scenarios depending upon the introduction or growth of varying technologies. These result from our profligate use of energy materials in the past, and are heightened by international political developments. Domestic political questions, such as big versus small business, or control versus deregulation, are also going to affect the scale of the technologies employed, as well as the technologies themselves. Technological developments give options which enable us to choose among different socioeconomic-political policies.

In this connection it is interesting to recall that at one time it was thought that technological developments made imperative large-scale enterprises. Logically this might seem so because of economies of scale, but the historical facts often depart from logic — or at least outmode the old logic. For example, the early electronic computers, built with vacuum tubes, were becoming larger and larger in size; this might have meant that only centralized, large-scale computing operations could be carried on because of the high cost of these monster automatons. But a scientific-technical revolution — building on developments in solid-state physics — produced the transistor. While increasing in capacity, computers decreased in size and cost. Instead of huge centralized computing facilities, we have a multiplicity of computer terminals which can be located everywhere. Furthermore, the growth of personal computers gives everyone access to the computational capabilities that were once reserved for experts. By 1985 it is estimated that there will be some 8.8 million home computers, and that these will allow for information exchange, playing games, and a wide variety of home activities, including control systems, which had been undreamt of hitherto [6].

Another example of past thinking that vast, vertically-integrated technological operations would be the wave of the future is afforded by the automotive industry. During the 1920's Henry Ford's River Rouge Plant provided what many thought was to be a model of the world to come. The plant was visited by engineers and businessmen from all over the world who marveled at its size and its accomplishments: raw materials flowed in at one end and the finished product rolled off the assembly line at the other end. But the fact is that nobody ever built another River Rouge Plant; instead, the automotive industry decentralized in response to a host of other factors.

CHOOSING AMONG AVAILABLE TECHNOLOGIES

The point I am trying to make is that technology does not make imperative massive productive units which we used to think were essential for technological growth. Instead, technology offers options — for centralization or decentralization, for large-scale or small-scale operations. If today we witness the growth of great conglomerate corporations, it is not because of any technological imperative. What technology has done is to provide the communication mechanisms enabling the exercise of decision-making powers and control over far-flung enterprises. But whether or not conglomerates will wax or wane depends upon political policies expressed in tax structures and anti-trust laws, rather than technology. Thus, instead of autonomous technology, we have technology giving us autonomy in our choices as to the paths we might follow to achieve certain social goals.

Likewise, changes in our environmental ethos affect the nature and direction of our technological developments. Attempts to clean up our air pollution force engineers to develop new technologies: catalytic converters for automobile exhausts, scrubbers for smokestacks, and the like. But then a change in the fuel supply, caused by external pressures, can bring about differences in the way in which we view the environmental risks. Thus, for example, when a pipeline across Alaska was proposed in the early 1970's, it was stalled because of concern about the effects of the pipeline technology on the Arctic environment. But with the onset of the Arab oil embargo in 1973, approval of the Alaskan pipeline was an almost immediate reaction. Of course, the opposition of the environmentalists did produce a more environmentally-conscious design and construction of the pipeline itself. But the point is that technology does interact with other societal factors.

SOCIAL CHANGES AND THEIR POTENTIAL IMPACT

What are some of the other changes in prospect which will affect society, our values, and hence our technological future? Some of these are global in nature, such as the ratio of population size to resources. During classical antiquity, there were probably fewer than 200 million people on the entire earth — and it had taken millennia to reach that figure. By 1650 it had gone up to one-half billion, and scientific and technical developments allowed it to double to one billion about 1830. But with industrialization, things began moving faster. By the beginning of this century the world's population had soared to over 1.5 billion. Then it took only 79 years to more than double to today's 4.4 billion, and it is expected to double again by the year 2013. Indeed, a projected increase of 2.26 billion between 1975 and 2000 means that the population rise during our present quarter century would equal the entire world population increase from the time of Christ to 1950.

In November 1978 we heard some heartening news from the U.S. Census Bureau, which reported a "perceptible decline" in the rate of increase of the world's population over the last ten years. In the period 1965-1970, the population grew 1.9% per year, dropping 5% to 1.88% per year in the 1975-1977 period. But that "good news" still means a growth of 80 million a year, so, instead of doubling every thirty-five years, the world's population will now double every thirty-seven years!

Providing food, clothing, and shelter for this unprecedentedly rapid growth in the world's population will put great demands upon our technology. New technologies will come to the fore and more intensive use will have to be made of old technologies. Our perceptions of risk — or our willingness to take risks currently regarded as unacceptable — might change in order to meet the enormity of the new needs created by this burgeoning population. Thus we might chance more environmental pollution through, say, the use of DDT in order to increase agricultural production, or we might decide to undergo the risks of increased radiation or a nuclear disaster in order to provide adequate energy. Tradeoffs will have to be made — unless we decide to remain indifferent to world population growth, but that, too, would involve changes in values and tradeoffs in political and other matters.

The burden upon resources from the growth of the world's population is evident. Even at current rates of consumption, a number of our resources are being exhausted. What will happen whan a growing world population with a growing expectation of material goods begins using up resources faster than ever? Obviously, great challenges will be posed to technology to provide substitutes, to eke out the existing supplies, to recycle, and to develop new techniques. At the same time, new styles of living and consuming will emerge.

The ratio of population to resources is not the only variable in our equation of the future; there is also the matter of *how* people live. Fundamental changes are occurring in the basic institution of the American family: cohabitation without marriage, the growing incidence of divorce, the rise of one-person households, and the decrease in family size. According to the U.S. Census Bureau [7], the 1980 census is expected to show that the traditional family household of mother, father, and one or more children now accounts for less than a third of the nation's households, the lowest percentage ever; that the number of unmarried couples sharing a household has more than doubled in ten years; and the number of husbandless women who are heading families has soared to more than 8 million, nearly 50% over 1970. What is this going to mean in terms of our housing, working, and living patterns — and what new demands will be made on our technology?

Furthermore, our population is growing older. As of July 1, 1979, the U.S. Census Bureau reported that the median age of all Americans was precisely 30.0. As the population grows more mature, that will surely involve some shift in perception of risks and willingness to make changes in lifestyles and patterns.

There will also have to be major improvements in productivity because, as the population grows older, the burden of providing social security for the older citizens will fall on fewer and fewer workers. In 1930, for example, there were 12.1 workers for every retired person; today it is 6.1 — or only half as many. By the year 2025, there will only be 3.5 workers per retired person. What does this require in terms of improvements in technology so that we can support all the additional people who have retired?

Will the changing population/resource ratio and the aging of the population put an end to our mass-consumption society? Most people will not want to lose the material goods and creature comforts which they have enjoyed — and those who have not yet enjoyed them will still demand them. Since we are not likely to change our value system overnight, we will probably try to maintain the same rate of consumption by employing substitutes or by various "technological fixes." In that case, great demands will be made on technology to provide these substitutions. Some of the risks which we now perceive and which might possibly slow down the production of consumer goods in the future might be diminished in the public's perception when compared with the loss of these consumer goods. Perhaps our perception of risks will change when the tradeoffs become plainer for all to see.

Lifestyles and leisure patterns have already begun to change as sociotechnical changes have encouraged more and more people to enter the working force. Since the 1970 census, some two-thirds of the 47.4 million families have incomes from both spouses [8]. Mostly this is in order to maintain the consumption patterns of our "consumer society," indicating the reluctance to change values.

Although today's Americans are spending more time in school than their parents and grandparents, we discover that the level of scholastic proficiency has been declining. Don't we need more and better education in order to function in a technological world? Just to keep a technological society going — to keep the TV sets repaired, the automobiles functioning, the plumbing working — we need more technically proficient people. Already there are changing educational emphases: toward vocationalism and continuing education programs.

Further educational changes will be forthcoming, for there are new means of communication and information dissemination. We have entered into the electronic age, and education is no longer confined to the classroom — if it ever was. But we have yet failed to make full educational use of our new electronic means — even though we know that a big chunk of the new leisure time given to Americans is apparently taken up by watching television and even though we know that by the time a child graduates from high school he has spent more time watching the television that he has in the classroom.

Our cities are in transition also. Private living conditions have improved. Nearly seven in ten American homes are single family units, with more than 75% of these in suburbs or rural areas or other areas outside the city centers. People are living in larger houses, and almost 30% of all owner-occupied homes have two or more complete bathrooms. As recently as 1940, just over half of the households had indoor plumbing; now 98% have it.

Nevertheless, in the midst of the improvement in private living conditions, the public setting — the urban center — is deteriorating [9]. Although we live in better houses and have more creature comforts, the dazzling spiral of rising affluence has also been accompanied by painful increases in crime, drug addiction, civil disorder, violence, and a steady deterioration of confidence in public authority [10].

It is expected that the "energy crisis" might bring about a reversal, or at least a slowdown, in the suburbanization of America, as well as bringing about a change in transportation means and habits. Urban mass transportation is getting renewed attention, and people are spending their vacations closer to home — in order to make certain that they have sufficient gas to get them to their destinations and back. Furthermore, there are signs that some middle-class people are returning from the suburbs and are revitalizing rundown urban areas.

While there are indications of a reversal of urban deterioration in the United States, the situation in the rest of the world appears to be getting worse. A World Bank study [11] points out that there will be more crowding taking place in the world's poorer cities, just exactly the ones which can least afford it. In 1950 only 29% of the world's population lived in urban areas, but it had increased to 39.3% in 1975, and it is expected to reach 51.5% by the year 2000. Some of the predictions boggle the imagination, such as the one that Mexico City may have a population of 30 million by the end of this century.

If the world's urban centers continue to grow in that fashion, not only will there be great strain on urban services — transportation, sewage, housing, medical facilities, and the like — but there will also have to be tremendous agricultural development to feed the growing number of people who are no longer engaged in primary food production on farms.

References p. 332.

CULTURAL AND TECHNOLOGICAL LAG

The above are but a few of the major social changes already occurring in this country and abroad. Technological changes will have to keep pace.

Back in the 1930's William Fielding Ogburn, the sociologist, popularized the notion of "cultural lag," wherein he postulated the view that technology changes rapidly but that our institutional mechanisms and society change slowly, lagging behind the changes in technology. We might now be faced with the opposite situation: a "technological lag." Society is changing rapidly, but is our technology keeping up?

Many engineers would not agree with my diagnosis. They claim that we already possess the technical knowledge to deal with many of the social changes I have been describing. Even if that is so, the fact is that this knowledge is not yet in place, in usable form to meet those changes. In that event, we certainly need social innovation in order to bring our existing technical knowledge into practice.

At this point another complication enters the picture, arising from the question of societal risk assessment. Man has always assessed the consequences of his technology. But throughout most of history he did so on a very limited and immediate basis. His major assessment criteria were, in the case of civilian technology, how to gain more profit and, in the case of military technology, how to kill or maim his enemies more effectively. In civilian technology, the values were those of the marketplace; in military technology, monetary costs were not the primary consideration, and throughout history nations have been lavish in military spending. Leaving aside the special case of the military, we find that we have lived in a society where economic considerations have determined much of the nature and direction of our technical activities. Now, however, we are adding a new dimension to cost-benefit analysis, for we are now beginning to talk about social costs and social benefits.

This change in emphasis derives from the concatenation of several circumstances. One derives from the very effectiveness of our technological efforts in the past. After all, for most of man's history he lived in a society of scarcity, so his attention was concentrated upon obtaining a sufficiency of material goods. The great outpouring of goods made possible by industrialization within the past two centuries has altered this. Now that we have reached the prospect — if not the complete fulfillment — of an abundant society, we can begin to turn our attention beyond the immediate satisfaction of our animal needs and creature comforts. We can begin to think of technology's impact upon society as a whole and upon the physical environment — as witness the development of the environmentalist movement in the past two decades.

It past ages, when men worked from sunup to sundown just to eke out a living from the soil, with the hope that they would have enough to subsist until the next day or until the next harvest, men simply did not have time to worry about how their activities affected the environment; children, also brought up in backbreaking work, simply did not have time to worry about how they related to their parents; and most human experiences were so demeaning in character for the vast body of mankind that there was little call to "share the experience." Our society of abundance has given us the opportunity to worry about more than meeting the minimum needs for sustaining life — and we are now free to worry about the use of our leisure time, the so-called "higher" things of life, the natural environment, and even about others.

M. Kranzberg

At the same time that technology was effecting this great change from an economy of scarcity to one of abundance, a great change was occurring in our democratic society, namely, the trend toward participatory democracy. As the American people increasingly comprehend the important role which technology plays in their daily lives, they are demanding greater control over technology, as over all other aspects of their lives.

Heretofore technical decisions were made on an economic basis, and the actual technology was left to the "experts," the scientific and technical elite, the corporate managers of American business, the military, the government. But now, even though the public might not be sufficiently informed to act directly on scientific-technical decisions, it increasingly wants to hold the decision-makers accountable for their actions. The people recognize that any activity affecting the public should not be controlled directly by its practitioners. Utility companies should not control Public Utility Commissions, bankers should not control the Security Exchange Commission, and scientists and engineers might not be the best ones to direct our scientific and technical applications. As Dan Greenberg, the science writer, has stated: "Never ask the barber if you need a haircut!"

Science and technology are too important to the public to be left to a small coterie of decision makers. Scientists and technologists simply do not possess the political wisdom to make these decisions, even though they possess the scientific and technical expertise upon which such decisions must be based. These decisions should be left to the political process — and that, in a democratic society, is exactly where they belong. However, there are many pitfalls and problems in the politics of technical decisions [12].

Up to now, participatory democracy has given great power to single-issue groups, and has manifested itself more through protest movements than through any national consensus. In terms of risk assessment, there are pressures to move to an entirely risk-free society. But, as Aaron Wildavsky has pointed out, "No risk is

References p. 332.

the highest risk of all" [13]. If we adopt an over-cautious attitude toward new technological developments, this might well paralyze our scientific and technical endeavor and end up leaving us worse off than we were before. A static society is impossible as well as undesirable, and efforts to reduce risk leave us more vulnerable than in the past.

But if a no-risk society is an impossible goal, most people would like to reduce the risks. But the assessment of risk is a difficult matter, not only in defining the risks, but also because of different value systems of different groups within America.

NEW ANSWERS TO THE QUESTION "HOW SAFE IS SAFE ENOUGH?"

It used to be said that the basic question of political science is: Who gets what and from whom? Now, with the entrance of societal risk assessment into the political process, that basic question will be posed in a new form: Who pays for the risks, and who receives the benefits?

After all, there is a hierarchy of risks. Some things might be more risky than others, and how are we going to decide? Besides, in some cases, the risk is borne by one group in the population while the benefits are borne by others. Thus, for example, in the case of coal-fired generators of electricity, the risks are borne by the miners and transporters of coal, but the electricity-using population, which might be many miles away, derives the benefits. Thus the problem is posed in a different fashion: What are the lives and health of the coal miners worth in terms of the availability of electric power for faraway city dwellers.

People also perceive risks differently, especially when the question of control over their own risks is involved. For example, the accident at Three Mile Island produced a great public outcry. The proponents of nuclear power point out, correctly, that no one was killed by the "catastrophe" at Three Mile Island and that no deaths have occurred in the operation of commercial reactors in this country; they compare this with the fact that some 50,000 Americans are killed in automobile accidents each year and that hundreds of thousands more are injured in them — yet there are no mass protests with banners proclaiming "Ban the Auto!"

There are several reasons for this difference in the perception of risks. One, of course, is historical, namely, the association of nuclear energy with its genesis in the atom bomb. A second might lie in our familiarity with the risks posed by the automobile; we have lived with automobiles for a long time, so we accept the risks just as we accept death and taxes. Nuclear energy is much newer, and we have not yet grown accustomed to the face of danger presented by it. But perhaps the reason there is more opposition to nuclear energy than there is to the automobile, which historically has been the greater killer, lies in the fact that we can, as individuals, choose not to drive automobiles, whereas, as individuals, we do not have a choice about whether or not our electricity comes to us through a nuclear generator. Yes, we choose to have electricity, but the choice as to the way in which that electricity will be generated does not rest upon the individual. Hence we have a feeling of helplessness about the dangers posed by nuclear reactors, while we do not possess that same sense of helplessness about the possibility of automotive accidents.

Of course, if we were utterly realistic in our assessments, we might feel just as helpless about the automobile. For one thing, because of the dearth of public transportation, our suburban pattern of living, and the dispersion of commerce and industry from residential areas, we are forced to use the automobile for much of our work and leisure activities. So we don't always have the choice we think we have about the automobile, although the fact that we say we are in the "driver's seat" gives us a sense of control over the automobile and our destinies.

The point is that the question of societal risk involves subjective judgments. Risks and benefits involve value notions and ethical judgments. This does not mean that the question of values is simply a subjective one, but it means that we have to define specifically what we mean by values and what the ethical bases of our judgments might be. In political terms it means that the basic question is: Who runs the risks and who gets the benefits?

Although it sometimes seems as though each pressure group in America is concerned only with its own narrow interests, so that we want someone else to run the risks while we receive all the benefits, in our saner and more rational moments we realize that this cannot be the case. We realize also that every action involves some risks. We know, for example, that most accidents occur in the home — and most frequently in the bathroom and the kitchen. Does this mean that we are going to stop taking baths or avoid cooking our meals? Of course not. Nor does it mean that we should simply accept all risks as a part of life. There are different risks, and some of them we can reduce by wise action taken in time.

The real danger lies not in the risks themselves, but in that we will be so paralyzed by the delays of the bureaucratic process in determining risks and by our own failure to achieve a consensus regarding the risks we are willing to take that we will not take any action whatsoever. It is no wonder that President Jimmy Carter finds a "crisis of confidence" in American society today.

Certainly, nagging doubts have arisen which challenge the boundless confidence of earlier generations of Americans: the recognition, following our involvement in Vietnam, that events around the world do not always turn out the way we want; the Watergate scandal which shook our faith in our constitutional system; the realization that some of our scientific and technical triumphs had begun to boomerang and pose harmful as well as beneficial results; and now the spiraling inflation which seems to challenge the work-hard-and-get-ahead ethic which fueled America's rise to industrial and technical preeminence throughout the world.

Archibald MacLeish, the great poet and Librarian of Congress, once said, "America was promises." Many people believe that the promise is over.

But I believe that America is still promises. There is much evidence that points toward the continuing vitality and dynamism of American society. America still remains preeminent in its science and technology, in its adherence to democratic ideals, in its concern for the individual, and in its moral strength. The nations of the rest of the world turn to us for moral leadership and material help. The real question is whether or not we possess the will to utilize our great strength and power to meet these problems head on by taking the risks which the future will demand of us.

References p. 332.

REFERENCES

1. D. J. Price, *Big Science, Little Science*, Columbia University Press, New York, *1963.*
2. P. Handler, "Basic Research in the United States," *Science, 204: 474-481, May 4, 1979.*
3. L. Winner, *Autonomous Technology: Technics Out-of-Control as a Theme in Political Thought,* MIT Press, Cambridge, Mass., *1977.*
4. J. Ellul, *The Technological Society,* Alfred A. Knopf, New York, *1964.*
5. D. L. Bazelon, "Risk and Responsibility," *Science, 205: 277-280, July 20, 1979.*
6. D. R. McGlynn, *Personal Computing: Home, Professional, and Small Business Applications,* John Wiley & Sons, New York, *1979.*
7. "Census '80 Information," *Bureau of the Census,* Document 01.11:05, Washington, *1979.*
8. U. S. Census Bureau, GPO: 1979-657-004/145.
9. K. R. Schneider, *On the Nature of Cities: Toward Enduring and Creative Human Environments,* Jossey-Bass, San Francisco, *1979.*
10. A. Campbell, P. E. Converse, and W. L. Rogers, *The Quality of American Life: Perceptions, Evaluations, and Satisfaction,* Russell Sage Foundation, New York, *1979.*
11. International Bank for Reconstruction and Development, *World Development Report, 1979,* Washington, D.C., *1979.*
12. D. Nelkin, ed., *Controversy: The Politics of Technical Decisions,* Sage Publications, Beverly Hills, *1979.*
13. A. Wildavsky, "No Risk Is the Greatest Risk of All," *American Scientist, 67: 32-38, Jan.-Feb. 1979.*

At this point, there were no questions or comments of Dr. Kranzberg. The chairman used the time to pursue discussions which had been curtailed on previous talks.

MISCELLANEOUS DISCUSSION

C. Starr *(Electric Power Research Institute)*

I don't know if Dorothy Nelkin is gone by now . . .

W. D. Rowe *(The American University)*

Her colleague is here.

C. Starr

I was very impressed with her paper, but I felt, and I mentioned this to her on the way out, that she had missed the point of the last question she got. We have in the country a Congress of the United States that is elected through a process which is reasonably democratic. We all have lived through the political machinery, we know what's involved, and the result has probably represented the cross section of public interest better than anything we've got. Congress has used every type of device in terms of hearings, committees, special studies, authorizing agencies in the government to do studys, asking the National Academies to tackle topics and inform it, to bring together experts of all kinds. I doubt if there's any expert voice in the country that hasn't had an opportunity to participate in a congressional discussion which has been of interest. Nelkin's litany of things that were being done for resolution of some of the public arguments on technical social issues, omitted the congressional approach and I would like to suggest that for discussion by this group. It seems to me that the objection and the criticism, and the so called loss of faith in the political institutions that are in existence, really is a reflection of the fact that there are organized groups who don't like the answers that are coming out of the democratic process. I'm wondering if what we are seeing is a challenge to the democratic process and the public interest rather than the protection of the public interest.

W. D. Rowe

Well, I open up to that I will also point out that maybe their dissatisfied with the lack of answers that are coming out and also point out too that we are talking about the fact that people are questioning their institutions. And I'm not so sure that this is just Congress, I wonder if it's not all institutions that are being challenged for some reason and whether this is a symptom or whether it's a cause, I don't know and that's opened for discussion.

A. Shantz *(U.S. General Accounting Office)*

I take exception to the last statement. I think the legislative procedure with the acts of the Congress is required to submit a statement to the respective house and senate government operations committee indicating the probable or potential impacts of the proposed legislation or ammendment and as you look at any of these statements it's quite obvious the Congress does not take that responsibility very seriously at all in major legislation, i.e. air, water, and land control — very significant legislation. The statements are one or two pages in length, they represent very little effort and to systematically determine just what the impact of the legislation will be. It seems obvious by inference that if the Congress intended to seriously come to grips with these issues they would not address it as they have superficially in those statements.

W. D. Rowe

Let me add one more word and that is that part of the cause of the problem might well be that most of our institutions have very short term outlooks. Congress is interested in the next election, so is the top part of the Executive Branch, and many of the problems we're looking at require long term solutions and long term planning. The short term outlook may be inconsistent with the long term needs and suddenly the one place where I thought there was some inertia for longevity, the Executive Branch suddenly now has a senior executive system which will automatically require turnover. Just a comment.

L. G. Lave *(Brookings Institution)*

I was a little puzzled by the Nelkin suggestion that what we ought to do is to fund the preparation of experts getting together with some of the people who were critical of certain technological innovations. There are two problems that I see there. One problem, is that the nature of the objections basically are one that experts can have very little to say on. That is, the nature of the concern is probably not focused on the safety of the nuclear reactor for example, so much as some generalized kinds of concerns about how perfectable systems are or how far in the future you want to look. And so, trying to get some expert together with a group is not likely to do anything to allay the kinds of fears that they have. This is a second kind of puzzle that I get out of it and perhaps this is a ray of hope. That is that it may be that what would happen is that in the course of some "neutral", whatever that means, experts getting together with a public group that there would be a greater amount of education that would go on where the group would more likely trust the expert of their choice than they might with some expert who'd been hired by a group unknown to them, including the Federal Government. That might lead them to much more opinion change than sort of listening to some sort of generalized hearings where they were not directly related. So there is at least a possibility for some conflict resolution in the suggestion. That seems to me to be not quite what that paper was suggesting initially.

W. Lowrance *(Stanford University)*

I have two comments. One to Lester. It seems to me that we are in an age in which it's no longer really that sophisticated to talk about the experts versus the public or the laypeople. Think about all sorts of aspects of the nuclear weapons proliferation efforts, the waste disposal, recombinent DNA for example, in which the experts have been on both sides. In fact many experts are torn internally, I am on all of these issues about many different aspects. You have to think in somewhat different terms. The second is that I find it very hard to talk about the public. Walter Lippman once wrote a book called *The Phantom Public* in which he said that there is no such thing and Carl Sandburg echoed that with a poem, "Your Public Series is a Myth." This provokes my classes and continues to provoke me to say, that one really shouldn't speak of the public interest unless one can defend exactly what one means by that phrase.

D. B. Straus *(American Arbitration Association)*

I would like to revert back, also, to Dr. Nelkin's remarks because they obviously struck a raw nerve in here. I agree totally with her that the processes and procedures that have thus far attempted to be developed have not worked very well. Where I think I disagree with her was her discarding the continuing, as I understood it, attempts at procedural efforts rather than perhaps some other efforts which did not come to clear. Just one last comment on this. From my perspective which is obviously a narrow one, it seems to me that one of the problems that we are confronting today is that we are heaping on democracy burdens which the democratic processes as we now know them are not equipped to handle. This does not make me feel that we should turn away from democracy, but that we should try to continue to bolster up the democratic processes in handling complex issues. I think for example this whole session is a good example of what I'm talking about. We have been wrestling for the past two days with some very, very complex issues. If we were to attempt to come out with some consensus or even some decisions which I think is different from consensus at the end of this meeting, I think we would feel a little bit inadequate, not because of the quality of the people here but because we have not yet found the right processes or procedures for intercommunicating in a group of this kind in a way that will move toward process. So that I end up really with a plea to those who are here to not discard the procedural innovations simply because they haven't worked in the past, but to look with some hope at least, because I think it is the only hope we have in trying to develop a better process.

W. D. Rowe

I think I would emphasize that to the extent that I think the process is what works. The question is when do you tell when the process has stopped. When you've gone far enough and can turn it off? I have only what I call a pragmatic definition, if you want to call it a definition of acceptable risk or whatever we're talking about, and that is when the people involved including the regulators are no longer anxious about it and the attention seems to go away. That may be after the due process has gone through or it may be because some other issue overtakes the one under discussion which is then discarded. There's no method, I think that has been established, only a process. Which process is what we can argue about.

M. A. Schneiderman *(National Cancer Institute)*

I'm a perennial optimist, and I want to underline these last remarks by giving an example of process that I think worked which I hope Dorothy would have mentioned in her talk. I gave her an opportunity to but she decided that she really didn't want to. I think that's what she did, I really don't know. I really shouldn't have said that. About two years ago, I guess it was, we at the Cancer Institute were very much involved with an issue that was upsetting people very deeply, the use of mamography for early diagnosis of breast cancer. We found that all the experts involved in this process were physicians. Now a physician who is involved in this process is very hard put to tell you that his process has flaws because he doesn't really think it has flaws. There were some people who were concerned with risks of radiation, they were concerned with whether these techniques were appropriate techniques to use on women of all ages as they were being used on women at age 30 where breast cancer is relatively rare. We wound up with what was essentially a 2 day consensus meeting. We had a panel of people, some of whom were experts and at least one of whom had lost a breast to breast cancer. In certain senses this made her a far greater expert than any of the experts. She had some personal acquaintance with some of these problems which none of the experts, including all the surgeons, had. There were a whole array of people, ethicists, physicians, radiologists, and surgeons. For two days information was presented back and forth, arguments, discussions, it broke down into smaller groups who then had further arguments and discussions and out of this came a set of recommendations paying attention in some way to almost all of the people who were involved in the discussions and the information givers. The experts were only looked upon as information givers in this. The chairman of this group was not a person involved in the problem himself, he was a very fine physician, a professor of medicine, but he was not involved in this particular issue. At the end of this two day process, proposals were made, recommendations were made for how mamography was to be used and some proposals were made for how the diagnosis and treatment for breast cancer, first stage surgery breast cancer, would come about. There were large numbers of participants in this who were not completely happy. In fact I can think of no one who was completely happy with this particular solution, yet this has become essentially public policy in the United States. It's become medical policy to use the techniques this way. The Cancer Institute, by the way, found itself in conflict with the American Cancer Society who were pressing these techniques much harder than we were. All of this has become, I think, public policy because in addition to the consensus meeting and the consensus activity, Joe and I talked about this at the break, there was something in the background. Any physician who doesn't follow these procedures and who has patients who develop some problem is going to be hauled before a court of law for not having followed the procedures and something untoward happens. The Tort things Judge Green was talking about are going to come into affect. But here was a process that I think has worked. We ask people what their measures of success are. My measure of success for a decision is how long it lasts. The good decisions last three years and poor decisions last two years.

W. D. Rowe

It's now law in Federal facilities and all Federal doctors to follow that, by the way.

M. A. Schneiderman

It's now within the law.

W. D. Rowe

It's the law. It's been signed off by the President in federal guidance. Chauncey have the last word.

C. Starr

I just wanted to mention the DNA situation as he said these issues don't go away. I think that's a bad description. The DNA issue and the risk benefit analysis, when first brought out, was cataclysmic, because the ultimate risk was wiping out the human race. The scientific community working with government agencies and through public discussion has come to a resolution of how to set up the levels of protection in experimentation as the DNA work goes on, as genetic experimentation goes on, and how it gets controlled, while the research in how to control and the degree of risk gets cleared up. I think that's an excellent solution. The risk hasn't been brought to zero but it's been made very, very small. The fact that the issue has to be continuously watched is a normal thing with any kind of a risk situation and I think that that is going to be the characteristic on all these major risks. Not a black or white, yes or no, you go ahead or you don't go ahead, but rather the degree and level of management of the risk. I think the risk management which you mentioned in words but haven't discussed very much, is in fact the resolution of these issues. And the public mechanism did work in the DNA situation. So, I think we have not examples of things that aren't resolved, but actually examples of problems that were resolved.

W. D. Rowe

Just as an addition to that — this is a particular case where further scientific effort has reduced the risk level especially the catastrophic risk aspect that Paul was referring too. And it actually never went through an actual public process in the sense of going through a legal process, it was done almost at a stage of experts and all of the experts themselves eventually agreed that the risks were low and there were no longer dissenters at that point. So that's important. This is a case where expert consensus did help the situation. So it's not necessarily one that would work in every place, but I think you're right Chauncey.

CONCLUDING REMARKS

Howard Raiffa

Harvard University
Cambridge, Massachusetts

First, if you'll pardon me, I'll start with a facetious remark. I thought a few minutes about the subtitle of this conference: "How Safe is Safe Enough?" My judgmental, but professional, assessment is .3624 *units*. Really, all the members in this conference realize that this question, in the abstract, does not make sense. It could make sense, of course, if suitably amplified. There is no absolute standard for safety — nor should there be. In any particular policy or decision choice where safety is a concern there is undoubtedly a myriad of other concerns. What are the full panoply of costs of benefits and of risks and what are their distributional impacts? If an action is contemplated, what are other possible contending action alternatives? What do we know about present uncertainties, the disputes about these uncertainties and how might the assessments of these uncertainties change over time? What about the precedent that will be established if such and such an action is taken? "How Safe is Safe Enough?" is a short-hand, catchy-sounding phrase that is merely a pitifully weak and misleading simplification of a very complex problem.

The trouble is that most people think the question (How Safe is Safe Enough?) deserves an answer. Most people do not think in terms of comparative analyses of alternatives; they do not think in terms of value tradeoffs; they do not think in terms of uncertainties; they do not think about how one policy choice may set up a dynamic sequence of adjustments and that indirect effects may far outweigh immediate direct effects.

By and large I was pleased with this conference. I gained some new knowledge and insights; and I heard several elegant statements and summaries of a lot of things I already knew ... But still I feel a bit uncomfortable, not unlike I feel when I read many dissertations or articles in public policy where the author brilliantly diagnoses a problem but offers no first-step proposals about how to ameliorate those problems. It's the well-known case of the missing last chapter. So with this conference! Where do we go next? Oh sure, there were some utopian suggestions for a new society, or proposals for getting rid of lawyers and replacing them with economists, but there was a paucity of proposals for helpful first steps. Now some one could say that you should not take first steps if you don't even know the direction you want to go. Granted, but I suspect a lot of us agree on what's wrong and we might even have a large consensus on a proper direction of change, but we have to be politically relevant and offer concrete, pragmatic suggestions. I missed hearing such proposals.

Our societal dialogues are becoming increasingly strident, more adversarial, more litigious, and I'm afraid less constructive. We're not as a society exploring and seeking compromises — we're not engaged in collaborative and collegial problem-solving exercises that seek joint gains for all — or for "almost all". It's damn difficult to find creative compromises that subtly exploit differences in perceptions of uncertainties and differences in value tradeoffs — especially when the protagonists in a controversy are behaving strategically with a zero-sum-game mentality.

The public is confused about technological controversies and well they should be. They get their snippets of information through the noisiest of communication channels. Our press is not designed to seek out a random sample of expert opinion but rather it is designed to let all sides have their say. They artificially amplify the weak signals out on the far-ends of the distribution of expert opinions and dampen the less interesting judgments in the middle. Consequently the avid reader of the press or listener of commercial TV believes that the experts differ much more sharply than really is the case. And, of course, sometimes the proponents at the tails of these distributions turn out to be right, which complicates matters. And, of course, sometimes so-called experts are not really expert: they are merely reporting their informal summary of the non-random sampling of opinions of other experts. So who should you believe? We desperately need better channels to get evidence from the scientific community of a less adversarial kind.

Thanks to social scientists like Paul Slovic and his associates we now know a lot about public perceptions of risk: What the public believes, and why they believe the way they do. Of course, more work has to be done here. But now what should we do with this information? How should perceptions and misperceptions about risk be factored into policy analysis? An analyst *should* be concerned about the full array of consequences of decisions and certainly the public's anxieties are very real and deserve attention even if these anxieties are based on infirm and dubious facts and on a perverse sampling of expert opinions. An analyst, in recommending policy, *should* realize that the public's perceptions of uncertainties will have a major effect on the political acceptability of proposals. So perceptions do enter and should enter into the evaluations of consequences and also into the assessments of probabilities of final outcomes. That's the easy part. But if public perceptions and reality dramatically differ, it is in societies' interest to inform the public and public opinion leaders about these differences. The ethical issus are deep here. Whose "reality" should we favor? How can we maintain a distinction between benign education and adversarial, self interested indoctrination. I wish more were said at this Conference about this dilemma.

I attended a meeting some time ago when Paul Slovic discussed his experiments with fault-tree analysis. He examined pathways that might cause an automobile to fail to start. Some of the principal headings were: battery charge insufficient, starting system defective, fuel system defective, ignition system defective, other engine problems, mischievous acts or vandalism. Well, the next morning after hearing Slovic's talk I hitched a ride from my motel to the Conference Center with another participant at the Conference, and guess what? His car wouldn't start. The cause was not on the fault tree diagram either! We got into the wrong car. Now there are two points to this story. First, it's extremely difficult to foresee all pathways. Second, and more important, over time we should expect to be surprised and learn from these surprises. If we amplify and adaptively correct our first efforts, these surprises should come less frequently.

Too many studies are done in an episodic fashion; they become targets for criticism and their authors become overly defensive and try to sell their analyses by claiming more than they should. The community of experts take up sides and one rarely hears constructive criticism, because the experts themselves become propagandists for their own viewpoints. And these viewpoints largely depend on the viewpoints they hear from their colleagues.

We know so little about so many potential risks and a natural impulse says: hold up until you learn far more. But next year we will only know just a trifle more and if we remain paralyzed, then our inactions will merely displace one risk with another risk. We should continue to worry about risk assessments — and learn how to do these better — but I believe the real improvements, for the time being, will come in the better *management* of risks. We need a more experimental societal approach, a more adaptive approach. We need to remain loose, flexible, and resilient.

SYMPOSIUM SUMMARY
THE SAFETY PROFESSION'S IMAGE OF HUMANITY

C. West Churchman

University of California, Berkeley, California

In the 1780's, Immanuel Kant was struggling to understand the basic principle underlying morality. First he stated it as a "categorical" imperative, meaning that it holds unconditionally: "you ought to do X," and no if's, and and's, or but's. The X you ought to do is to act so that you can will the principle of your action to hold universally, i.e., for everyone in every situation. In order to clarify the meaning of this categorical imperative, Kant gave us an alternative version: "so act as to treat humanity, either in yourself or in another, never as means only but as an end withal."

This second version of Kant's moral law is both beautiful and puzzling, beautiful because it views everyone sharing in each other's humanity, puzzling because it leaves the meaning of "humanity" a mystery.

To me, it has become a matter of intense interest to try to understand how the various professions interpret "humanity in yourself or another." I'm sorry to conclude that many professions - including the oldest - tend very strongly to regard "humanity" to be some kind of purposeful machine. Thus the "medical model" is primarily interested in fixing up deficiencies in the human body, or in preventing them, much as a good mechanic cures a carburetor or puts one in your car which has a long life. Many educators regard students as beings with large and rather empty information tanks that need filling. Nutritionists view each of us as an intake machine that needs a carefully designed mixture of nutritional ingredients.

Other professions - like the law and accounting - view "humanity" as being basically foolish, with an occasional inclination to be dishonest and with an overriding anxiety about what's going to happen to each of us.

Neither of these attitudes seems very laudatory, and certainly neither captures what Kant had in mind. Of course, the professions can claim that they are in the business of taking care of the weaknesses and foibles of people, and not their basic

References p. 346

humanity. But then one can't help wondering what happens to this "basic" humanity when we're being treated either like machines or fools. It seems plausible to say that if I'm being treated as a machine or a fool, then my humanity is being treated as a means only.

Rather than argue further on this point, I want in the "summary" of this conference to examine the evidence that has been created at the conference on how the "safety profession" views humanity. My evidence consists of sentences uttered by the conferees that talk about people, what they do, behave and value. The following quotations are not exhaustive, of course. I've selected those that seem to bring out a "humanity-image" most clearly.

> With so much more to lose than our parents or grandparents, it is only natural that we [the people] should be more concerned with protecting what we have.[1]

> Many people do not differentiate between process and outcome.[2]

> The individual who desires regulation to lessen the risk of an untoward event over which he has almost complete control is asking the government to take on the role of his keeper. Most people do not desire to have anyone assume this role.[3]

> . . . a sample of members of the League of Women Voters estimated that motor vehicles cause 28,000 deaths per year in the United States; a sample of students estimated 10,500 deaths. In fact, there are about 50,000 deaths each year caused by automobile accidents.[4]

> . . . people's perception of probabilities to a professional really reaches absurdities.[5]

> People are now questioning whether we really need to do much about our chemical environment. They argue that no great harm has yet shown itself from our living in an environment in which synthetic chemicals are much more common.[6]

> And I am not concerned only because such a data gathering is likely to be inefficient. What is "efficient" is not necessarily "good." What is good lies in the class of truth and beauty. It lies in the eye of the beholder, with the beholder's vision modified by the place and time, the morality and ethic in which he lives, and to which he subscribes.[7]

> Man has always assessed the consequences of his technology. But throughout most of history he did so on a very limited and immediate basis. His major assessment criteria were, in the case of civilian technology, how to gain more profit and, in the case of military technology, how to kill or maim his enemies more effectively. In civilian technology, the values were those of the marketplace; in military technology, monetary costs were not the primary con-

sideration, and throughout history nations have been lavish in military spending. Leaving aside the special case of the military we find that we have lived in a society where economic considerations have determined much of the nature and direction of our technical activities. Now, however, we are adding a new dimension to cost-benefit analysis, for we are now beginning to talk about social costs and social benefits.[8]

The aesthetics of high standard mountaineering are such that a proposed route is only felt to be worthwhile if there is considerable uncertainty as to its outcome.[9]

The fears of risk are our fears, the people making and taking risks are ourselves and our neighbors. When we intellectualize ourselves away from these ambiguities, our work becomes sterile, our subjects ciphers. When we tackle them directly, our involvement makes critical interpretation impossible and broader interpretations irrelevant. Unable to see inside the problem, we trivialize it. Unable to see outside the problem, we become part of it.[10]

A particularly pernicious aspect of heuristics is that people are typically very confident in judgments based on them.[11]

Technical consensus may narrow the range of choices, but procedures that bypass underlying value concerns will have little effect on the resolution of disputes.[12]

C. W. Churchman

(I apologize for not having quoted each speaker, but in some cases the author's style did not produce clear "people" descriptions.)

I judge that the safety profession has to some extent come to realize that the experts are also human, i.e., the experts are people even when they make their expert judgments. They also seem to be aware of a change of people's attitudes towards safety, e.g., that the humanity within us often seeks risk rather than avoiding it. Sometimes our humanity is overly cautious. Sometimes our humanity leads

References p. 346

us to believe that we no longer control the technologies we invented, but there may be a will to put us back in the driver's seat.

I felt some confusion about Ron Howard's paper because I thought he was proposing an ethical procedure for assessing the economic value of a life. But to me both characters in the black pill example were immoral in Kant's sense: the person who made the offer and the person who accepted it (to accept the offer of deliberately running a risk of death for dollars gain is treating humanity in yourself as a means only, whether or not you agree, or perhaps especially if you agree). But I may well have misinterpreted Howard's intent.

In summary, there is reason to be cheerful about the future of the safety profession because it seems to be on the pathway of not treating people as children who should be told how to behave safely, but rather as humans all sharing in the quality of each other's lives, and of realizing that risk is an essential in the design of a good life.

REFERENCES

1. Lester B. Lave, "Economic Tools for Risk Reduction," This Volume, p. 117.
2. Ibid, p. 118.
3. Ibid, p. 119.
4. Raphael G. Kasper, "Perceptions of Risk and Their Effects on Decision Making," This Volume, p. 74.
5. Chauncey Starr, discussion comments, This Volume, p. 4.
6. Marvin A. Schneiderman, "The Uncertain Risks We Run: Hazardous Materials," This Volume, p. 22.
7. Ibid, pp. 37-38.
8. Melvin Kranzberg, "Prospects for Change," This Volume, p. 328.
9. Michael Thompson, "Risk: The Three Faces of Everest," This Volume, p. 278.
10. William C. Clark, "Witches, Floods and Wonder Drugs: Historical Prospectives on Risk Management," This Volume, p. 288.
11. Paul Slovic, Baruch Fishoff and Sarah Lichtenstein, "Facts and Fears: Understanding Perceived Risk," This Volume, p. 185.
12. Dorothy Nelkin and Michael Pollak, "Problems and Procedures in the Regulation of Technological Risk," This Volume, p. 243.

PARTICIPANTS

Agnew, W. G.
General Motors Research Laboratories
Warren, Michigan

Albers, W. A., Jr.
General Motors Research Laboratories
Warren, Michigan

Altshuler, B.
New York University
New York, New York

Amann, C. A.
General Motors Research Laboratories
Warren, Michigan

Amin. H.
General Motors Research Laboratories
Warren, Michigan

Ancker-Johnson, B.
Environmental Activities Staff, GMC
Warren, Michigan

Anderson, E. V.
Johnson and Higgins
New York, New York

Anderson, B.
Environmental Protection Agency
Washington, D. C.

Atkinson, T. R.
Financial Staff, GMC
New York, New York

Babcock, L. R.
University of Illinois
Chicago, Illinois

Barnes, G. J.
Environmental Activities Staff, GMC
Warren, Michigan

Bauer, M. H.
Industry-Govt. Relations Staff, GMC
Warren, Michigan

Bidwell, J. B.
General Motors Research Laboratories
Warren, Michigan

Bird, C. G.
General Motors Research Laboratories
Warren, Michigan

Bollen, K. A.
General Motors Research Laboratories
Warren, Michigan

Bordley, R. F.
General Motors Research Laboratories
Warren, Michigan

Boroush, M.
University of Michigan
Ann Arbor, Michigan

Bowditch, F. W.
Environmental Activities Staff, GMC
Warren, Michigan

Bruce-Briggs, B.
New Class Study
Fort Lee, New Jersey

Burke, F. E.
University of Waterloo
Waterloo, Ontario, Canada

Butterworth, A. V.
General Motors Research Laboratories
Warren, Michigan

Buzan, L. R.
General Motors Research Laboratories
Warren, Michigan

Caplan, J. D.
General Motors Research Laboratories
Warren, Michigan

Chapman, W. C.
Industry-Govt. Relations Staff, GMC
Washington, D. C.

Chenea, P. F.
General Motors Research Laboratories
Warren, Michigan

Chock, D. P.
General Motors Research Laboratories
Warren, Michigan

Churchman, C. W.
Univ. of California, Berkeley
Berkeley, California

Clark, W. C.
Int'l. Inst. for Applied System Anal.
Laxenburg, Austria

Cole, R.
　　Battelle Institute
　　Seattle, Washington

Colucci, J. M.
　　General Motors Research Laboratories
　　Warren, Michigan

Crouch, E. A. C.
　　Harvard University
　　Cambridge, Massachusetts

Curran, A. S.
　　Westchester Cty., Dept. of Health
　　White Plains, New York

Dierkes, M.
　　Int'l. Inst. for Environ. & Society
　　Berlin, West Germany

Dorf, R.
　　University of California
　　Davis, California

Drescher, S.
　　Financial Staff, GMC
　　New York, New York

Elder, C. J.
　　Environmental Activities Staff, GMC
　　Warren, Michigan

Evans, L.
　　General Motors Research Laboratories
　　Warren, Michigan

Evans, J.
　　Harvard University
　　Cambridge, Massachusetts

Everett, R. L.
　　Environmental Activities Staff, GMC
　　Warren, Michigan

Ferreira, J., Jr.
　　Massachusetts Inst. of Technology
　　Cambridge, Massachusetts

Fischhoff, B.
　　Decision Research
　　Eugene, Oregon

Fisher, T. M.
　　Environmental Activities Staff, GMC
　　Warren, Michigan

Fraser, J.
　　Purdue University
　　West Lafayette, Indiana

Gagliardi, A. V.
　　Public Relations Staff, GMC
　　Detroit, Michigan

Gallasch, H. F., Jr.
　　General Motors Research Laboratories
　　Warren, Michigan

Gallopoulos, N. E.
　　General Motors Research Laboratories
　　Warren, Michigan

Gardels, K. D.
　　General Motors Research Laboratories
　　Warren, Michigan

Gough, M.
　　Office of Technology Assessment
　　Washington, D. C.

Green, H. P.
　　George Washington University
　　Washington, D. C.

Griesmeyer, J. M.
　　UCLA
　　Los Angeles, California

Hamilton, L.
　　Brookhaven National Laboratory
　　Upton, New York

Hammond, J. D.
　　Pennsylvania State University
　　University Park, Pennsylvania

Hare, L.
　　National Science Foundation
　　Washington, D. C.

Heuss, J. M.
　　General Motors Research Laboratories
　　Warren, Michigan

Hildrew, J. C.
　　Mobil Oil Research & Devel. Group
　　New York, New York

Holt, J. L.
　　Exxon Company, U.S.A.
　　Houston, Texas

Holzwarth, J. C.
　　General Motors Research Laboratories
　　Warren, Michigan

Horowitz, A. D.
　　General Motors Research Laboratories
　　Warren, Michigan

Howard, R. A.
Stanford University
Stanford, California

Howell, L. J.
General Motors Research Laboratories
Warren, Michigan

Huntsman, J.
Applied Decision Analysis
Menlo Park, California

Jaksch, J.
Los Alamos Scientific Labs
Los Alamos, New Mexico

Jenkins, C. L.
Industry-Govt. Relations Staff, GMC
Detroit, Michigan

Johnson, W. R.
Environmental Activities Staff, GMC
Warren, Michigan

Joksch, H.
Center for the Environment & Man
Hartford, Connecticut

Jones, S. T.
General Motors Research Laboratories
Warren, Michigan

Kachman, N. C.
Environmental Activities Staff, GMC
Warren, Michigan

Kamal, M. M.
General Motors Research Laboratories
Warren, Michigan

Kasper, R. G.
National Research Council
Washington, D. C.

Kingman, R. T., Jr.
Public Relations Staff, GMC
Detroit, Michigan

Klimisch, R. L.
General Motors Research Laboratories
Warren, Michigan

Kneese, A.
Resources for the Future
Washington, D. C.

Kranzberg, M.
Georgia Institute of Technology
Atlanta, Georgia

Lave, L. B.
The Brookings Institution
Washington, D. C.

Lawless, E.
Midwest Research Institute
Kansas City, Missouri

Linde, E.
U.S. Environmental Protection Agency
Washington, D. C.

Lorenzen, T. J.
General Motors Research Laboratories
Warren, Michigan

Lowrance, W.
Stanford University
Stanford, California

Luckey, M.
Ford Motor Company
Dearborn, Michigan

Marks, C.
Engineering Staff, GMC
Warren, Michigan

Martens, S. W.
Environmental Activities Staff, GMC
Warren, Michigan

Martin, D. E.
Environmental Activities Staff, GMC
Warren, Michigan

Maxey, M.
University of Detroit
Detroit, Michigan

McCoubrey, A. O.
National Bureau of Standards
Washington, D. C.

McDonald, R. J.
General Motors Research Laboratories
Warren, Michigan

McDonald, G. C.
General Motors Research Laboratories
Warren, Michigan

McLean, D.
University of Maryland
College Park, Maryland

Morgan, M. G.
Carnegie-Mellon University
Pittsburgh, Pennsylvania

Moskowitz, H.
Purdue University
West Lafayette, Indiana

Muench, N. L.
General Motors Research Laboratories
Warren, Michigan

Mullen, T. J.
Public Relations Staff, GMC
Detroit, Michigan

Naylor, M. E.
Transportation Systems Center, GMC
Warren, Michigan

Nelkin, D.
Cornell University
Ithaca, New York

North, D. W.
Decisions Focus, Inc.
Palo Alto, California

Okrent, D.
University of California
Los Angeles, California

Owen, D.
SRI, International
Menlo Park, California

Page, Toby
Environmental Protection Agency
Washington, D.C.

Parker, F.
Tenneco, Inc.
Houston, Texas

Pate, M. E.
Stanford University
Stanford, California

Patton, G.
American Petroleum Institute
Washington, D. C.

Peterson, A. V., Jr.
University of Washington
Seattle, Washington

Phillips, B. A.
General Motors Research Laboratories
Warren, Michigan

Plott, C. R.
California Institute of Technology
Pasadena, California

Pollak, M.
Cornell University
Ithaca, New York

Pollock, S.
University of Michigan
Ann Arbor, Michigan

Potter, D. S.
Public Affairs Group, GMC
Detroit, Michigan

Raiffa, H.
Harvard University
Cambridge, Massachusetts

Repa, B. S.
General Motors Research Laboratories
Warren, Michigan

Richter, B. J.
Enviro Control, Inc.
Rockville, Maryland

Rodriguez, R. N.
General Motors Research Laboratories
Warren, Michigan

Rosenthal, A. B.
Customer Relations & Serv. Staff, GMC
Warren, Michigan

Rothery, R. W.
General Motors Research Laboratories
Warren, Michigan

Rowe, W. D.
The American University
Washington, D. C.

Runkle, D. L.
Chevrolet Motor Division
Warren, Michigan

Sabey, B.
Transport & Road Research Lab.
Crowthorne, Berkshire, England

Sapre, A. R.
General Motors Research Laboratories
Warren, Michigan

Schmidt, D. E.
General Motors Research Laboratories
Warren, Michigan

Schneiderman, M.
National Cancer Institute
Bethesda, Maryland

Schulze, W.
  University of Wyoming
  Laramie, Wyoming

Schwing, R. C.
  General Motors Research Laboratories
  Warren, Michigan

Scott, G. H.
  Proctor and Gamble Company
  Cincinnati, Ohio

Scott, S. L.
  Environmental Activities Staff, GMC
  Warren, Michigan

Shantz, A.
  U.S. General Accting Office
  Washington, D. C.

Shariq, S. Z.
  Environmental Protection Agency
  Research Triangle Park, N.C.

Siegla, D. C.
  General Motors Research Laboratories
  Warren, Michigan

Slovic, P.
  Decision Research
  Eugene, Oregon

Smith, G. W.
  General Motors Research Laboratories
  Warren, Michigan

Spreitzer, W. M.
  General Motors Research Laboratories
  Warren, Michigan

Stairs, G.
  Duke University
  Durham, North Carolina

Starr, C.
  Electric Power Research Inst.
  Palo Alto, California

Starr, L.
  Celanese Corporation
  New York, New York

Stebar, R. F.
  General Motors Research Laboratories
  Warren, Michigan

Straus, D. B.
  American Arbitration Association
  New York, New York

Swanson, S. M.
  American Petroleum Institute
  Washington, D. C.

Thomas, K.
  Troy, Michigan

Thompson, M.
  Inst. for Policy & Mgmt. Research
  Bath, Avon, England

Tobin, R. J.
  State Univ. of New York
  Buffalo, New York

Tuesday, C. S.
  General Motors Research Laboratories
  Warren, Michigan

Tummala, V. M. R.
  University of Detroit
  Detroit, Michigan

Upson, B. T.
  Financial Staff, GMC
  Detroit, Michigan

van Schayk, C.
  Motor Vehicle Manufacturers Assn.
  Detroit, Michigan

Viano, D. C.
  General Motors Research Laboratories
  Warren, Michigan

von Buseck, C. R.
  General Motors Research Laboratories
  Warren, Michigan

von Gizycki, R.
  Battelle Institute
  Frankfurt AM Main, Germany

Vostal, J. J.
  General Motors Research Laboratories
  Warren, Michigan

Wallace, W. A.
  Rensselaer Polytechnic Inst.
  Troy, New York

Wasielewski, P. F.
  General Motors Research Laboratories
  Warren, Michigan

Weaver, P.
  Ford Motor Company
  Dearborn, Michigan

Whipple, C.
  Electric Power Research Inst.
  Palo Alto, California

Whyte, A.
  University of Toronto
  Toronto, Ontario, Canada

Wiggins, J. H.
  J. H. Wiggins Company
  Redondo Beach, California

Williams, R. L.
  General Motors Research Laboratories
  Warren, Michigan

Wilson, R. A.
  Environmental Activities Staff, GMC
  Warren, Michigan

Wolff, G. T.
  General Motors Research Laboratories
  Warren, Michigan

Wolsko, T.
  U.S. DOE, Argonne Natl. Laboratory
  Argonne, Illinois

Yesley, M. S.
  The Rand Corporation
  Santa Monica, California

PROPER NAME INDEX

Albers, Walter A. Jr., vi
Altshuler, Bernard, 82
Anderson, E. V., 13, 66, 107, 177, 315
Arrow, Kenneth, 219
Ashby, Eric, 22
Auden, W. H., 304

Backman, J. E., 165
Balzhiser, R., 135
Bazelon, David L., 323
Bentham, Jeremy, 220
Bonington, Chris, 276
Boyland, Eric John, 38
Breslow, Lester, 39
Brooks, Harvey, 306, 307
Bruce-Briggs, B., 110
Burke, F. E., 106, 142, 252, 313
Burton, Ian, 294

Calabresi, G., 132, 139
Callahan, Daniel, 140
Carson, Rachel, 290
Carter, Jimmy, 122, 331
Casey, Albert, 76
Chapman, W. C., 177
Churchman, C. West, 343-346
Clark, William C., 39, 126, 248, 268, 287-318
Cohen, B., 130-132
Coleman, W. T. Jr., 130
Crouch, E. A. C., 111
Crozier, Michael, 309
Curran, A. S., 109

Drescher, S., 250
Dubos, Rene, 304
Duerr, H. P., 289

Edgeworth, F, 221
Edison, T. A., 320
Einstein, Albert, 320
Ellul, Jacques, 321
Enstrom, James, 41

Ferrari, J. R., 165
Ferreira, J., 177, 208, 215
Feyerabend, P., 306
Fischhoff, Baruch, 137, 181-216
Ford, Gerald, 122
Ford, Henry, 324
Friedman, Milton, 268
Fuchs, Victor, 129

Green, Harold P., 255-269
Greenberg, Dan, 329
Grosse, R. N., 130

Hammond, J. D., 147-178
Hand, Learned, 257
Handler, Philip, 321
Harris, Marvin, 289, 293
Haston, Dougal, 278
Hayakawa, S. I., 222
Hillary, E., 278
Hofflander, A. E., 166
Holt, J. L., 253
Horatius, 281
Howard, Ronald A., 89-112, 124, 126, 230, 268, 346
Hoyle, Fred, 279
Hunt, John, 277
Huntsman, J., 111
Hutt, P. B., 293
Hynes, M., 187

Inhaber, H., 135, 136
Inquisitor Alonso Salazar y Frias, 292, 293, 308
Institor, 290, 304

Joksch, H., 67, 126

Kahneman, D., 161, 163, 214
Kant, Immanuel, 219, 343
Kasper, Raphael G., 71-84
Kates, Robert, 188, 294
Keeny, R. L., 208
Kefauver, E., 290, 300
Kipling, R., 280
Kneese, Allen, 85-87, 125, 144, 145
Kranzberg, Melvin, 319-332
Kulp, C. A., 148
Kunreuther, H., 150, 174, 175

Lave, Lester B., 115-128, 207, 216, 225, 268, 334
Lawless, E., 292
Lederberg, Joshua, 301
Lichtenstein, Sarah, 137, 181-216
Lindblom, C. E. , 305, 307
Linstone, H., 139
Lippman, Walter, 335
Loftus, K. H., 298
Lowrance, William W., 5-17, 176

MacDonald, Gordon, 11, 15
MacLeish, Archibald, 331

Majone, G., 307, 308, 309
Markowitz, H. M., 165, 169
McLean, D., 112, 317, 318
Michael, D. N., 309
Midgley, Mary, 139
Midlefort, H. C. E., 293
Mill, John, 220
Mishan, E. J., 223, 225
Morgan, M. Granger, 110, 307
Muskie, Edmond, 188

Nelkin, Dorothy, 189, 233-253, 307, 309,
  333
Nietzsche, Friedrich, 221
Nozick, Robert, 231

O'Day, J., 133
Ogburn, William Fielding, 328
Okrent, David, 81, 267, 268
Owen, D. , 269

Page, Talbot, 40
Parker, F., 144
Pate, E., 108, 314
Patey, Tom, 282
Patton, G., 66, 251, 253
Peltzman, Samuel, 117, 127
Pigou, A. C., 220
Plott, C. R., 215
Pollak, Michael, 233-253
Pollock, S., 69
Pope Innocent VII, 290
Popper, K. , 291

Quade, E. S., 307

Raiffa, Howard, 271, 339-341
Rand, Ayn, 223
Rasmussen, Norman, 75
Ravetz, Jerome, 306
Rawls, John, 221
Rethans, A. , 188
Rice, Dorothy, 225
Rosen, S., 135, 218
Rowe, William D., 38, 109, 179, 180, 249,
  268, 314, 315, 333

Sabey, Barbara E., 43-70
Sandburg, Carl, 335
Schantz, A., 334
Schelling, Thomas, 121
Schmidt, Alexander, 188
Schneiderman, Marvin A., 19-41, 144, 249,
  336
Schulze, William D., 217-232
Schweig, B., 150, 161
Schwing, Richard C., 129-145
Seskin, Eugene, 225
Sharpe, W. F., 169
Shilling, N., 166
Slesin, L., 208
Slovic, Paul, 76, 137, 181-216, 340,
Smith, R. S., 218
Starr, Chauncey, 1, 15-17, 40, 78, 80, 83,
  124, 143, 205, 206, 214, 225, 315, 316,
  333, 337
Stockton, Frank Richard, 302
Straus, D. B., 68, 143, 248, 315, 335
Summers, M., 289
Svenson, O., 188
Swanson, S. M., 83

Taylor, Harold, 43-70
Thaller, R., 135, 218
Thompson, Michael, 112, 273-286, 309
Tobin, R. J., 125, 252
Trevor-Roper, H. R., 289, 294
Tversky, A. , 161, 163, 214

VanMarcke, E., 187

Wallace, W. A., 177
Weinberg, Alvin, 305, 307
Weinstein, N. D., 188
Whillans, Don, 278, 279
White, Gilbert, 294
Wiggins, John H., 83, 108, 111
Wildavsky, A., 290, 307, 329
Williams, J. D., 133
Wilson, R., 130, 208
Winner, Langdon, 319, 321, 323

Young, Winthrop, 279

Zeckhauser, R., 144

SUBJECT INDEX

Abnormal risk, 257
Acceptability, 6, 85, 180
Acceptable, 45, 256
Accident data, highway, 44
Accident savings
  road environment, 64
  vehicle safety, 65
Accident severity, distribution of impacts,
  55
Accidents
  automobile, 74
  causation, 58
  contributory factors, 49-53
  discrete small-scale, 9
  falls, 107
  multiplicity of causes, 49-53
  reduction potential, 58,59
  statistics, 100
Acquittal, 291
Acrylonitrile, 24,27
Acts of God, 9
Actuarial capabilities, 148
Adaptive design, 306
Adverse conditions, 49,52
Advisory models, 235-241
Aesthetic
  framework, 278
  perspective, 273-286
  view, 273-286
Agriculture, 298
Air pollution, 145
Airplanes, 9
Alienation, 253
American Cancer Society, 336
American Civil Liberties Union, 143
Amish, 14
Animal studies, extrapolation from, 33
Answers, Utopian, 339
Anthropological view, 273-286, 295
Anthropology, 280
Anticipation, 22
Apathy, 252
Apprehension, perceived risk of, 48
Army Corps of Engineers, 122
Arrogance, 77, 81
Arsenic, 24, 31
Asbestos, 10, 24, 27
Atomic bomb, 75
Atomic Energy Act, 263-265
Atomic Energy Commission, 86
Attitude change, 80

Attitudes, 60
Automobile
  accidents, 330, 344
  regulations, 127,128
  safety, 218
Autonomous technology, 321, 324
Availability heuristic, 183

Barclay Bank, 276, 280
Base camp, 282
Bayesian, 153
Beauty, 344
Behavior change, 120
Behavior modes, 115
Bellagio Conference on Science and
  Technology, 307
Benefit-cost, 45, 72, 86, 122, 218, 224, 226,
  230, 231
Benefits, 61, 63, 196, 203, 255
  judgments, 203
  of relieving apprehension, 77
Benthamite, 220
Benzene, 12, 25, 28
Bill of Rights, 91
Biological system perspective, 287-318
Biorhythm, 60
Black pill, 92, 94, 108
Blameworthiness, 49-54
British Council for Science and Society, 242
British Everest Expedition, 276
Browns Ferry, 188
Bureaucracies, public, 233, 234
Burgerdialog, 242

Calculations, 73
Cambridge Citizens Review Board, 244
Cambridge City Council, 240
Canadian Burger Commission, 239, 243,
  252
Cancer
  cervix, 38
  environmental, 38
  lung, 39
  rate, 40, 41
  risk estimation, 31
  stomach, 38
  viruses, 38
Carbon dioxide, 11, 15-17
Carcinogens, 23, 35

Careerism, 293
Catastrophic, 115, 149, 181, 207, 208, 210
Catchphrases, 281
Categorical imperative, 219, 343
Cause and effect, 73
Certainty, highway risk, 45
Certainty-Uncertainty dimension, v
Certified Public Scientist, 235
Chance of death, 92
Change, 319-332
Chemical Food Additives Amendments, 261
Chemicals, 22, 23
Chest injury, 55, 56
Children, 346
Christian, 222
Chromium, 25, 28
Chronic effects, 187
Church, 289, 290
Cigarette smoking, 22, 39, 41
Civil law, 44
Claim forcasting, 148, 149
Claim
  frequency, 154
  severity, 154
  stability, 152, 154
Clean Air Act Amendments of 1970, 22, 121
Coal, 134
Collectivism, 282-284
Comfort compromised, 76
Committee on Government Operations, 187
Common good, 290
Common law, 256
Community, 7
Compensated risk, 225, 230
Concerns of people, 116
Concluding remarks, 339-341
Concorde SST, 78
Conflict, 46, 233, 243
Conflict resolution, 235, 247, 334
Confrontation, 79
Congress, 79
Congressional approach, 333, 334
Conseil d'Information sur l'Energie Electro-nucleaire, 241, 245
Consensus, 78, 143, 236-238, 246, 253, 335,
Consent, 91
Consumer Product Safety Commission, 118
Consumerists, 117
Consumers desire, 119
Consumption, 92
Context of risk, 75
Context or culture, 273-286
Continuing risks, 100

Contradictions, 115
Contradictory catchphrases, 281
Contradictory rationalities, 273
Contrary evidence, 189
Contributing factors, 67
Coping strategies, 287
Cost minimizing, 144
Cost-effectiveness, 37, 45, 90, 126, 133, 344
Costly business, 284
Costs, 61,63
  highway accidents, 47
  pain, grief, and suffering, 47
  property damage, 48
  of regulation, 143
Countermeasures, 54, 58
Credibility, 153
Criminal law, 44
Crisis of confidence, 331
Cultural change, 319
Cultural lag, 328
Culture or context, 273-286

DC-10, 13, 72, 76, 79, 83, 187
DDT, 240, 325
DNA, 11, 76, 181, 208, 234, 240, 250, 293, 335, 337
Dams, 9
Darwinian, 223
Death
  judged seriousness, 194
  decisions, 89
  rates, 20, 21
  frequency, 184, 190
Decide, who shall, 179-269
Decision maker perspective, 71-84, 89-112
Decision making, 2, 76, 79, 86, 180, 182, 256
  corporate, 72
  framework, 89
  government, 72
  individual, 72, 92
  procedures, 234
  under uncertainty, 148
Decision theory, 10
Degenerative disease, 8
Degree of control, 75
Delaney Clause, 79, 122, 258
Delayed risk, 75
Democracy, 180
Democratic process, 335
Denial, 188
Dependency, 138
Desire for certainty, 188
Diethylstilbestrol, 12

Directions and perspectives, 271-346
Disaster
  multipliers, 193
  potential, 192
Discipline, creation of, 2
Disclosure. 3
Discount, 148
Discount rate, 139
Discounting, 225
Diseases of cleanliness, 297
Dispute closure, 250
Divergent opinions, 189
Dose-response, 33, 35
Dread, 199, 210
Driver error, 49-51
Drug
  disaster, 313
  safety, 299, 303
  testing, 300
Drunk driving, 60, 69

Earthquakes, 4,9
Economic
  perspective, 115-128
  considerations, 274, 345
  criteria, 135
  theory, 117, 225
  tools, 115-128
Economists, 117, 126, 217
Educators, 343
Efficacy, 90
Efficiency, 37, 126, 133, 344
Efficient positions, 167-173
Efficient underwriting frontier, 171
Egalitarian, 221, 226, 231
Electric Power Research Institute, 78, 80
Electricite de France, 239
Electronic age, 327
Elitist, 127, 222, 227, 235-238, 241
Embargo, 10
Emission control devices, 120
Empirical analysis, 10
End state ethics, 230
Ends and means, 343
Energy policy, 134
Engineer perspective, 129-145
Engineers, 130, 180, 303
Enlightenment, 289
Environment
  chemical, 20
  physical, 20
  damage, 135
Environmental Defense Fund, 241
Environmental Protection Agency, 125,
  140, 262, 263

Epidemic, 7
Epidemiology, 32, 33
Ethical principles, 91
Ethical systems, 219
Ethics, 89-112, 217-232, 244
Ethologists, 139
Ethylene dibromide, 26, 30
Everest, 276
Evidence, 214
Executive branch, 334
Experience, 73
Experimental design, in regulation, 301
Expert, 250, 334, 335
  as human, 345
  judgment, 189
  opinion, 137
Expertise
  distribution of, 245
  institutionalized, 289
Exposed number, 199
Exposed persons, 31
Exposure, 210
Exposure
  degree of, 202
  estimation, 31
  intensity, 32
Externality, 121
  corrections, 120

Factor analysis, 199, 201
Facts, 140, 181-216, 308
Facts and values, 74
Familiarity, 199, 210
Fault tree
  radiation, 182
  starting a car, 185, 186, 341
Fear of risk, 288
Fears, 77, 181-216
Federal Insecticide, Fungicide, and
  Rodenticide Act, 115
Federal Trade Commission, 118
Financial risk, 278-280
Finite resources, 85, 129-133
Fire insurance, 157
First collision, 121, 128
Fixed budget, 225
Fixed resources, 85
Flammable Fabrics Act, 261
Flammable fabrics, 261
Flood control, 295
Floods, 9, 187, 287-318,
Flying club, 109
Food
  additives, 12
  contaminants, 12

safety, 12
Food and Drug Administration, 12, 188,
    261, 263, 300, 304
Food, Drug, and Cosmetic Act, 122
Force, use of, 91
Ford Foundation/MITRE Report, 11, 15-16
Fraud, 111
Freedom, 61
    individual, 109, 256
French Declaration d'Utilite Publique, 240
Fuel savings, 133
Future directions, 61, 63, 173
Future generations, 13
Future risks, 138

Garden of Eden, 320
General Accounting Office, 140
General Motors, 140
Generally Recognized as Safe, 310
German Atom Law, 240
German Minister of Economics, 244
German Ministry of Science and
Technology, 239
God
    act of, 90
    wrath of, 90
Good science, 306
Government, role of, 119
Greater Himalayan Range, 274
Gross national product, 131
Groupe d'Information Nucleaire, 241
Growth industry, 289
Guilt, 291

Hastings Center, 250
Hazardous materials, 19, 24
Hazards
    buying, 99
    catastrophic, 49
    classification, 8
    low-level delayed, 10
    management, 181, 182
    modification, 100
    over-rated, 55
    overview, 197
    selling, 99
    under-rated, 55
    unquantifiable, 19
Health programs, 130
Health status, 116
Herring gull, 139, 140
Highway, 9
    actual risk, 45
    codes, 44

fatalities, 46
history of, 44
injuries, 46
journey, 44
perceived risk, 45
regulation, 44
risk, 43-70
Highway Tranportation Safety Act, 123
Highway risk perspective, 43-70
Himalayan climbing, 273-286
Hindu-Buddhist cycle, 278
Hindus, 274-286
Hip dislocation, 55, 57
Historical perspective, 287-318, 319-332
History, 344
Holistic, 140, 314
Homogeneity, 153
House Commerce Committee, 262
Human
    as machine or fool, 344
    error, 8, 9, 11, 49-51
    error interactions, 66
    goals, 90
    judgement, 181
    nature, 140
    value of life, 89, 94, 97, 102, 218, 225
Humanity, 343-346
Humility, 11
Hurricanes, 9
Hysteresis, 284

Imagined, 71-84
Immorality, 133
Impact timing, 137
Impairment, 49, 51
Imported fuel, 137
Incentives, 176
Indians, souls of, 75, 84
Individual
    risk accepting, 273
    risk avoiding, 273
    exposure, 19
Individualism, 282-284
Industrial Revolution, 320
Industrial societies, 295
Inefficiency, 344
Infectious disease, 8
Influenza, 7
Information models, 237-242
Initiators, 35
Injury tolerance, 55
Innocent parties, 121
Inquisition, 289, 291, 308
Insecticides, 296

Institute for Nuclear Power Operations, 81
Institute for Reactor Safety, 244
Institution flexibility, 288
Institutions, 115
Insurability, criteria, 149
Insurance
    brokers, 162
    captive, 174
    contract, 148
    non-claims information, 152
    institution perspective, 147-178
Intellectual risk, 278-280
Interest rate, 87
Interlock device, 118
Internal Revenue Service, 174
Interstate highway, 133
Intervener funding, 251-253

Joint Committee on Atomic Energy, 265
Judged risk study design, 195
Judgement underwriting, 161
Judgmental bias, 183
Judgmental heuristics, 183

Kamikaze pilot, 281
Kant's Categorical Imperative, 219, 343
Kantian, 221
King and king equivalents, 91
Known risks, 43-70
Kranzberg's First Law, 322

Lady or the Tiger, 302
Laissez-faire, 116
Lamaist Buddhists, 274-286
Large numbers, 153
Law, corporate, 105
Lawyers, 126
Lay
    estimates, 137, 192
    public, view, 7
Lead effects, 73
League of Women Voters, 81, 183, 189, 344
Legal
    criteria, 255
    perspective, 255-269
    position, 92
    procedures, 268
Legislation, 257
Length of life, 92
Lexicographic preferences, 229
Liability, 105, 156, 256, 269
    limited, 105
    third party, 105

Libertarian, 217, 222, 227, 229
Liberty infringement, 49
Life
    decisions, 89
    expectancy, 116
    extending programs, 130
    saving programs, 130
    span, 144
    table, 95
Limits of choice, 246
Limits of regulation, 299
Limits to growth debate, 139
Liquid natural gas, 75, 83, 187, 208
Logit model, 33, 34
Longevity, weighted value, 143
Longevity gain, 144
    cost per year, 132
Loss exposure, 147
Loss spreading, 148
Love canal, 78

Mackenzie Valley Pipeline, 239
Malleus, 289-292
Mamography, 336
Man, planning animal, 36
Man-days lost, 136
Marginal drugs, 300
Market economy, 116
Mathematical model, 33
Measures of risk, calculated and perceived,
    72
Mediation, 248
Medical model, 343
Medical profession, 44
Metabolic activation , 33
Mexico City population, 327
Micromort, definition, 99
Miscellaneous discussion, 333-338
Modern industrial society, 285
Morality, 133
Mountaineering, 277
Multi-stage model, 35
Murder, 105
Myths, 76

Napalm, 75
National Academy of Sciences, 299, 321
National Association of Insurance
    Commissioners, 166
National Cancer Institute, 336
National Earthquake Hazards Reduction
Act, 13
National Highway Traffic Safety
Administration, 117, 120, 121

National Institute of Health, 240
National Occupational Health Survey, 31
National Research Council, 13, 15
National Science Foundation, 321
National characteristics, 63
Natural disasters, 9
Nature of risk, 5-17
Negligence, 44, 256
Negotiation, 243, 244
Newspaper coverage, 185
Nickle, 25, 29
Nietzschean, 217, 222
Noise, 10
Novelty, 75
Nuclear, 8, 134
     power, 77, 78, 181, 189, 207, 241, 249
     power characteristics, 196
     proliferation, 15
     reactors, 106, 330
Nuclear Electric Insurance,Ltd., 81
Nuclear Regulatory Commission, 187,
     263-265
Nuclear Safety Analysis Center, 80
Nutritionists view, 343

Oak Ridge National Laboratories, 240
Objective view, 71-84
     highway risk, 48
     truth, 6
Occupational Safety and Health Act, 11, 12,
     86
Office of Management of Budget, 122
Office of Technology Assessment, 140
Oil, 134
One hit model, 33, 34
Optimal programs, 144
Optimist, 281
Outcome, 118
Overconfidence, 185
Overstated risk, 32, 33
Oystercatcher, 139, 140

Pandemic, 7
Paretian, 222
Pareto superiority, 223
Participants List, 347-352
Participatory, 236-241
Pedestrian error, 49-51
People problem, 287
Perceived
     domestic risk, 137
     national risk, 137
     risk, 181-216
Perception, 4, 36, 68, 71-84, 183, 344

Perception
     altering, 77
     determinants, 189-210
     discrepancies between calculated and
          perceived, 74
     discrepancies between real and perceived,
          74
     faulty, 181
     highway risk, 45
     multiple fatality, 48
     quantification, 189-210
     versus reality, 340
Perchlorethylene, 26, 30
Personal habits, 22
Perspective
     aesthetic, 273-286
     biological system, 287-318
     decision maker, 71-84, 89-112
     engineer, 129-145
     highway risk, 43-70
     historical, 287-318, 319-332
     insurance institution, 147-178
     psychologist, 181-216
     toxicologist, 19-41
Pessimist, 281
Pesticides, 13, 79, 86
Physical risk, 278-280
Plutonium, 13, 15
Police department, 143
Policy making, 76
Polio, 296, 297
Political science question, 324, 329
Pollution control, 78
Population change, 325
Portfolio theory, 147, 167-170, 174
Posterity, 22
Poverty, 274
Power generation costs, 135
Power plants, 9
Precedents, 256
Preface, v
Premium distribution, 155
Premiums to surplus, 166-170
Price Anderson Act, 106
Primary rural road, 133
Primary safety, 63
Private Passenger Automobile Physical
     Damage Insurance, 157-159
Private Passenger Automobile Liability,
157-159
Private risk, 217
Private safety, 228, 229
Probability of death, 225
Probit model, 33, 34
Procedural acceptability, 243

Process, 80, 118, 336
Process and outcome, 344
Professional engineering, 303
Program evaluation, 87
Promises, 331
Promoters, 35
Promoting agents, 19
Proof, 308
Propaganda, 71, 77, 78
Prospects for change, 319-332
Prosperity, 116
Psychologist perspective, 181-216
Psychologists, 117
Public, 180, 334, 335
    airline safety, 218
    fears, 79
    insurance funds, 177
    interest, 300, 306
    interest groups, 116
    opinion, 118
    participation recommendations, 248
    policy, 108
    policy decisions, 255
    safety, 228, 229
Public Utility Commissions, 329
Purchasing decision, 89
Pure premium, 153
Purposeful machine, 343

Quality
    control, 306
    of dying, 143
    of life, 323
    of life years, 144
    of living, 143
Quantification, 258
Quantitative, versus qualitative, 267, 268
Quantitative methods, 73

Radiation, 10, 182
Rare events, Highway, 44
Rawlsian, 217, 221
Real, 71-84
Reasonable, 256
Recalls, Automotive, 72, 79
Referenda, 241
Reformation, 289
Regulation, 12
    desired, 344
    experiments, 316, 317
    future direction, 123
    government, 44
    limits, 299
    mitigated gains, 117

procedures, 233-253
reform, 124
Regulators, 115, 116
    disappointing success, 118
    success, 118
Regulatory
    agencies, 117, 259
    budget, 122
    choice criteria, 123
    efficacy, 300
    frameworks, 122,123
Reinsurance, 150
Relative risk, 40
Renaissance, 289
Research needs, 61, 63
Resiliency, 287-318
Resource management, 294, 297
Retired persons, 326
Retrospective study, 288
Right to health and safety, 129
Rights and responsibilities, 13
Risk
    adjustments to, 118, 205
    avoider, 282
    avoiding, 280
    benefit, 10, 11, 202, 257, 300, 330
    characteristics, 195, 198, 200, 205
    commodity, 217
    compensated, 217
    control of, 119
    creating projects, 104
    decisions, 10
    definition of, 6, 82
    dimensions, 199
    false perception, 48
    free technology, 247
    imposed by another, 90
    imposition, 103
    involuntary, 91
    issues in society, 101
    job, 102
    long range, 244
    marketplace, 102
    measurement, 92
    narrowing, 281-284
    people problem, 287
    political questions, 234
    politicized issue, 233
    reduction, 115-128
    reduction, value of, 99
    remedies, 60
    return , 167-173
    spectrum, 90
    spreading, 147-178, 281-284
    taking, 149, 274, 276, 280

teaching of, 5
technical questions, 234
transfer of, 148
uncompensated, 217, 225, 227, 228, 230
Risk assessment
    confusions, 3
    flexibility, 314
Risk assessors, 290
Risk management, 10, 12
    adaptive, 302
    feasibility, 287
    government, 13
    individual, 13
    personal, 280
    policy, 306
    scale, 309
Riskless society, 147
Risky business, 275
River Rouge Plant, 324
Road design, 49, 54
Road engineering benefits, 58
Role of law, 255-269
Russian roulette, 225

Saccharin, 34, 118
Safe Drinking Water Act, 11, 22
Safety
    automotive, 79
    definition, 261
    margin, 54
    profession, 343-346
    regulations, 260
Science, 7
Science Court, 235
Science Research Council, 279
Scientific
    analysis, 258
    expertise, 234
    revolution, 320
    truth, 235
Scientists role, 82
Seat belt, 49, 55
Second collision, 121, 128
Secondary safety, 63
Security Exchange Commission, 329
Self-interest, 110
Senate Committee on Labor and Public
    Welfare, 261
Sherpas, 276
Ships, 9
Side effects, 317, 318
Signal value, 209
Skylab, 72
Slave, 108

Sobriety test, 128
Social
    change, 325
    conflict, 181
    context, 281
    costs, 3, 345
    preferences, 217
    psychology, 289
    value, 246, 257
    welfare, 226
Societal dialogues, 339
Sociopolitical change, 319
Sociopolitical disruption, 10
Soft-fail strategies, 287
Speed limits, 60, 133, 258
Sports, 9
Spruce budworm, 296
Standard of living, 91
Statistics, highway, 46
Stopping rule, 291
Strategies, appropriate, 273
Stress, 10
Subjective, 71-84
    judgements, 181
    risk, 54
    versus objective, v, 71-84
Subjectivity, 7
Subranormal stimuli, 140
Supreme Court, 12
Surityship, 148
Susceptibility of animals and humans, 33
Swine influenza, 7
Symposium
    dimensions, v
    objective, v
    questions, 3
    summary, 343-346
Synthetic chemicals, 22

Technical Review Board, 235
Technological
    age, 320
    change, 323
    determinism, 319
    fixes, 326
    lag, 319, 328
    society, 140
Technological Magistrature, 235
Technology
    adaptive mechanisms, 315, 316
    autonomy, 319
    choices, 324
    failure, 2, 9
    forcing laws, 121

impacts on society, 72
new , 2, 234
risks, 191
Ten Commandments, 219
Terminally ill, 13
Terrorism, 10
Teton dam, 13, 78, 187
Thalidomide, 313, 314
Thermal oxide reprocessing plant, 239
Three Mile Island, 7, 72, 77, 78, 83, 187,
    189, 209, 315, 330
Threshold, injury, 55
Tibet trade, 275
Time horizon, 138
Tolerance level, 55
Tools, 36
Tornadoes, 9
Toxic Substances Control Act, 11, 22, 86,
    262, 263
Toxic
    chemicals, 87
    non-cancer effects, 35
    substances, 262
Toxicologist perspective, 19-41
Trade-off, 129-145, 217, 228, 303, 325
    journey time, 45
    lives for lives, 130
    lives for money, 133
    lives for time, 133
    multidimensionsal, 134
    of risks and benefits, 72
    with efficiency, 123
Trans science, 305, 307
Transitivity, 219
Trichlorethylene, 26, 29
Truisms, 2
Truth, 344
Two hit model, 33, 34
Type I-Type II Errors, 152, 160

U. S. Weather Bureau, 7
Unacceptable, 256
Uncertain risks, 19
Uncertainty, v, 11, 72, 74
    inherent, 140
Uncompensated risk, 225, 227, 228, 230
Underestimates, 76
Understated risk, 32, 33
Underwriting, 147-178
    capacity, 165
    cyclical nature, 163
    data, 156

decision diagram, 151
error, 160
judgement, 161
pessimistic and conservative, 164
portfolio, 166, 170
uncertainty, 154
United Auto Workers, 140
United States Census Bureau, 326
United States Forest Service, 240
University of Colorado, 143
Unreasonable risk, 256
    definition, 262
Uranium, 13
Uselessness, 277, 280
Utilitarian, 217, 220, 226
Utility functions, 180, 221

Value
    appraisal, 10, 11
    judgment, 250
    of a friend, 108
    of life, 89, 94, 97, 102, 225
    of safety, 117, 217-232
    systems, 3
    trade-offs, 340
Values, 140, 325, 330
Vehicle defects, 49, 53
Vehicle safety measures, 58
Vinyl chloride, 10
Voluntariness, 205
Voluntary, 75, 281

Watershed, 290
Wealth, 92
Welfare, maximize, 274
Wetenschapswinkels, 245
White pill, 97, 98
Wisconsin Supreme Court, 257
Witchcraft, 91, 287-318, 317, 318
Wonder drugs, 287-318
Worker exposure, 31, 35
Workplace, 9, 31, 32, 135
World Bank Study, 327
World view, 283

X rays, 187, 196

Yak route, 278

Zero risk, 300